Advanced Sciences and Technologies for Security Applications

Founded by H. J. Koelsch

Editor-in-Chief: B. Javidi, UConn, Storrs, CT, USA

Editorial Board: E. Carapezza, DARPA, USA
J.P. Huignard, Thales, France
N. Nasrabadi, US Army
H. Tiziani, Stuttgart, Germany
T. Tschudi, Darmstadt, Germany
E. Watson, US Air Force, USA
T. Yatagai, Tsukuba, Japan

T0137776

Advanced Sciences and Technologies
for Security Applications

The series Advanced Sciences and Technologies for Security Applications focuses on research monographs in the areas of

—Recognition and identification (including optical imaging, biometrics, authentication, verification, and smart surveillance systems)

—Biological and chemical threat detection (including biosensors, aerosols, materials detection and forensics),

and

—Secure information systems (including encryption, and optical and photonic systems).

The series is intended to give an overview at the highest research level at the frontier of research in the physical sciences.

The editors encourage prospective authors to correspond with them in advance of submitting a manuscript. Submission of manuscripts should be made to the Editor-in-Chief or one of the Editors.

Editor-in-Chief

Bahram Javidi
Distinguished Professor of Electrical and Computer Engineering
University of Connecticut
Electrical & Computer Engineering Dept.
371 Fairfield Road, Unit 1157
Storrs, CT 06269-1157, USA
E-mail: bahram@engr.uconn.edu

Editorial Board

Edward Carapezza
Program Manager, DARPA Advanced Technology
Office
3701 N. Fairfax Drive
Arlington, VA 22203-1714, USA
E-mail: ecarapezza@darpa.mil

Jean-Pierre Huignard
Thales Research & Technology
Domaine de Corbeville
91404 Orsay cedex, France
E-mail: jean-pierre.huignard@thalesgroup.com

Nasser Nasrabadi
Department of the Army
US Army Research Laboratory
ATTN: AMSRL-SE-SE
2800 Powder Mill Road
Adelphi, MD 20783-1197, USA
E-mail: nnasraba@arl.army.mil

Hans Tiziani
Universität Stuttgart
Institut für Technische Optik
Pfaffenwaldring 9
70569 Stuttgart, Germany
E-mail: tiziani@ito.uni-stuttgart.de

Theo Tschudi
Technische Universität Darmstadt
Institut für Angewandte Physik
AG Licht- und Teilchenoptik
Hochschulstraße 6
D-64289 Darmstadt
E-mail: Theo.Tschudi@physik.tu-darmstadt.de

Edward A. Watson
AFRL/SNJ Bldg 620
2241 Avionics Circle, Ste 2
WPAFB, OH 45433-7304
E-mail: Edward.Watson@wpafb.af.mil

Toyohiko Yatagai
Institute of Applied Physics
University of Tsukuba
Tsukuba Ibaraki 305, Japan
E-mail: yatagai@bukko.bk.tsukuba.ac.jp

Bahram Javidi
Editor

Optical and Digital Techniques for Information Security

With 120 Figures

 Springer

Bahram Javidi
Distinguished Professor of Electrical and Computer Engineering
University of Connecticut
Electrical & Computer Engineering Dept.
371 Fairfield Road, Unit 1157
Storrs, CT 06269-1157, USA
E-mail: bahram@engr.uconn.edu

Library of Congress Cataloging-in-Publication Data
Javidi, Bahram.
    Optical and digital techniques for information security / Bahram Javidi.
        p. cm.
    Includes bibliographical references and index.

    1. Optical data processing.   2. Information storage and retrieval systems—Security
measures.   3. Computer security.   I. Title.

TA1630.J38   2004
005.8—dc22   2003066025

ISBN-13: 978-1-4419-1920-5   e-ISBN: 978-0-387-25096-0   Printed on acid-free paper.

Printed in the United States of America.   (BPR/SBA)

9 8 7 6 5 4 3 2 1

springeronline.com

Contents

For my sister Nika

Preface

There are wide-ranging implications in information security beyond national defense. Securing our information has implications for virtually all aspects of our lives, including protecting the privacy of our financial transactions and medical records, facilitating all operations of government, maintaining the integrity of national borders, securing important facilities, ensuring the safety of our food and commercial products, protecting the safety of our aviation system—even safeguarding the integrity of our very identity against theft. Information security is a vital element in all of these activities, particularly as information collection and distribution become ever more connected through electronic information delivery systems and commerce.

This book encompasses results of research investigation and technologies that can be used to secure, protect, verify, and authenticate objects and information from theft, counterfeiting, and manipulation by unauthorized persons and agencies. The book has drawn on the diverse expertise in optical sciences and engineering, digital image processing, imaging systems, information processing, mathematical algorithms, quantum optics, computer-based information systems, sensors, detectors, and biometrics to report novel technologies that can be applied to information-security issues.

The book is unique because it has diverse contributions from the field of optics, which is a new emerging technology for security, and digital techniques that are very accessible and can be interfaced with optics to produce highly effective security systems.

The book has contributions from internationally recognized outstanding leaders in the field. It has compiled recent critical breakthroughs from academic and industry research of the leaders in the field. The book's technical focus is in the following areas of authentication and verification technologies: secure optical data storage, biometrics recognition, secure ID cards, data encryption, public-key cryptography, quantum cryptography, digital watermarking and data embedding, digital holography and its application to secure data storage and communication, and secure displays.

The book's objective is to focus on the state-of-the-art optical and digital technologies at the science and engineering levels with significant impact on information security. It includes sensors, systems processing algorithms, and their applications aimed at information security. The book provides examples, tests, and experiments on real-world applications to clarify theoretical concepts. A bibliography for each chapter is also included to aid the reader.

The intended audience is electrical, electronics, signal-processing, computer, optical, and communication engineers; computer scientists; imaging scientists; applied physicists; mathematicians; scientists in industry and defense technology; and upper-level undergraduate and graduate students in these disciplines.

I would like to thank the contributors, many of whom I have known for many years and are my friends, for their fine contributions and hard work. I also thank Mr. Hans Koelsch for his encouragement and support and Ms. Margaret Mitchell for her valuable assistance. I hope that this book will be a useful tool to further appreciate and understand a very important field.

Storrs, Connecticut   Bahram Javidi

Contributors

Pedro Andrés
Departamento de Óptica
Universidad de Valencia
E-46100 Burjassot
Spain

Hans I. Bjelkhagen
Center for Modern Optics
De Montfort University
Leicester LE1 9BH
UK

Rajarathnam Chandramouli
Department of Electrical and Computer
    Engineering
Stevens Institute of Technology
Hoboken, NJ 12345
USA

R. Chellappa
Center for Automation Research
University of Maryland at College Park
College Park, MD 20740
USA

Vicent Climent
Departament de Ciències Experimentals
Universitat Jaume I
E-12080 Castellón
Spain

N. Cuntoor
Center for Automation Research
University of Maryland at College Park
College Park, MD 20740
USA

Damien Delannay
Communications and Remote Sensing
    Laboratory
Université catholique de Louvain
Louvain, Belgium

Edward J. Delp
School of Electrical and Computer
    Engineering
Purdue University
1285 Electrical Engineering Building
West Lafayette, IN 47907-1285
USA

J.D. Franson
Johns Hopkins University
Applied Physics Laboratory
Laurel, MD 20723
USA

B.C. Jacobs
Johns Hopkins University
Applied Physics Laboratory
Laurel, MD 20723
USA

Bahram Javidi
Electrical and Computer Engineering
    Department
U-157
University of Connecticut
Storrs, CT 06269
USA

A. Kale
Center for Automation Research
University of Maryland at College Park
College Park, MD 20740
USA

Sherif Kishk
Electrical and Computer Engineering
  Department
U-157
University of Connecticut
Storrs, CT 06269
USA

Jesús Lancis
Departament de Ciències Experimentals
Universitat Jaume I
E-12080 Castellón
Spain

Franck Leprévost
Université Joseph Fourier
UFR de Mathématiques
100, Rue des Maths
BP 74, F-38402 Saint-Martin d'Hères
  Cedex
France
and
Centre Universitaire du Luxumbourg
  162 A
Avenue de la Faïencerie
L-1511 Luxembourg, Luxembourg

Eugene T. Lin
School of Electrical and Computer
  Engineering
Purdue University
1285 Electrical Engineering Building
West Lafayette, IN 47907-1285
USA

Benoit Macq
Communications and Remote Sensing
  Laboratory
Université catholique de Louvain
Louvain, Belgium

Lisa M. Marvel
U.S. Army Research Laboratory
APG, MD 21005
USA

Osamu Matoba
Department of Computer and Systems
  Engineering
Kobe University
Rokkadai 1–1
Nada, Kobe 657-8501
Japan

Nasir Memon
Department of Computer and Information
  Science
Polytechnic University
Brooklyn, NY 11201
USA

Thomas J. Naughton
Department of Computer Science
National University of Ireland
Maynooth, Kildare
Ireland

Takanori Nomura
Department of Opto-Mechatronics
Wakayama University
930 Sakaedani, Wakayama 640-8510
Japan

A.N. Rajagopalan
Department of Electrical Engineering
Indian Institute of Technology
Madras, Chennai, 600 036
India

Joseph Rosen
 Department of Electrical and Computer
  Engineering
Ben-Gurion University of the Negev
P.O. Box 653
Beer-Sheva 84105
Israel

Toru Sasaki
Optics Technology Research Center
Canon Inc.
23–10 Kiyohara-Kogyodanchi
Utsunomiya, Tochigi 321–3231
Japan

Enrique Tajahuerce
Departament de Ciències Experimentals
Universitat Jaume I
E-12080 Castellón
Spain

Jun Tanida
Graduate School of Information Science
  and Technology
Osaka University
2-1 Yamadaoka
Suita, Osaka 565–0871
Japan

B. Yegnanarayana
Department of Computer Science and
  Engineering
Indian Institute of Technology
Madras, Chennai, 600 036
India

1

Quantum Cryptography

J.D. Franson and B.C. Jacobs

Summary. Quantum cryptography is a new method for the transmission of secure information whose security is based on the fundamental laws of nature. The basic concepts of quantum cryptography are described along with a discussion of several different methods of implementation. Quantum cryptography systems operating in free space as well as in optical fibers are described. The possibility of future developments, including improved single-photon detectors and quantum repeaters is considered.

Quantum cryptography (Bennett and Brassard, 1984) is an optical technique for the transmission of secure communications. Unlike conventional methods for cryptography, the security of the transmitted information is guaranteed by the laws of nature. The secret key is transmitted in the form of single photons, after which the information itself is encoded and transmitted over an open communications link. Systems for quantum cryptography, also known as quantum key distribution, have now been demonstrated in optical fibers and in free-space communications links.

Quantum cryptography is motivated by the fact that potential vulnerabilities may exist in conventional forms of cryptography. For example, some forms of secure communications are based on the exchange of a secret key that can then be used to encode information at one location and decode it at another location. Vulnerabilities in systems of that kind can arise if the secret key is compromised in some way, either during its transmission or while being stored for future use. Public-key cryptography systems do not rely on the exchange of a secret key but are based instead on the assumed difficulty in performing certain mathematical operations, such as factoring large integers. Systems of that kind may be compromised by the development of quantum computers, which are expected to be able to factor large numbers in an efficient way. Considerable progress has recently been made on an optical approach to quantum computing that makes use of techniques similar to those used in quantum cryptography (Knill, Laflamme, and Milburn, 2001; Pittman, Jacobs, and Franson, 2001, 2002).

Quantum cryptography avoids potential vulnerabilities of that kind by making use of the uncertainty principle of quantum mechanics. Any measurement on a quantum system will produce unavoidable changes in the state of the system. In the case of quantum cryptography, any attempt by an eavesdropper to gain information from the single photons carrying the secret key will produce a change in the state of the photons. Those changes can be identified by the system in the form of an increased error rate. If the raw error rate is below a certain threshold, an eavesdropper cannot have obtained any significant information regarding the secret key, which is then used to encode the information to be transmitted. If, on the other hand, the error rate is above that threshold, the system rejects the secret key, and no information is encoded or transmitted until the error source has been eliminated. Any information that is transmitted in this way is known to be secure, regardless of the technical capabilities of a potential eavesdropper.

The remainder of this chapter will provide an introductory review of the current state of quantum cryptography systems. A more technical review can be found in a recent article by Gisin et al. (2002). The chapter begins with an intuitive discussion of quantum measurements and their uncertainties. The implementation of quantum cryptography systems using the polarizations of single photons is then described, followed by a discussion of systems based on optical interferometers. Free-space quantum cryptography systems that may allow secure communications within a global network of satellites and ground stations will also be discussed. The chapter concludes with a discussion of possible future developments in quantum cryptography, including increased range in optical fibers using quantum repeaters.

1.1 Quantum Measurements and Uncertainties

The uncertainty introduced as a result of a quantum-mechanical measurement can be illustrated by considering a measurement of the polarization of a single photon, as illustrated in Figures 1.1 and 1.2. Polarization measurements of this kind also form the basis of the quantum cryptography systems discussed in the next section.

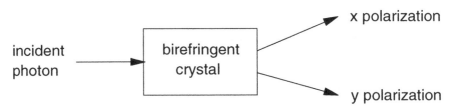

Fig. 1.1. Measurement of the polarization of a single photon using a birefringent crystal.

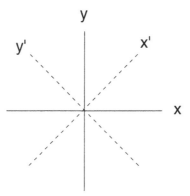

Fig. 1.2. Coordinate frames $x - y$ and $x' - y'$ used for the polarizations of single photons.

The polarization of a beam of light can be measured by passing it through a birefringent crystal, as illustrated in Figure 1.1. If a classical beam of light is passed through such an analyzer, some fraction of the intensity will be transmitted into the upper path corresponding to an x polarization, while the remainder will be transmitted into the lower path corresponding to the y polarization component. A measurement of the two amplitudes (including their relative phase) can determine the state of polarization completely.

The situation is very different, however, if a single photon is incident on a birefringent crystal. The photon must exit from either the upper path or the lower path, and the detection of the photon in one path or the other provides only a single bit of information. As a result, it is not possible to completely determine the state of polarization of a single photon. Furthermore, a measurement of the polarization of a single photon will in general change its state of polarization. This is illustrated in Figure 1.2, which shows two coordinate frames, $x - y$ and $x' - y'$, that are rotated through a 45° angle with respect to each other. If the incident photon was linearly polarized along the x' axis, for example, then after it emerges from the birefringent crystal its polarization will have been changed to either x or y, depending on which path it is in. This is a simple example of the quantum-mechanical uncertainty principle, which states that the measurement of one variable will produce unavoidable uncertainties in other variables. In this example, the initial polarization of the photon was known with certainty in the $x' - y'$ basis, but after the measurement its polarization becomes completely uncertain in that basis.

The inability to completely determine the state of polarization of a single photon forms the basis for one kind of quantum cryptography system. The secret key is transmitted using photons polarized along one of the four axes shown in Figure 1.2, as will be described in more detail in the next section. An eavesdropper can attempt to determine the polarizations of the photons as they pass by using a polarization analyzer and single-photon detector as in Figure 1.1. The detection process destroys the original photon, and the best that an eavesdropper can do is to generate a replacement photon whose polarization is chosen to be the same as the result of the measurement on the original photon. The eavesdropper does not know which coordinate frame to choose for the measurements and will choose the wrong coordinate frame 50% of the time. (For example, the initial photon may have been polarized in the $x' - y'$ basis but the measurement was made in the $x - y$ basis.) When the eavesdropper chooses the wrong coordinate frame, the polarization of the replacement photon will not be the same as that of the incident photon. The errors that result from this process can be identified by the quantum cryptography system, which is thus secure against such an attack, usually referred to as an intercept-and-resend attack.

One might ask whether or not it is possible for the eavesdropper to copy the initial photon, producing two or more identical photons with the same polarization. If so, polarization measurements could be made on the copies to determine their state of polarization, while the original photon could be transmitted onto the quantum cryptography system's receiver with no change in its polarization. It can be shown that the quantum state of a single photon cannot be copied in this way, which is an example of the so-called no-cloning theorem (Wooters and Zurek, 1982).

1.2 Polarization Implementations

The first proposal for quantum cryptography (Bennett and Brassard, 1984) was based on the impossibility of completely determining the state of polarization of a single photon, as described above. The operation of these systems will be described below, while systems based on various optical interferometers will be described in the following section. Polarization-based systems can be implemented using either linear polarizations or a combination of linear and circular polarizations. Here we will describe the use of linear polarizations since it is more straightforward and is the approach that we use in our systems.

Once again, the goal of the system is to transmit a secret key in the form of single photons. The system consists of a transmitter and a receiver connected by an optical fiber or free-space path. The first step in the process is for the transmitter to randomly choose one of the two coordinate frames, $x - y$ or $x' - y'$, shown in Figure 1.2. A single photon whose polarization is randomly chosen along one of the two axes of that coordinate frame is then transmitted through the fiber toward the receiver; the transmitted photon is thus linearly polarized along either the x, y, x', or y' directions. The transmitter does not reveal its choice of polarization basis at this time.

At the receiver, the coordinate frame for the polarization measurement is also chosen at random since the receiver does not know the basis chosen by the transmitter. As a result, the two bases will only be the same 50% of the time.

After the receiver has measured the polarization of the transmitted photon in its basis, the transmitter and receiver openly compare their choice of bases (but not the polarization of the photon). If they happened to choose the same basis, then the results of that event are accepted: an x or x' polarization is taken to represent a bit 0 in the secret key, while a y or y' polarization is taken to represent a bit 1 in the secret key. When they happen to choose different bases, that event is simply discarded. By repeating this process, the transmitter and receiver can accumulate a string of bits that are the same at both ends of the system, and that can subsequently be used as the secret key.

It is important to note that the polarization basis chosen by the transmitter is not revealed until after the photon has passed by the position of any potential eavesdropper. As a result, an eavesdropper does not know the correct basis to use for a measurement of the polarization and will choose the incorrect basis 50% of the time. An intercept-and-resend attack will thus change the polarization of the photon 50% of the time, which will cause the receiver to measure an incorrect polarization 25% of the time. The transmitter and receiver continuously monitor the error rate of the system by sacrificing a small fraction of the bits and openly comparing their values (those bits are not used in the secret key). In this way, they can put an upper bound on the fraction of the bits that an eavesdropper may have learned.

A number of error sources can contribute to an incorrect measurement of the polarization at the receiver even in the absence of an eavesdropper. In our systems, the raw error rate is typically on the order of 0.5%. These errors can be identified and corrected in a secure manner by putting the raw key data into the form of a matrix and then openly comparing the parity of the rows and columns of the matrix. As long as one bit from each row or column is discarded after such a comparison, the eavesdropper does not gain any significant information regarding the values of the remaining bits. By repeating this process, the probability of an error in the remaining bits can be made negligibly small.

Quantum cryptography systems must also deal with the possibility that an eavesdropper may have learned the values of a small number of bits. This can result from the fact that some finite error threshold has to be accepted when attempting to determine whether or not an eavesdropper is present, which would allow for an eavesdropper to intercept a small fraction of the bits. In addition, many systems do not use a true source of single photons. Instead, a classical pulse of light is attenuated to the point where the probability that it will contain even a single photon is small. The probability of there being two photons in the same pulse is even smaller, but the two-photon events provide some opportunity for an eavesdropper to gain information regarding the polarization of one photon while allowing the second photon to be transmitted to the receiver without any change in its polarization. Despite these possibilities, the information available to an eavesdropper can be made exponentially small using a process known as privacy amplification (Bennett, Brassard, and Robert, 1988).

The basic idea in privacy amplification is to replace the original bits in the secret key with the parity of a block of N bits. In order to know the value of the parity, an eavesdropper would have to know the value of all of the N bits, which is extremely unlikely. The information available to an eavesdropper can be made exponentially small by repeating this process a sufficiently large number of times while discarding a small fraction of the original bits.

The first experimental demonstration of quantum cryptography was reported by Bennett and his colleagues in 1992 (Bennett et al., 1992). This was an optical-bench experiment in which photons from an attenuated laser beam were transmitted over a distance of 32 cm. Quantum cryptography can be performed over useful distances in optical fibers, but that is complicated by the fact that the state of polarization of the photons will change as they propagate through a long fiber due to birefringence and other effects. The first experiments with optical fibers were performed around 1993 (Franson and Ilves, 1993, 1994a, 1994b; Muller, Breguet, and Gisin, 1993; Townsend, 1994). We demonstrated a fully automatic system in which two Pockels cells were used to provide real-time compensation for any polarization changes (Franson and Jacobs, 1995). By applying suitable voltages to the two Pockels cells, the transmitter could produce an arbitrary state of polarization. A feedback loop was used to generate four sets of voltages, each of which would produce one of the four polarization states x, x', y, or y' at the output of the fiber, as required for the operation of the system. A small computer at each end of the system automatically performed the necessary error correction and privacy amplification. Photographs of the transmitter and receiver are shown in Figures 1.3 and 1.4.

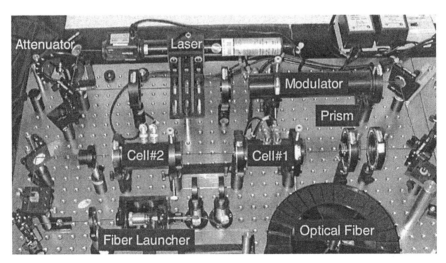

Fig. 1.3. Photograph of the transmitter end of a polarization-based quantum cryptography system developed at the Applied Physics Laboratory.

Fig. 1.4. Photograph of the receiver end of a polarization-based quantum cryptography system.

All-fiber versions of a polarization-based quantum cryptography system have also been developed (Townsend, 1994). The advantages of polarization-based systems include the simplicity of polarization measurements and their relative stability compared with some of the interferometric systems described in the next section. Most free-space systems make use of photon polarizations since there is very little change in polarization for photons propagating through the atmosphere.

1.3 Interferometer Implementations

Quantum cryptography systems based on single-photon polarizations must be able to compensate for changes in the state of polarization during propagation, as described above. This requirement can be avoided in systems that make use of interferometers, in which case the information in the secret key is transmitted in the form of the relative phase between two different single-photon wave packets. The separation in time of the two wave packets is very short, so systems of this kind are relatively insensitive to any time-dependent phase shifts.

Interferometric approaches to quantum cryptography originated with the two-photon interferometer shown in Figure 1.5, which was suggested by one of the authors (Franson, 1989, 1991a). In this nonclassical device, a source emits two correlated photons, each of which travels toward two separated Mach–Zehnder interferometers. The photons are known to have been emitted at the same time, but the time at which they were emitted is totally uncertain in the quantum-mechanical sense. Photon pairs of this kind can be created by passing a laser beam through a nonlinear crystal, which splits individual laser photons into two photons of lower energy in a process known as parametric down-conversion (Shih et al., 1993).

The two interferometers are assumed to have a long path L and a short path S, and the difference in path lengths is chosen to be much larger than the coherence length of the photons. As a result, no interference would be expected classically, and the photons have an equal probability of emerging

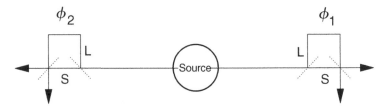

Fig. 1.5. Two-photon interferometer with nonclassical correlations between the output paths chosen by the photons.

from either exit path from the interferometers. It was predicted (Franson, 1989), however, that the random choices of exit paths would be totally correlated if the phase shifts in the two interferometers were the same. Roughly speaking, one photon decides at random which exit path to take, after which the other photon takes the corresponding path. This effect violates Bell's inequality (Bell, 1964), which means that any classical interpretation of these experiments would require the instantaneous transmission of information from one location to the other. It should be emphasized, however, that this does not allow the transmission of messages faster than the speed of light since the choice of exit paths is random and cannot be controlled in order to transmit any information. These effects have now been verified in numerous experiments (Kwiat et al., 1990, 1993; Ou et al., 1990; Brendel et al., 1991, 1992; Franson, 1991; Rarity and Tapster et al., 1992; Shih et al., 1993; Shih and Sergienko, 1994; Strekalov et al., 1996; Tittel et al., 1998) over distances of up to 11 km in optical fibers (Tittel et al., 1998, 1999; Ribordy et al., 2001).

The origin of this effect can be intuitively understood from the interference of quantum-mechanical probability amplitudes. In quantum mechanics, all of the probability amplitudes for the various ways in which an effect can occur must be added together, after which the total probability amplitude is squared to obtain the probability itself. In the operation of these interferometers, only events in which both photons arrive at the detectors at the same time are accepted, while events in which they arrive at different times are ignored. Since the photons were emitted at the same time, they both must have traveled along the longer paths or they both must have traveled along the shorter paths in order for them to arrive at the same time. We will let A_{1L} denote the probability amplitude for photon 1 (the one on the right) to travel along the longer path and arrive at the upper detector, while the probability amplitude for it to travel along the shorter path to the upper detector will be denoted by A_{1S}. The corresponding probability amplitudes for photon 2 will be denoted by A_{2L} and A_{2S}. The total probability P for both photons to arrive at the upper detectors at the same time is then given by

$$P = |A_{1L}A_{2L} + A_{1S}A_{2S}|^2 = A^2|e^{i\Phi_1}e^{i\Phi_2} + 1|^2 \qquad (1.1)$$

Here A is the absolute value of the probability amplitudes, which are assumed to have equal magnitudes, while ϕ_1 and ϕ_2 are the phase shifts along the two longer paths through the interferometers. This can be rewritten in the form

$$P = \alpha \cos^2[(\phi_1 + \phi_2)/2], \qquad (1.2)$$

where α is a constant. It can be seen that interference between the long–long events and the short–short events produces a sinusoidal dependence on the sum of the two phase shifts in the interferometers.

Although it is not possible to transmit any information using this effect, the correlations between the output ports chosen by the two photons can be used to establish a secret key and to implement a quantum cryptography system (Ekert et al., 1992). For example, we can take a photon leaving the upper port of the interferometer to represent a bit 0, while a photon leaving the lower port of the interferometer will represent a bit 1. The operation of the system is completely analogous to that of the polarization-based systems described above, where the phase shifts ϕ_1 and ϕ_2 replace the polarization angles θ_1 and θ_2 of the photons in the earlier approach. Systems of this kind have recently been experimentally demonstrated (Ribordy et al., 2001).

Charles Bennett (Bennett, 1992) realized that similar effects could be obtained if a single photon were transmitted through two interferometers in series, as illustrated in Figure 1.6. This allows the source of correlated pairs of photons to be replaced with an attenuated classical pulse, which is easier to implement. In this arrangement, photons arriving at the detectors will show interference between the long–short and short–long paths through the two interferometers if the timing window is chosen properly. Because of its relative insensitivity to changes in the state of polarization, quantum cryptography systems of this kind have been implemented by a number of groups (Townsend, Rarity, and Tapster, 1993; Townsend, 1994; Marand and Townsend, 1995, Hughes et al., 1996, 2000). Secret-key transmission rates on the order of 10^6 bits per second are achievable over distances of roughly 50 km in optical fibers.

Although the information is not carried in the polarizations of the photons, it is still necessary to control the state of polarization within the two interferometers. In addition, it can be difficult to maintain the difference in path lengths between the two interferometers and to stabilize the phase. These difficulties are largely avoided in a clever variation on the interferometer scheme

Fig. 1.6. A single photon transmitted through two interferometers in series, which has some of the same properties as the two-photon interferometer of Figure 1.5.

of Figure 1.6 that was introduced by Gisin and his colleagues, which they refer to as a "plug-and-play" system (Muller et al., 1997; Zbinden et al., 1997). In a plug-and-play system, the photons are first transmitted through the fiber in one direction, reflected off a Faraday mirror, and then transmitted back through the fiber in the opposite direction. The Faraday mirror changes the state of polarization to the orthogonal polarization, so any birefringence experienced during the first trip through the fiber is canceled by that experienced during the return trip. A single interferometer at the transmitter end performs the function of both interferometers in Figure 1.6, since the photons pass through it in both directions, while the receiver end simply applies a phase shift to one of the two wave packets. Systems of this kind are straightforward to implement and have been developed by several groups (Bethune and Risk, 2000), although they can be sensitive to back-reflection during the first pass through the fiber.

1.4 Free-Space Systems

The range of fiber-based quantum cryptography systems is limited by detector noise and the loss of photons in an optical fiber. Optical amplifiers cannot be used because they would destroy the quantum-mechanical coherence of the single photons; noise-free amplification would effectively copy the photons and is ruled out by the no-cloning theorem discussed above. In addition to the attenuation in the raw data rate, error correction and privacy amplification become increasingly inefficient as the signal-to-noise ratio decreases. Once the bit error rate exceeds 25% due to detector dark counts, it is no longer possible to rule out the possibility of the presence of an eavesdropper. As a result, there is a cutoff range beyond which the secure throughput of a fiber-based system abruptly drops to zero (Lutkenhaus, 2000; Gilbert, Hamrick, and Thayer, 2001).

Quantum cryptography in free space offers a potential solution to the limited range in optical fibers since a network of satellites and ground stations could provide global coverage. The most obvious challenge in building such a system is the enormous number of photons that are present in the environment, which at first may seem to make it impossible to transmit and receive single photons as required for quantum cryptography. The background rate due to ambient photons can be greatly reduced, however, by using a combination of short timing windows, narrow-band filters, and a small acceptance angle.

We demonstrated the first free-space quantum cryptography system in 1995 (Franson and Jacobs, 1996) using the apparatus shown in Figure 1.7. The system was capable of secure communications under daylight conditions over a path length of approximately 75 meters, which was limited by the diameter of the optics that were used. Optical fibers from our polarization-based experiment led into an outdoor unit that launched the photons into a collimated beam using a beam expander. The direction of the beam was controlled by a computer using a motorized mirror mount. The photons were reflected from a retro-reflector located on the roof of an adjacent building and then collected using a small telescope, which coupled them back into an optical fiber. A narrow bandwidth was maintained using a computer-controlled étalon (filter). The system was fully operational and performed nearly as well as our fiber-based system. Other groups have subsequently demonstrated the feasibility of free-space quantum cryptography over ranges up to 2 km (Buttler et al., 2000; Rarity, Tapster, and Gorman, 2001).

There are a large number of trade-offs in the design of a free-space system capable of secure communications with a satellite. For example, the transmission through the atmosphere and the efficiency of the single-photon detectors must both be taken into account when choosing the optimal wavelength. This is illustrated in Figure 1.8, which shows an estimate of the combined detection probability as a function of wavelength, including the detector efficiency, atmospheric transmission, and diffraction effects for half-meter optics. These studies suggest that free-space quantum cryptography could be used to transmit secure information to and from a satellite at useful data rates.

1.5 Future Development

Practical applications of quantum cryptography would benefit from improved performance with regard to data rate and maximum range. Some expected areas of future development are described in this section.

The rate of secure key transmission is currently limited by the dead time of single-photon detectors. The most commonly used detectors in the visi-

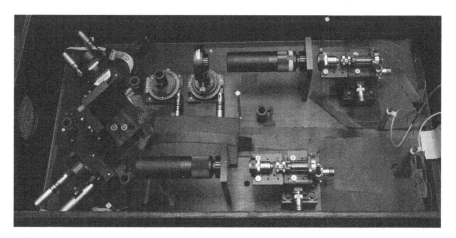

Fig. 1.7. Part of the apparatus used in the first demonstration of quantum cryptography in free space under daylight conditions.

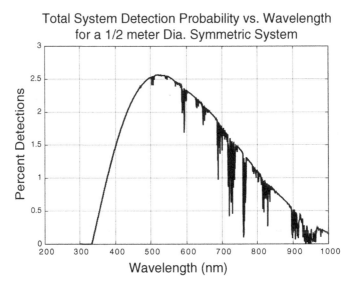

Fig. 1.8. Calculated overall single-photon detection probability for a satellite link with half-meter optics.

ble spectrum are silicon avalanche photodiodes, while operation in the infrared requires the use of InGaAs. Detectors of this kind are currently limited to roughly 5×10^6 photons per second by the time required to quench the avalanche. A new generation of superconducting single-photon detectors may extend the detection rate into the GHz region (Gol'tsman et al., 2001).

The data rate is also limited by the use of highly attenuated classical light pulses rather than true single photons. In order to keep the probability of two-photon events to an acceptably low level, the average number of photons per pulse must be correspondingly low. True single-photon sources are under development by a number of groups, and a quantum cryptography system operating with true single photons has been demonstrated (Beveratos, 2002).

Although optical amplifiers cannot be used, the range of quantum cryptography systems in optical fibers could be extended indefinitely using quantum repeaters (Kok, Williams, and Dowling, 2002). These nonclassical devices overcome the losses in an optical fiber by using quantum teleportation (Bennett et al., 1993) to transmit the photons from one location to another. Quantum teleportation with a maximum efficiency of 50% has been experimentally demonstrated, while more efficient methods should allow 100% efficiency (Franson et al., 2002). The techniques required for quantum repeaters are similar to those required for quantum computing but on a smaller scale. One of the requirements for a quantum repeater is the ability to perform logic operations using single photons. We have recently demonstrated several types of quantum logic operations using linear optical elements (Pittman, Jacobs, and Franson, 2002), and further work in this area may lead to the development of quantum repeaters.

1.6 Summary

Quantum cryptography is an optical technique for the transmission of secure information where the security of the information is guaranteed by the uncertainty principle of quantum mechanics. The need for quantum cryptography may increase if the security of public-key cryptography systems is threatened by the development of quantum computers, for example. In quantum cryptography, a secret key is transmitted in the form of single photons while the message itself is encoded, transmitted over an open communications link, and then decoded. The security of the system is ensured by the fact that an eavesdropper would produce unavoidable changes in the state of the photons that can be detected by the system in the form of an increased error rate. Quantum cryptography has been demonstrated in optical fibers and in free space, and systems can be implemented using the polarizations of single photons or various interferometric techniques. Future developments may include satellite systems, the use of true single-photon sources rather than weak classical pulses, increased transmission rates, and quantum repeaters to extend the range in optical fibers.

Acknowledgments

This work was supported in part by the Office of Naval Research.

References

[1]   Bell, J.S., 1964, *Physics* **1**, 195.
[2]   Bennett, C.H., 1992, *Phys. Rev. Lett.* **68**, 3121.
[3]   Bennett, C.H. and G. Brassard, 1984, In *Proceedings of the International Conference on Computers, Systems, and Signal Processing*, Bangalore, India, December 10–12, pp. 175–179.
[4]   Bennett, C.H., G. Brassard, and J.-M. Robert, 1988, *SIAM J. Comput.* **17**, 210.
[5]   Bennett, C.H., F. Bessette, G. Brassard, L. Salvail, and J. Smolin, 1992, *J. Cryptol.* **5**, 3.
[6]   Bennett, C.H., G. Brassard, C. Crepeau, R. Jozsa, A. Peres, and W.K. Wooters, 1993, *Phys. Rev. Lett.* **70**, 1895.
[7]   Bethune, D. and W. Risk, 2000, *IEEE J. Quantum Electron.* **36**, 340.
[8]   Beveratos, A., R. Brouri, T. Gacoin, A. Villing, J.-P. Poizat, and P. Grangier, 2002, Preprint quant-ph/0206136.
[9]   Brendel, J., J.E. Mohler, and W. Martienssen, 1991, *Phys. Rev. Lett.* **66**, 1142.
[10] Brendel, J., J.E. Mohler, and W. Martienssen, 1992, *Europhys. Lett.* **20**, 575.

[11] Buttler, W.T., R.J. Hughes, S.K. Lamoreaux, G.L. Morgan, J.E. Nordholt, and C.G. Peterson, 2000, *Phys. Rev. Lett.* **84**, 5652.

[12] Ekert, A.K., J.G. Rarity, P.R. Tapster, and G.M. Palma, 1992, *Phys. Rev. Lett.* **69**, 1293.

[13] Franson, J.D., 1989, *Phys. Rev. Lett.* **62**, 2205.

[14] Franson, J.D., 1991a, *Phys. Rev. Lett.* **67**, 290.

[15] Franson, J.D., 1991b, *Phys. Rev. A* **44**, 4552.

[16] Franson, J.D. and H. Ilves, 1993, *Opt. Soc. Am. Tech. Dig.* **3**, 266.

[17] Franson, J.D. and H. Ilves, 1994a, *Appl. Opt.* **33**, 2949.

[18] Franson, J.D. and H. Ilves, 1994b, *J. Mod. Opt.* **41**, 2391.

[19] Franson, J.D. and B.C. Jacobs, 1995, *Electron. Lett.* **31**, 232.

[20] Franson, J.D. and B.C. Jacobs, 1996, *Opt. Lett.* **21**, 1854.

[21] Franson, J.D., M.M. Donegan, M.J. Fitch, B.C. Jacobs, and T.B. Pittman, 2002, *Phys. Rev. Lett.* **89**, 137901.

[22] Gilbert, G., M. Hamrick, and F.J. Thayer, 2001, Preprint quant-ph/0108013.

[23] Gisin, N., G. Ribordy, W. Tittel, and H. Zbinden, 2002, *Rev. Mod. Phys.* **74**, 145.

[24] Gol'tsman, G.N., O. Okunev, G. Chulkova, A. Lipatov, A. Semenov, K. Smirnov, B. Voronov, A. Dzardanov, C. Williams, and R. Sobolewski, 2001, *Appl. Phys. Lett.* **79**, 705.

[25] Hughes, R., G.G. Luther, G.L. Morgan, and C. Simmons, 1996, *Lect. Notes Comput. Sci.* **1109**, 329.

[26] Hughes, R., G. Morgan, and C. Peterson, 2000, *J. Mod. Opt.* **47**, 533.

[27] Knill, E., R. Laflamme, and G.J. Milburn, 2001, *Nature* **409**, 36.

[28] Kok, P., C.P. Williams, and J.P. Dowling, 2002, Preprint quant-ph/0203134.

[29] Kwiat, P.G., W.A. Vareka, C.K. Hong, H. Nathel, and R.Y. Chiao, 1990, *Phys. Rev. A* **41**, 2910.

[30] Kwiat, P.G., A.M. Steinberg, and R.Y. Chiao, 1993, *Phys. Rev. A* **47**, 2472.

[31] Lutkenhaus, N., 2000, *Phys. Rev. A* **61**, 052304.

[32] Marand, C. and P.D. Townsend, 1995, *Opt. Lett.* **20**, 1695.

[33] Muller, A., J. Breguet, and N. Gisin, 1993, *Europhy. Lett.* **23**, 383.

[34] Muller, A., T. Herzog, B. Huttner, W. Tittel, H. Zbinden, and N. Gisin, 1997, *Appl. Phys. Lett.* **70**, 793.

[35] Ou, Z.Y., X.Y. Zou, L.J. Wang, and L. Mandel, 1990, *Phys. Rev. Lett.* **65**, 321.

[36] Pittman, T.B., B.C. Jacobs, and J.D. Franson, 2001, *Phys. Rev. A* **64**, 062311.

[37] Pittman, T.B., B.C. Jacobs, and J.D. Franson, 2002, *Phys. Rev. Lett.* **88**, 257902.

[38] Rarity, J.G. and P.R. Tapster, 1992, *Phys. Rev. A* **45**, 2052.

[39] Rarity, J.G., P.R. Tapster, and P.M. Gorman, 2001, *J. Mod. Opt.* **48**, 1887.

[40] Ribordy, G., J. Brendel, J.-D. Gautier, N. Gisin, and H. Zbinden, 2001, *Phys. Rev. A* **63**, 012309.

[41] Shih, Y.H., A.V. Sergienko, and M.H. Rubin, 1993, *Phys. Rev. A* **47**, 1288.

[42] Shih, Y.H. and A.V. Sergienko, 1994, *Phys. Lett. A* **191**, 201.

[43] Strekalov, ,D.V., T.B. Pittman, A.V. Sergienko, and Y.H. Shih, 1996, *Phys. Rev. A* **54**, R1.

[44] Tittel, W., J. Brendel, and N. Gisin, 1998, *Phys. Rev. A* **57**, 3229.

[45] Tittel, W., J. Brendel, and N. Gisin, 1999, *Phys. Rev. Lett.* **81**, 3563.

[46] Tittel, W., J. Brendel, N. Gisin, and H. Zbinden, 1999, *Phys. Rev. A* **59**, 4150.

[47] Townsend, P.D., 1994, *Electron. Lett.* **30**, 809.

[48] Townsend, P.D., 1998, *IEEE Photonics Technol. Lett.* **10**, 1048.

[49] Townsend, P.D., J.G. Rarity, and P.R. Tapster, 1993a, *Electron. Lett.* **29**, 634.

[50] Townsend, P.D., J.G. Rarity, and P.R. Tapster, 1993b, *Electron. Lett.* **29**, 1291.

[51] Wooters, W.K. and W.H. Zurek, 1982, *Nature* **299**, 802.

[52] Zbinden, H., J.-D. Gautier, N. Gisin, B. Huttner, A. Muller, and W. Tittel, 1997, *Electron. Lett.* **33**, 586.

New OVDs for Personalized Documents Based on Color Holography and Lippmann Photography

Hans I. Bjelkhagen

Summary. Optical variable devices (OVDs), such as holograms, are now common in the field of document security. Up until now mass-produced embossed holograms or other types of mass-produced OVDs are used not only for banknotes but also for personalized documents, such as passports, identification cards, travel documents, driving licenses, credit cards, etc. This means that identical OVDs are used on documents issued to individuals. Today, there is a need for a higher degree of security on such documents and this chapter covers new techniques to make personalized OVDs.

The recent introduction of color holography offers a possibility to apply full color volume reflection holograms in the field of document security. A presentation of the technique and recording materials used in color holography is provided. Another technique, interferential photography of Lippmann photography, represents a new type of OVD, which belongs to the interference security image structures. In this type of photography, color is recorded in a photosensitive film as a black-and-white interference structure. The technique offers additional advantages over holographic labels for unique security applications. The application of the Lippmann OVD for document security and counterfeit-resistant purposes is presented here.

Optical variable devices (OVDs), such as holograms, are now common in the field of document and product security [1]. The OVDs offer improved anticounterfeiting and antiforgery characteristics over earlier devices. OVDs based on light diffraction are called diffractive optically variable image devices (DOVIDs). OVDs based on light interference are known as interference security image structures (ISISs). DOVIDs in the form of embossed rainbow holograms have been used on credit cards since the early 1980s. Later, mass-produced DOVIDs (embossed holograms, kinegrams, or other similar diffractive devices) were applied to banknotes and also to personalized documents such as passports, ID cards, and drivers licenses, for example. However, recently there has been an interest in making new types of OVDs that are unique to a particular document. New OVDs based on holographic and other optical techniques have been developed. Recently, ISISs have appeared on documents (e.g., a monochrome 2-D reflection hologram is used on the German passport, having a unique hologram recorded for each passport).

In this chapter, new optical devices of the ISIS type are introduced. The first one uses full-color reflection holography, a technique that requires more sophisticated equipment and special recording materials to produce OVDs. This type of hologram is more difficult to record and copy than the embossed mass-produced holograms of today. However, using color images in holograms is not a new idea. The embossed holograms, based on the rainbow holography technique, can be recorded so that the final image displays different colors, which change depending on illumination and observation directions. The recording of a color hologram requires at least three different laser wavelengths using special recording materials. For example, a unique color hologram for each passport is a possibility. In the future, mass-produced identical OVDs will most likely no longer be of interest for personalized documents.

The other optical device described here is based on a very old photographic technique invented to record the first color photographs. It is known as a Lippmann photograph and represents a new type of OVD (Lippmann OVD) that which belongs to the ISIS group. The main advantage of this new device is that it can be individually made, providing a unique label for each document, which the issuer of the document can produce in-house. This type of OVD is only intended for personalized security documents.

2.1 Color Holography

In the field of embossed holography, transmission color holograms of the rainbow type have been used in which color-separated photographs or movie or video recordings have been used to produce a 3-D holographic stereogram. For example, Microsoft color security holograms of this type have been attached to software products. However, holographic color images of the rainbow type can provide a correct color image only at a fixed position along a horizontal line in front of the hologram. For security holograms, it may not be a disadvantage that there is a color change when the illumination and observation directions change. However, this type of embossed color hologram is really not much different from any other type of mass-produced rainbow hologram and can be copied in the same way as other types of embossed holograms are copied.

2.1.1 Principle of Color Reflection Holography

Over many years, holographic techniques have been used to produce holograms with different colors, referred to as *pseudocolor or multicolor holograms*. Described in this chapter is *color holography* (full-color or true-color holography). Its applications in the field of document security are of particular interest here. Here, we are concerned with the volume type of hologram, where the recorded information is stored as a refractive-index distribution in

Table 2.1. Characteristics of the Slavich color emulsion.

Silver halide material	PFG-03C
Emulsion thickness	7 μm
Grain size	12–20 nm
Resolution	$\sim 10,000$ lp/mm
Blue sensitivity	~ 1.0–$1.5 \cdot 10^{-3}$ J/cm^2
Green sensitivity	~ 1.2–$1.6 \cdot 10^{-3}$ J/cm^2
Red sensitivity	~ 0.8–$1.2 \cdot 10^{-3}$ J/cm^2
Color sensitivity peaked at:	633 nm and 530 nm

the emulsion. The reflection type of volume hologram is sometimes referred to as a *Lippmann hologram*. A special holographic technique is required to record color 3-D images of objects. The color rendition has to be as close as possible to the color of the real object. This is very important for applications of color holography for display purposes, in museums, and so on. Perfect color rendition in security holograms may be less important. In this case, the complicated recording process and the requirements of special panchromatic recording materials are more important as well as special copying techniques.

Not until panchromatic ultrafine-grain silver halide emulsions were introduced in Russia in the early 1990s was it possible to record high-quality color holograms in a single emulsion layer [2]. Reflection holography can offer full-parallax, large-field-of-view 3-D color images. To be able to record high-quality color reflection holograms, it is necessary to use extremely low-light-scattering recording materials. This means, for example, the use of ultrafine-grain silver halide emulsions (grain size about 10 nm or less). Currently, the only producer of a commercial holographic panchromatic ultrafine-grain silver halide material (glass plates and film) is the Micron branch of the Slavich photographic company located outside Moscow [3]. Some characteristics of the Slavich PFG-03c material are presented in Table 2.1.

By using suitable processing chemistry for the PFG-03c emulsion, it has been possible to obtain high-quality color holograms. These holograms are recorded in a single-layer emulsion, which greatly simplifies the recording process as compared with many earlier techniques.

Another type of panchromatic recording material used for color holography is the photopolymer film from E.I. du Pont de Nemours & Co [4]. In particular, this type of material is suitable for the mass production of color holograms. Although it is less sensitive than the ultrafine-grain silver halide emulsion, it has its special advantages of easy handling and dry processing (only UV-curing and baking). Since this film is used in the field of document security, the panchromatic polymer material is only supplied to specially selected and approved optical security companies. Holograms can be recorded manually, but in order to produce large quantities of holograms, special machines are required. For hologram replication, the scanning technique can provide the highest production rate. In this case, three scanning laser lines are needed, which can be adjusted in such a way that all three simultaneously can scan the film. After the exposure is finished, the film has to be exposed to strong white or UV light. After that, the hologram is treated in an oven in order to increase the brightness of the image.

2.1.2 Laser Wavelengths for Recording Color Holograms

Color reproduction in most imaging techniques is based on Maxwell's three-color RGB principle. It may seem that the main aim of choosing the recording wavelengths for color holograms would be to cover as large an area of the chromaticity diagram as possible. However, there are many other considerations that must be taken into account when choosing the optimal wavelengths for color holograms. One of these important considerations is the question of whether three wavelengths are really sufficient for color holography. For perfect color rendition, four or more wavelengths may be needed. However, in the field of document security, perfect color rendition may not be necessary since document holograms in most cases will not be correctly illuminated. The wavelength selection problem has been discussed in several papers (e.g., [5], [6]).

2.1.3 Setup for Recording Color Holograms

A typical reflection hologram recording setup is illustrated in Figure 2.1. The different laser beams necessary for the exposure of the object pass through

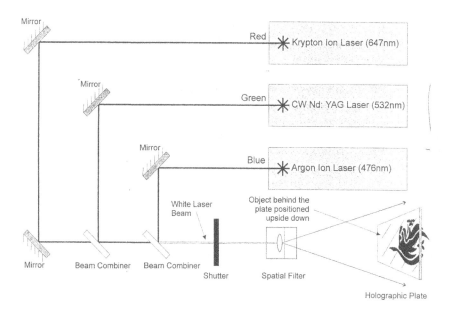

Fig. 2.1. The setup for recording color reflection holograms.

the same beam expander and spatial filter. A single-beam Denisyuk arrangement is used (i.e., the object is illuminated through the recording holographic plate). The light reflected from the object constitutes the object beam of the hologram. The reference beam is formed by the three expanded laser beams. This "white" laser beam illuminates both the holographic plate and the object itself through the plate. Each of the three primary laser wavelengths forms its individual interference pattern in the emulsion, all of which are recorded simultaneously during the exposure. In this way, three holographic images (a red, a green, and a blue image) are superimposed upon one another in the emulsion.

Three laser wavelengths are employed for the recording: 476 nm, provided by an argon ion laser, 532 nm, provided by a cw frequency-doubled Nd:YAG laser, and 647 nm, provided by a krypton laser. Two dichroic filters are used to combine the three laser beams. The "white" laser beam goes through a spatial filter, illuminating the object through the holographic plate.

By using the dichroic filter beam combination technique, it is possible to perform simultaneous exposure recording, which makes it possible to control independently the RGB ratio and the overall exposure energy in the emulsion. The RGB ratio can be varied by individually changing the output power of the lasers, while the overall exposure energy is controlled solely by the exposure time.

Color reflection holograms can also be produced using copying techniques so that a projected real image can be obtained; however, they are normally associated with a restricted field of view. For many display purposes, the very large field of view obtainable in a Denisyuk color hologram is often more attractive.

2.1.4 Processing of Color Holograms

The dry processing of color holograms recorded on photopolymer materials has already been described. The process is simple and very suitable for machine processing using, for example, a baking scroll oven. The processing of silver halide emulsions is more difficult and critical. The Slavich emulsion is rather soft, and it is important to harden the emulsion *before* the development and bleaching takes place. Emulsion shrinkage and other emulsion distortions caused by the active solutions used for the processing must be avoided. In particular, when recording master color holograms intended for photopolymer replication, shrinkage control is extremely important. The processing steps are summarized in Table 2.2.

It is very important to employ a suitable bleach bath to convert the developed silver hologram into a phase hologram. The bleach must create an almost stain-free clear emulsion so as not to affect the color image. In addition, no emulsion shrinkage can be permitted, as it would change the colors of the image. Washing and drying must also be done so that no shrinkage occurs.

Table 2.2. Color holography processing steps and duration.

1. Tan in a formaldehyde solution	6 min
2. Short rinse	5 sec
3. Develop in the CWC2 developer	3 min
4. Wash	5 min
5. Bleach in the PBU-amidol bleach	~ 5 min
6. Wash	10 min
7. Soak in acetic acid bath (printout prevention)	1 min
8. Short rinse	1 min
9. Wash in distilled water with wetting agent added	1 min
10. Air dry	

A suitable spotlight to reconstruct color holograms is a 12-volt 50-watt halogen lamp. The selection of a lamp for the reconstruction of color holograms is much more important than the selection of lamps for monochrome holograms. The color balance for the recording of a color hologram must be adjusted with the type of spotlight that is going to be used for the display of the finished hologram in mind. Figure 2.2 shows a typical normalized spectrum obtained from a white area of the color-test target hologram.

This means that the diffraction efficiency of each color component is obtained assuming a flat spectrum of the illuminating source. One should note the high diffraction efficiency in blue, which is needed to compensate for the rather low blue-light emission of the halogen spotlight. The noise level, mainly in the blue part of the spectrum, is visible and low. The three peaks are exactly

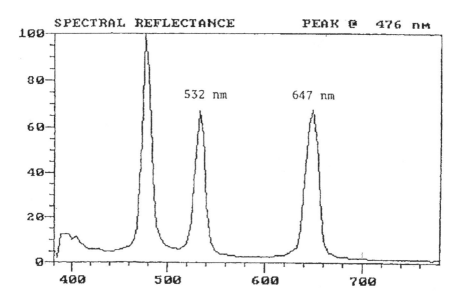

Fig. 2.2. Normalized spectrum from a white area of a color test-target hologram.

Fig. 2.3. Photograph of a color hologram.

at the recording wavelengths (i.e., 647, 532, and 476 nm). A color reproduction of a color hologram by the author is presented in Figure 2.3.

2.1.5 Computer-Generated Color Reflection Holograms

Today it is not possible to obtain a computer-generated hologram (CGH) with the same huge information content as the laser-recorded ones of real objects. What may become possible in the future is a technique to compute and record the interference pattern that is stored in a Denisyuk color hologram, which upon illumination can produce an image like the laser-recorded ones of real objects. Therefore, the best compromise today is to use holographic stereograms, which can provide high-quality computer-generated images. A holographic stereogram is created by using a series of 2-D photographic images or 2-D images displayed on an LCD screen, from which the hologram is recorded. In order to obtain a high-quality holographic image with a large field of view, many 2-D images are needed. There has been rapid progress in color CGHs of the reflection type over the last few years. Remarkable results have been achieved by Klug et al. [7] at Zebra Imaging Inc. in the USA. A new technique to record full-parallax color reflection CGHs has been developed. Color holograms can be produced having both vertical and horizontal parallaxes with a 100° field of view. The generation of a holographic hard copy of either digitized images or computer graphics models is based on the following technique. The "object" subbeam is directed through a sequence of digital images on a liquid-crystal screen. Each resulting exposure, about two millimeters square, is called a "hogel." The full-color hogels are the holographic building blocks of a finished CGH image recorded on DuPont panchromatic photopolymer film by simultaneous RGB laser exposures.

2.1.6 Applications of Color Holograms in the Field of Document Security

Currently, the applications of color reflection holograms are limited. One reason is that the cost of producing color holograms is higher than the cost of monochrome hologram production. There is still more research and development work needed to make the technique easier to employ and less expensive to increase its use for security applications. The first company producing color reflection holograms is DAI Nippon Printing Co., Ltd. (DNP) in Japan [8,9]. In 1998, as the first company in the world and still the only one, DNP launched the mass-produced color holograms recorded in DuPont's panchromatic photopolymer material. DNP has several patents on the mass-production process [10]. These mass-produced holograms, called True Image[TM] were mainly manufactured as decorative holograms. Later, DNP introduced security color holograms under the name of Secure Image[TM], which happened in 2001. There are two types of Secure Image holograms: a transparent one and one that includes a reflective layer. Overlaminating the transparent type to a photo on an ID card will prevent card counterfeiting and ID photo tampering.

One example of the security application is a DNP hologram produced for the company Venture 21 (R&B 21 Group) in Japan. The holograms were for authentication of important rated trading cards. A photograph of this security hologram is shown in Figure 2.4.

In addition to the security feature, DNP can also include unique design elements of full-color–full-3-D impression in their holograms. Due to the easy recognition of DNP full-color holograms, it is easy to detect counterfeit holograms, such as embossed or monochrome volume holograms.

2.2 Lippmann Photography

After the invention of black-and-white photography in the 19th century, a lot of research was devoted to the possibility of recording natural color images. In

Fig. 2.4. DAI Nippon color security hologram for trading cards (*Venture 21*, R&B 21 Group in Japan). (Reprinted with permission from DAI Nippon.)

1891, Gabriel Lippmann invented a technique known as interferential photography, or interference color photography, for directly recording the first color photographs. Lippmann's work on what also has become known as *Lippmann photography* was published in several papers [11–13]. In this type of photography, color is recorded on a photosensitive film as a black-and-white interference structure. The direct color-recording technique was extremely interesting from a scientific point of view; however, it was not very effective for color photography since the technique was complicated. The fact that the color photographs could not be copied was another contributing factor that prevented Lippmann photography from becoming a practical photographic color-recording method at that time. However, 100-year-old Lippmann photographs are very beautiful, having extremely high resolution and good color contrast. The fact that the colors of the early Lippmann photographs are well-preserved indicates that their archival properties are very good. Gabriel Lippmann was awarded the Nobel Prize in Physics in 1908 for his invention. The limitations of the Lippmann technique have now become important advantages of the new type of document-security device, the *Lippmann OVD*, presented here.

2.2.1 Principle of Lippmann Photography

The original principle of Lippmann photography is shown in Figure 2.5. The demand for very high resolving power for recording Lippmann photographs

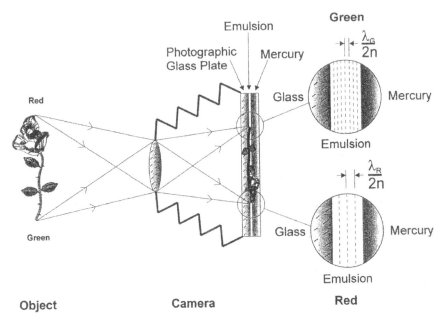

Fig. 2.5. Principle of Lippmann photography.

means that the recording material had a rather low light sensitivity. The photosensitive emulsion coated on Lippmann plates was brought in contact with a surface of high reflectivity. Lippmann used mercury in contact with the emulsion. This mercury mirror reflects the light back into the emulsion, which then interferes with the light coming from the other side of the emulsion. Standing waves of the interfering light produce a very fine fringe pattern throughout the emulsion with a periodic spacing of $\lambda/(2n)$ that is recorded (λ is the wavelength of light in air and n is the refractive index of the emulsion). The color information concerning the object is recorded in this way. For example, a large separation between the fringes in the emulsion indicates that the recorded wavelength is located at the red end of the spectrum. More closely spaced fringes indicate a shorter wavelength, such as green or blue. This description is correct only when monochromatic colors are recorded. A polychrome recording is more complex, and the interference pattern is located only in a very thin volume close to the emulsion surface.

When the developed Lippmann photograph is viewed in white light, different parts of the recorded image produce different colors due to the separation of the recorded fringes in the emulsion. The light is reflected from the fringes, creating different colors corresponding to the original ones that produced them during the recording. It is obvious that there is a demand for extremely high resolving power of the recording material to record fringes separated on the order of half the wavelength of the light. It is also clear that the processing of these plates is critically important, as any change in separation between the recorded fringes would affect color reproduction. To observe the correct colors, the illumination and observation must be at normal incidence. If the angle changes, the color of the image will change. This change of color, known as iridescence, is a very important feature of the Lippmann photograph as a document-security device. The Lippmann image is recorded as a Bragg structure in the film. Modern holography shows similarities to Lippmann photography. In both cases, an interference pattern is recorded in a high-resolution emulsion. The Bragg diffraction regime applies to both categories. The fundamental difference is that, in the Lippmann case, there is no phase recording involved; the recorded interference structure is a result of phase-locking of the light by the reflecting mirror. In holography, the phase information is actually recorded, being encoded as an interference pattern created between the light reflected from the object and a coherent reference beam.

2.2.2 Modern Lippmann Photography

Progress in the development of high-resolution panchromatic photosensitive recording materials has opened up new possibilities to investigate the old Lippmann photography technique again. New and improved recording materials combined with special processing techniques can make Lippmann photography an interesting imaging technique for the optical document-security market. In particular, the ultrahigh image resolution, the Bragg sensitivity of the image, the negative/positive image switch, the archival quality, the copying problem, and the possibility of individually recorded OVDs make them suitable for security applications. For this application, a special recording–processing device must be designed and manufactured in which a roll of the photosensitive film, laminated with a reflecting foil, is exposed. The principle of the system is shown in Figure 2.6.

Special illuminating lamps, such as strong halogen spotlights, are needed to record the security documents with a reasonably short exposure time. Strong flash light can also be considered. There are two modern recording materials suitable for Lippmann photography: panchromatic ultrafine-grain silver halide materials and panchromatic photopolymer materials. Silver halide materials require wet processing. Some photopolymer materials require only dry processing. This fact makes photopolymers suitable for Lippmann OVDs. A special type of panchromatic photopolymer material from E.I. du Pont de Nemours & Co. is required for Lippmann photography.

2.2.3 Brief Description of the Recording Technique

A Lippmann photograph can be recorded on a photopolymer material in the following way. The photosensitive polymer layer has to be rather thin, on the order of only a few micrometers. The photopolymer layer must be coated on

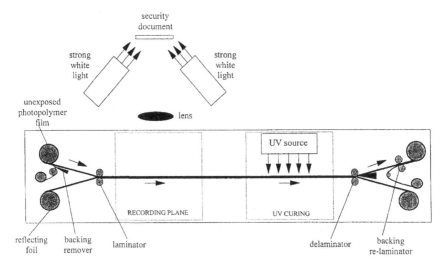

Fig. 2.6. Schematic of Lippmann OVD recording machine.

a flexible transparent base, and a special type of reflecting foil has to be laminated onto the photosensitive polymer layer in perfect contact with it. The polymer side is facing the camera lens during exposure. Experimental materials manufactured by DuPont were used for the initial recording experiments [14]. The polymer film laminated to the reflecting foil must be exposed in a special camera. If the recording material is not perfectly isochromatic, a correction filter may be needed in front of the camera lens to obtain the correct color balance.

After being exposed to the image-forming information, the film must be processed. The reflecting foil is detached from the photopolymer film, and the photopolymer layer is exposed to strong white light or UV light for developing. After that, the photograph is put in an oven for a certain time to increase the brightness of the image. The whole processing technique of the photopolymer film is completely dry. Based on this fact, an automatic recording and processing machine for the recording of Lippmann security labels can be developed. After being processed, the transparent photopolymer label is laminated to its corresponding security document. The polymer film contains no dyes or any fading chemicals, which means that the archival stability is expected to be very high. The photograph is simply a piece of plastic material with the information recorded in it as an optical phase structure (refractive index variations within the photopolymer layer).

2.2.4 Security Applications of Lippmann Photographs

The Lippmann OVD is still under development. No commercial applications exist. Sample documents with Lippmann OVDs have been produced. The application of a Lippmann photograph on a security document can be performed in the following way. The first step is the actual photographic recording of the document information (e.g., printed text, codes, signature, color photograph), similar to conventional photography, using a recording device. After being exposed, the film must be processed (dry processing for photopolymer materials, wet processing for silver halide materials). An automatic processing machine is required for this purpose. Actually, both the recording and the processing can be performed on the same piece of equipment. After being processed, and if desired, the recorded images can be laminated to a black backing plastic foil. Then the backed or unbacked Lippmann OVDs are laminated to their corresponding security documents.

Referring to Figure 2.7, in which a Lippmann photograph (1) is attached to a security document (2), the color of the image in the Lippmann photograph varies depending on the angle of observation. Perpendicular observation (3) gives the correct color image, oblique observation (4) shifting the colors toward shorter wavelengths. This feature, known as the Bragg sensitivity, makes it impossible to replace a Lippmann color photograph with a conventional color photograph; the switch would be easily detected. In addition, a security document with a Lippmann photograph cannot be scanned or copied in a color copier or scanner. The Bragg selectivity of the Lippmann OVD makes it a unique type of photograph, very different from modern conventional color photographs.

An interesting feature of the Lippmann photograph is that it can record latent images often used on security documents. A latent image is a pattern of parallel intaglio lines containing a foreground and a background. Latent images are concealed when observed at a normal angle of view and are increasingly perceptible with the acuteness of the viewing angle. Despite the fact that the Lippmann photograph is recorded at normal incidence, the latent image will be recorded in the photograph and is visible when the photograph is illuminated and observed at oblique angles.

In Figure 2.8, a sample U.S. passport is featured with a Lippmann OVD in the upper right-hand corner. In this case, only a black-and-white negative image is visible. The negative Lippmann OVD is also seen in Figure 2.9a. When observed under perpendicular diffuse light, the color image in the Lippmann OVD is revealed, as shown in Figure 2.9b. Although the Lippmann OVD can be seen in any diffused light, it is convenient to have a small illuminating device for fast verification of the document. It is an illuminated diffuser mounted upside-down in a stand that can be placed on a table. Such a device is shown in Figure 2.10.

2.2.5 Advantages of the Lippmann OVD

Currently, holograms are common in the field of document security, where mass-produced embossed holograms are attached to many types of security documents and credit cards. In almost every case where holograms are used,

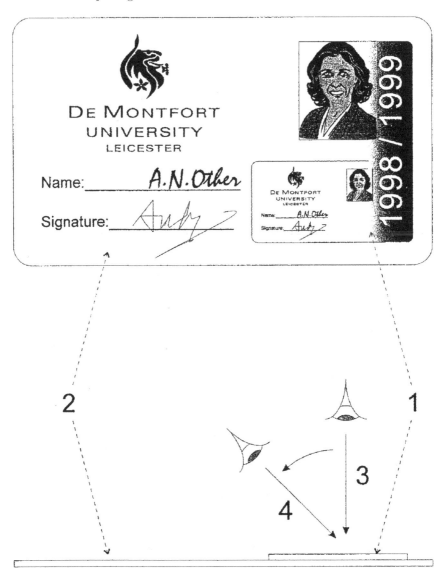

Fig. 2.7. A security document with a Lippmann OVD. **1**: Lippmann OVD; **2**: ID card; **3**: Perpendicular observation (correct color image); **4**: Oblique observation (image changing color).

Fig. 2.8. Lippmann OVD attached to a sample U.S. passport.

(a)   (b)

Fig. 2.9. The Lippmann OVD on the U.S. passport. (**a**) Negative image of the passport page; (**b**) color image of the passport page.

exactly the same hologram image is attached to a large quantity of security documents of the same type (e.g., the embossed dove hologram on the VISA credit cards). Since holograms are difficult to manufacture and lasers are required for the actual recording of a hologram, the use of holograms has been a valuable security device over many years. Nowadays, however, it is possible to copy holograms, and there are examples of illegally copied security holograms reported. Nevertheless, a hologram is a very valuable OVD for mass-produced security instruments such as banknotes, cheques, vehicle stickers, and product labels, for example. Lippmann photography offers a new type of optical security device that is unique and can be individually produced for each security

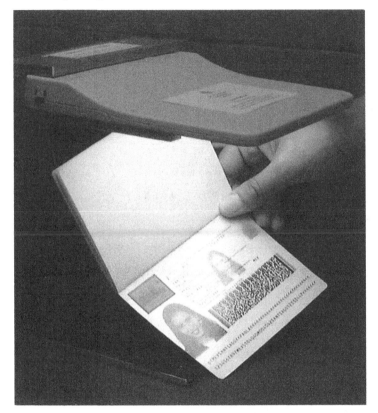

Fig. 2.10. Illuminating device for Lippmann OVDs.

document issued. Some of the advantages of a Lippmann OVD as a security device are:

1. Automatic recording and processing equipment for Lippmann OVDs can be manufactured to be used by security document producers and institutions issuing security documents.
2. The recording is rather simple to perform; no specially equipped laboratory is required.
3. The access to the recording photosensitive film (e.g., the special photopolymer materials) can be strictly controlled by the manufacturer of the film. Only approved producers of security documents and institutions issuing such documents can order the material from the film manufacturer.
4. The Lippmann OVD has a very high archival stability.
5. The Lippmann OVD is Bragg-sensitive, which means it changes its color depending on the angle of illumination and observation.
6. The Lippmann OVD can record latent images, which can be used as security devices on documents.

7. The Lippmann OVD cannot be copied by conventional color photography, nor can it be copied on color copy machines.
8. The Lippmann OVD of the phase type is completely transparent. The visibility of the image depends on the illumination. It can be laminated to a security document in such a way that printed information or other information can be visible through the Lippmann film.
9. The Lippmann OVD can be laminated to a light-absorbing material (e.g., black plastic foil), which means that it is not possible to see through the Lippmann film.
10. Since the resolution of the Lippmann OVD is extremely high, a reduced image of the security document can be laminated to the document, occupying only a limited area of it. In this case, magnifying techniques may be necessary to be able to read all the recorded information in the Lippmann image.

There are many potential applications of Lippmann photography for security and counterfeit-resistant purposes. For example, a Lippmann OVD of a passport page can be recorded, including specific information about the individual, the signature, and the conventional color photograph. Then, a reduced-size Lippmann image is laminated to the page at an appropriate place. The color shift of the Lippmann OVD indicates that it is a genuine Lippmann photograph and not a conventional photograph. In addition, all the information recorded in this OVD can be compared with the corresponding information in the document itself. It is a very difficult process to go through if someone wants to tamper with Lippmann-protected documents.

An alternative is to use a transparent Lippmann OVD as an over-laminated structure on a security document. For example, the transparent Lippmann OVD can be laminated over a signature or other important information of the security document.

The authenticity of a Lippmann OVD is easy to verify simply by looking at it. The most obvious feature is the switch between a black-and-white negative image and the positive color image. However, it is also possible to make automatic inspection equipment that can check the iridescence of the image or compare the high-resolution information recorded on the document itself with the corresponding information stored in the Lippmann OVD.

In principle, Lippmann photography is not limited to the visible part of the spectrum. It is possible to record a Lippmann security label in IR or UV, that can be "seen" only under infrared or ultraviolet light. In visible light, it would look like only a transparent foil laminated to the document. However, the IR version may show a red image under oblique illumination and observation.

2.2.6 How Secure Is the Lippmann OVD?

As mentioned in the introduction, a Lippmann photograph is almost impossible to copy. However, since Lippmann OVDs are unique to each document, there is never an interest in copying a Lippmann OVD. The only concern here is when a tampered or falsified security document has to be provided with a fake Lippmann OVD. Any normal color photograph is impossible to use, it will be detected at once. Instead, in this case it is necessary to record a Lippmann photograph of the tampered or falsified security document. Of course, it is possible for somebody familiar with Lippmann photography to record such an image, assuming at the same time that he or she can get access to the recording equipment and the unique recording material for the Lippmann security process (e.g., the special photopolymer material). Since the Lippmann process is extremely difficult to get working, it will take a very long time to learn how to do it if one has no knowledge or practice about the technique.

With regard to the current holograms used as security devices, a person familiar with holography can also produce a hologram or copy holograms provided he has access to the material and equipment. In this case, it is more attractive to copy holograms since large quantities of fake documents containing the same hologram can be made when the hologram has been copied.

Currently, the new optical security Lippmann process is being developed. It is protected by a U.S. patent, and European patents are pending [15]. A more detailed paper on this application was published in SPIE's Optical Engineering journal in the January 1999 special issue on optical security [16].

References

[1]   R.L. van Renesse, ed., *Optical Document Security*, 2nd ed. (Artech House, Boston and London, 1998).

[2]   H.I. Bjelkhagen, T.H. Jeong, and D. Vukiĕvić, "Color reflection holograms recorded in a panchromatic ultrahigh-resolution single-layer silver halide emulsion," *J. Imaging Sci. Technol.* **40**, 134–146 (1996).

[3]   Slavich Joint Stock Co., Micron Branch Co., 2 pl. Mendeleeva, 152140 Pereslavl-Zalessky, Russia.

[4]   S.H. Stevenson, "DuPont multicolor holographic recording film," in *Practical Holography XI and Holographic Materials III*, S.A. Benton and T.J. Trout, eds., Proceedings of SPIE **3011**, pp. 231–241 (SPIE, Bellingham, WA, 1997).

[5]   M.S. Peercy and L. Hesselink, "Wavelength selection for true-color holography," *Appl. Opt.* **33**, 6811–6817 (1994).

[6]   T. Kubota, E. Takabayashi, T. Kashiwagi, M. Watanabe, and K. Ueda, "Color reflection holography using four recording wavelengths," in *Practical Holography XV and Holographic Materials VII*, S.A. Benton, S.H. Stevenson, and T.J. Trout, eds., Proceedings of SPIE **4296**, pp. 126–133 (SPIE, Bellingham, WA, 2001).

[7]   M. Klug, "Display applications of large scale digital holography," in *Holography: A Tribute to Yuri Denisyuk and Emmett Leith*, H.J. Caulfield, ed., Proceedings of SPIE **4737**, 142–149 (SPIE, Bellingham, WA, 2002).

[8]   M. Watanabe, T. Matsuyama, D. Kodama, and T. Hotta, "Mass-produced color graphic arts holograms," in *Practical Holography XIII*, S.A. Benton, ed., Proceedings of SPIE **3637**, 204–212 (SPIE, Bellingham, WA, 1999).

[9]   D. Kodama, M. Watanabe, K. Ueda, "Mastering process for color graphics arts holograms," in *Practical Holography XV and Holographic Materials VII*, S.A. Benton, S.H. Stevenson, and T.J. Trout, eds., Proceedings of SPIE **4296**, 198–205 (SPIE, Bellingham, WA, 2001).

[10] Dai Nippon. U.S. patents on the mass-production process: 5798850, 5755919, 5843598, 5993600, 5504593.

[11] G. Lippmann, "La photographie des couleurs," *C.R. Acad. Sci.* **112**, 274–275 (1891).

[12] G. Lippmann, "La photographie des couleurs [deuxième note]," *C.R. Acad. Sci.* **114**, 961–962 (1892).

[13] G. Lippmann, "Sur la théorie de la photographie des couleurs simples et composées par la méthode interférentielle," *J. Phys.* **3** (3), 97–107 (1894).

[14] H.I. Bjelkhagen, "Lippmann photographs recorded in DuPont color photopolymer material," in *Practical Holography XI and Holographic Materials III*, S.A. Benton and T.J. Trout, eds., Proceedings of SPIE **3011**, 358–366, (SPIE, Bellingham, WA, 1997).

[15] H.I. Bjelkhagen, "Secure photographic method and apparatus," U.S. patent No. 5,972,546, October 26, 1999.

[16] H.I. Bjelkhagen, "New optical security device based on one-hundred-year-old photographic technique," *Opt. Eng.* **38**, 55–61 (1999).

3

Distortion- and Noise-Robust Digital Watermarking Using Input and Fourier-Plane Phase Encoding

Sherif Kishk and Bahram Javidi

Summary. In this chapter, we propose a technique for information hiding using input and Fourier-plane phase encoding. The proposed method uses a weighed double-phase-encoded hidden image embedded within a host image as the transmitted image. We develop an analytical presentation for the system performance using the statistical properties of double-phase encoding. The peak signal-to-noise ratio (PSNR) metric is used as a measure for the degradation in the quality of the host image and the recovered hidden image. We test, analytically, the distortion of the hidden image due to the host image and the effect of occlusion of the pixels of the transmitted image (that is, the host image containing the hidden image). Moreover, we discuss the effect of using only the real part of the transmitted image to recover the hidden image. Computer simulations are presented to test the system performance against these types of distortions. Simulations illustrate the system's ability to recover the hidden image under distortions and the robustness of the hidden image against removal trials.

3.1 Introduction

Digital watermarking and information hiding [1] can be considered as a method for protecting data from unauthorized distribution. Digital watermarking has many applications; for example, it can be used in ownership affirmation by adding a watermark to the image to be protected using a code that is known only by the author, and if someone claims the ownership of the image, he or she should be able to recover the hidden image. In this application, the watermark should be robust to intended destruction and removal trials. Another application of digital watermarking is in copy-prevention systems by developing a copying machine that detects the watermark and rejects any copying process if the document is not authentic. There are also so many other applications for digital watermarking, such as identification card verification and fraud detection. Several information-hiding and watermarking techniques have been proposed [2–5].

In digital image watermarking, an image is embedded within another image, referred to as the host, such that the host image does not suffer from severe degradation. An information-hiding system should satisfy a number of conditions. For example, the embedded data should be robust against modification trials, signal-processing operations, and removal attacks, and the embedded data should be hidden from the human eye.

Information-hiding systems can be categorized as either spatial-domain systems or Fourier-domain systems, according to the domain in which the image is embedded. Spatial-domain systems are easy to implement but suffer from the degradation in the host-image quality, and they are not robust against signal-processing operations such as compression and filtering. Fourier-domain filtering is more robust to signal-processing operations, but the hidden image may be easier to remove from the watermarked image.

In this chapter, we use double-phase encoding [6] to transform the watermarked image and embed it into the host image. This method uses two random phase codes—one in the input plane and the other in the Fourier plane—to convert the input image into a white-sense stationary noise. Because of this property, it is impossible to recover the hidden image using one of the phase-recovery algorithms [7]. Also, it was shown that double-phase encoding is robust to different types of noise and distortion [8,9,10].

We introduce a spatial-domain information-hiding [11] technique using double-phase encoding. This method will fix some of the problems of spatial-domain data-embedding algorithms by controlling the amplitude of the embedded image to reduce the undesirable effects on the host image. Also, because the double-phase-encoded hidden image is white-noiselike, any trial to remove the embedded image will result in damage to the host image. Moreover, it is difficult to recover the hidden image without knowing the phase keys because double-phase encoding is robust to blind-retrieval trials [6].

The chapter is organized as follows. Double-phase encoding and its properties are discussed in Section 3.2. In Section 3.3, we discuss the proposed information-hiding algorithm. In Section 3.4, we discuss the proposed system under several types of distortion. Computer simulations are presented in Section 3.5. Finally, our conclusion is presented in Section 3.6.

3.2 Double-Phase Encoding

Let the image to be encoded, $f(x,y)$, have a size of $M \times N$ pixels. Let (x,y) denote the spatial coordinates and (v,w) denote the coordinates in the Fourier domain. $f(x,y)$ is normalized to have a maximum value of one, and $\psi(x,y)$ is the double-phase-encoded image. Let $p(x,y)$ and $b(v,w)$ be two independent white sequences uniformly distributed from zero to one. The double-phase-encoded image, $\psi(x,y)$, is given by

$$\psi(x,y) = \{f(x,y)\exp[j2\pi p(x,y)]\} \otimes h(x,y), \qquad (3.1)$$

where $h(x, y)$ is the impulse response of $H(v, w) = \exp[j2\pi b(v, w)]$, and the symbol $\otimes$ stands for convolution.

To decode a double-phase-encoded image, the Fourier transform of the encrypted image is multiplied by $\exp[-j2\pi b(v, w)]$, then inverse Fourier-transformed, which produces $f(x, y)\exp[j2\pi p(x, y)]$. Finally, $f(x, y)$ can be recovered by multiplying it by $\exp[-j2\pi p(x, y)]$ or by just computing the magnitude of $f(x, y)\exp[j2\pi p(x, y)]$ if $f(x, y)$ is positive.

3.2.1 Properties of Double-Phase Encoding

In this section, we will discuss some of the properties of double-phase encoding that are useful for our analysis.

The encoded image is white and stationary. This property can be proved by calculating the autocorrelation function for $\psi(x, y)$; that is, $E[\psi^*(x, y)\psi(x + \tau, y + \beta)]$, where E stands for the expected value, and τ and β are shifts in the spatial domain.

Using Eq. (3.1), we can write

$$\psi(x, y) = \sum_{\eta=0}^{N-1}\sum_{\xi=0}^{M-1} f(\eta, \xi)\exp[j2\pi p(\eta, \xi)]h(x - \eta, y - \xi), \qquad (3.2)$$

and from Appendix A we obtain the autocorrelation of $\psi(x, y)$,

$$E[\psi^*(x, y)\psi(x + \tau, y + \beta)] = \frac{1}{N \times M}\left[\sum_{\eta=0}^{N-1}\sum_{\xi=0}^{M-1}|f(\eta, \xi)|^2\right]\delta(\tau, \beta), \qquad (3.3)$$

where δ is the Kronecker delta function.

It is evident from Eq. (3.3) that a double-phase-encoded image is white with zero mean and variance of

$$\sigma_\psi^2 = \frac{1}{N \times M}\left[\sum_{\eta=0}^{N-1}\sum_{\xi=0}^{M-1}|f(\eta, \xi)|^2\right]. \qquad (3.4)$$

3.3 Data Hiding Using Double-Phase Encoding

Using the same notation adopted in the previous section, we denote the host image by $C(x, y)$ and the watermarked image (the host containing the hidden image) by $I(x, y)$. Using this notation, the watermarked image will be given by

$$I(x, y) = \alpha\psi(x, y) + C(x, y), \qquad (3.5)$$

where α is an arbitrary constant chosen to assure the invisibility of the hidden image and the robustness of the hidden image against distortions. Figure 3.1

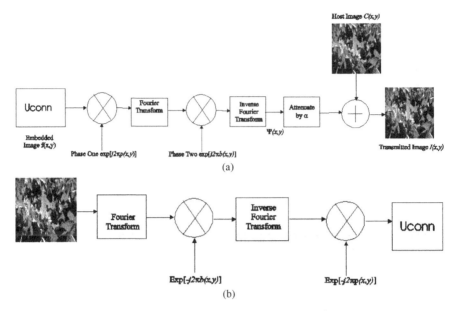

Fig. 3.1. Block diagram of the proposed image-hiding technique using double-phase encoding: (**a**) transmitter; (**b**) receiver.

shows a block diagram for the coder and the decoder of the proposed double-phase-encoding information-hiding system.

It is clear that $I(x, y)$ is complex. The real part is the host image plus the real part of the encoded hidden image. The imaginary part is the imaginary part of the encoded hidden image.

To recover the hidden image, we obtain the Fourier transform of $I(x, y)$, multiply it by $\exp[-j2\pi b(v, w)]$, then inverse Fourier transform and multiply by $\exp[-j2\pi p(x, y)]$. The decoded hidden image, $\tilde{f}(x, y)$, will have the form

$$\tilde{f}(x, y) = \alpha f(x, y) + \text{IFT}\{\hat{C}(v, w) \exp[-j2\pi b(v, w)]\} \exp[-j2\pi p(x, y)], \quad (3.6)$$

where $\hat{C}(v, w)$ is the Fourier transform of $C(x, y)$, and IFT stands for the inverse Fourier transform. The second term of the right-hand side of Eq. (3.6) can be written as

$$\Delta f(x, y) = \text{IFT}\{\hat{C}(v, w) \exp[-j2\pi b(v, w)]\} \exp[-j2\pi p(x, y)]$$
$$= \exp[-j2\pi p(x, y)] \frac{1}{N \times M} \sum_{v=0}^{N-1} \sum_{w=0}^{M-1} \hat{C}(v, w)$$
$$\times \exp\left[j2\pi \left(-b(v, w) + \frac{vx}{N} + \frac{wy}{M}\right)\right]. \quad (3.7)$$

Then, for large N and M, and using the central limit theorem [12–14], $\Delta(x, y)$ is an approximately zero-mean complex Gaussian random variable with variance given by

$$\sigma^2 = \frac{1}{N \times M} \left[\sum_{\eta=0}^{N-1} \sum_{\xi=0}^{M-1} |\hat{C}(\eta, \xi)|^2 \right], \tag{3.8}$$

provided that Lindeberg's condition is satisfied [13,14]. The one-dimensional version of Lindeberg's condition is

$$\lim_{N \to \infty} \frac{\sum_{v=0}^{N-1} \int_{\tau L(N)}^{\infty} x^2 dF_v(x)}{L(N)} = 0, \tag{3.9}$$

for any $\tau > 0$, where $L(N) = \sum_{v=0}^{N} E|\hat{C}(v)|^2$, and $F_v(x)$ is the probability distribution function of

$$\hat{C}(v) \exp \left\{ 2j\pi \left[-b(v) + \left(\frac{vx}{N} \right) \right] \right\}.$$

As a result, we can claim that a double-phase-encoded real image can be approximated by a Gaussian process. Also, a real image after double-phase decoding can be approximated by a Gaussian process.

Also, if the central limit theorem is applied to the real and imaginary parts of Eq. (3.7), direct calculation shows that the resultant Gaussian processes are uncorrelated. Using this result, $\tilde{f}(x, y)$ can be represented as

$$\tilde{f}(x, y) = \tilde{f}^{R}(x, y) + \tilde{f}^{I}(x, y), \tag{3.10}$$
$$\tilde{f}(x, y) = \{\alpha \times f(x, y) + \text{Re}(\Delta f(x, y))\} + \text{Im}(\Delta f(x, y)), \tag{3.11}$$

where $\tilde{f}^{R}(x, y)$ and $\tilde{f}^{I}(x, y)$ are the real and imaginary parts of $\tilde{f}(x, y)$. From Eq. (3.11), we can say that the error in the recovered signal is a complex Gaussian random variable with the real part having a variance of $\sigma^2/2$, where σ^2 is given by Eq. (3.8), and the imaginary part is a Gaussian random variable with the same variance.

3.4 Effects of Distortions on the Proposed Information-Hiding System

We will discuss several types of distortions that could appear in the proposed information-hiding system. First, we will discuss the distortion caused by the host image to the recovered hidden image and the distortion caused by the hidden image to the host image. Moreover, we conducted some simulation experiments to test the performance of the proposed system under several types of distortion by using only the real part of the watermarked image to decode the hidden image. Tests on the occluded watermarked image are also discussed.

3.4.1 Distortion Caused by the Host Image

The recovered image is the original image plus $\Delta f(x, y)$ [Eqs. (3.6) and (3.7)].
This term is approximated by a complex Gaussian noise with variance given
by Eq. (3.8). If this approximation is valid, the error caused by the host image
to the recovered hidden image will be modeled by an additive white Gaussian
noise having a variance given by Eq. (3.8).

We note in Eq. (3.6) that the desired image in the recovered image,
$\alpha f(x, y)$, is real. As a result, when finding the final output, we can use the
real part of Eq. (3.6) and discard the imaginary part, which will reduce the
noise due to the host image by half since the variance of the real part of the
error is half the variance of the total error.

3.4.2 Distortion Caused by Occlusion of some of the Watermarked Image Pixels

Occlusion of the watermarked image can be done as a trial to remove the hidden image. The occlusion of parts of the watermarked image can be modeled
as a binary function W, whose value $W(x, y)$ at position x, y is one if the
pixel is covered and zero otherwise. As a result, we can write the occluded
watermarked image as

$$I(x, y) = \{\alpha\psi(x, y) + C(x, y)\} \times \{1 - W(x, y)\}. \tag{3.12}$$

Hence,

$$I(x, y) = \alpha\{1 - W(x, y)\}\psi(x, y) + \{1 - W(x, y)\}C(x, y). \tag{3.13}$$

As shown in Appendix B, the double-phase encoding is a unitary process; in
other words, it preserves the energy of images. Thus, the effect of occlusion
of the image will be to reduce the energy of the decoded hidden image by
a factor of $\{1 - W(x, y)\}^2$. On the other hand, the variance of the resultant
additive noise in the decoded image will be reduced by the same factor. In
the next section, computer simulations are performed to illustrate this type
of distortion.

3.4.3 Effect of Using only the Real Part of the Watermarked Image to Recover the Hidden Image

On some occasions, the watermarked image bandwidth or the print size of the
watermarked image is of interest. In this case, transmitting the real part of
the watermarked image will save half of the bandwidth and half of the print
size of the watermarked image.

In this section, we consider the double-phase-encoded image $\psi(x, y)$, which
can be represented in the form of magnitude and phase as

$$\psi(x, y) = A(x, y) \exp[j\phi(x, y)], \tag{3.14}$$

where $A(x, y)$ and $\phi(x, y)$ are the magnitude and phase, respectively.

Using Eq. (3.14), the watermarked image can be written as

$$I(x, y) = A(x, y) \cos[\phi(x, y)] + C(x, y) + j A(x, y) \sin[\phi(x, y)]. \qquad (3.15)$$

The real part of Eq. (3.15), $I_R(x, y)$, is

$$I_R(x, y) = \frac{A(x, y)}{2} \{\exp[j\phi(x, y)] + \exp[-j\phi(x, y)]\} + C(x, y). \qquad (3.16)$$

This leads to

$$I_R(x, y) = \left\{ \frac{A(x, y)}{2} \exp[j\phi(x, y)] + C(x, y) \right\} + \frac{A(x, y)}{2} \exp[-j\phi(x, y)]. \qquad (3.17)$$

The first term in the brackets in Eq. (3.17) is the same as before, except for some constant attenuation. The second term is an image that is double-phase-encoded by the receiver. The effect of taking the real part of the watermarked image on the final recovered image is attenuation to the recovered image, and the variance of the additive white noise is increased due to the second term of Eq. (3.17).

3.4.4 Computer Simulations

We have conducted several computer simulations to illustrate the robustness of the algorithm. A popular image-error measurement metric is the peak signal-to-noise ratio (PSNR). It is based on the sum of the squared differences between corresponding pixels of two images. The exact formula is given below:

$$\text{PSNR} = 20 \log \left[\frac{2^n - 1}{\left(\frac{1}{N \times M} \sum_{i=0}^{N-1} \sum_{j=0}^{M-1} (P_{i,j} - Q_{i,j})^2 \right)^{1/2}} \right]. \qquad (3.18)$$

Here $P_{i,j}$ is a pixel in row i, column j of image P, and $Q_{i,j}$ is a pixel in row i, column j of image Q. N is the number of rows, and M is the number of columns, and n is the number of bits representing a pixel. The PSNR decreases as the difference between P and Q increases. PSNR is an indicator of image quality.

In the first experiment, we test the validity of applying the central limit theorem to $\Delta f(x, y)$. Figures 3.2(a) and 3.2(b) show the histogram for the 1-D slice of the real part and the imaginary part of the $\Delta f(x, y)$ compared with the histogram of an actual Gaussian distribution having a zero mean and a variance equal to $\sigma^2/2$, where σ^2 is given by Eq. (3.8). Also, we will be using the kurtosis as a measure of how similar the resulting histogram is to a normal distribution. The kurtosis is used as a measure of how outlier-prone

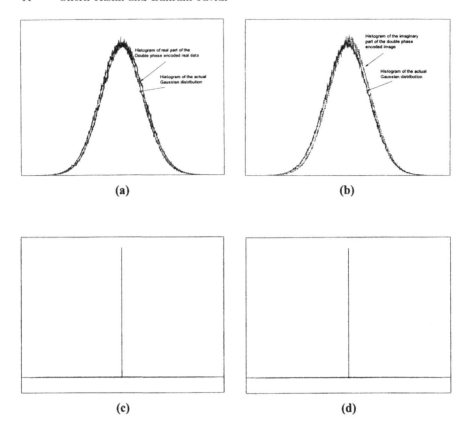

Fig. 3.2. (**a**) Histogram of the real part of the double-phase-encoded image compared with the histogram of an actual Gaussian distribution. (**b**) Histogram of the imaginary part of the double-phase-encoded image compared with the histogram of an actual Gaussian distribution. (**c**) Autocorrelation of the 1-D slice of the real part of the double-phase-encoded image. (**d**) Autocorrelation of the 1-D slice of the imaginary part of the double-phase-encoded image.

a distribution is. The kurtosis, k, of a random vector x is the fourth central moment divided by the fourth power of the standard deviation,

$$k = \frac{E|(x - \bar{x})^4|}{\sigma^4}, \tag{3.19}$$

where $\bar{x}$ is the mean of x, σ is the standard deviation of x, and $E[.]$ is the expected value. It is known that the kurtosis of a normal distribution equals 3. The experiments show that the kurtosis of the real part of the double-phase-encoded image is 2.9944 and that of the imaginary part 3.0034.

Fig. 3.3. Output without distortion when using $\alpha = 0.5$. (**a**) Original hidden image; (**b**) original host image; (**c**) watermarked image, PSNR = 23.7; (**d**) the recovered hidden image, PSNR = 22.0.

Figures 3.2(c) and 3.2(d) show the autocorrelation of the one-dimensional version of the real part and the imaginary part of the double-phase-encoded image.

In the following experiment, we test the system performance without any kind of added distortion to observe the effect of the hidden image on the host image and vice versa. We conduct this experiment using α [see Eq. (3.5)] equal to 0.5. Figure 3.3 shows the results for this experiment. The resultant PSNR for the watermarked image is 23.7, and the PSNR for the recovered hidden image is 22.0. Figure 3.4 illustrates how the value of α affects the PSNR for both the recovered hidden image and the watermarked image. There is no

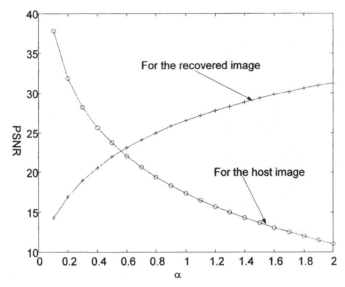

Fig. 3.4. The PSNR for the host image (circles) and for the recovered image (pluses) for different α.

optimum value for α, but the choice of α depends on the relative importance of keeping the appearance of the watermarked image unchanged and the ability to recover the hidden image under severe distortion. For comparison purposes, the value of α is set to 0.5 and the same hidden image and host image will be used in the following simulations.

In the second set of experiments, we will check the effect of occluding some pixels of the watermarked image. The reason behind this experiment is to investigate the robustness of the watermarked image against removal trials and to find out the ability to recover the embedded image from an occluded watermarked image. In the following experiment, we occlude 25%, 50%, and 75% of the watermarked image pixels. Figure 3.5 shows the occluded watermarked images and the corresponding recovered hidden images. As we can see from the figure, even with 75% occlusion, the embedded image can be recovered. The same experiment is repeated, but with a nonuniform occlusion, to demonstrate that the location of the occluded pixels has no effect on the recovered image quality. The scratches shown in Figure 3.6 cover 52% of the total number of pixels in the watermarked image. Table 3.1 summarizes the PSNR for the watermarked image and the recovered hidden image for the experiments shown in Figures 3.4 and 3.5.

Next, experiments are performed to test the effect of using only the real part of the watermarked image. Figure 3.7 shows the watermarked image and the recovered hidden image when only the real part of the watermarked image is used to recover the hidden image. The resultant PSNR for this experiment is 24.3 for the watermarked image and 18 for the recovered hidden image. In

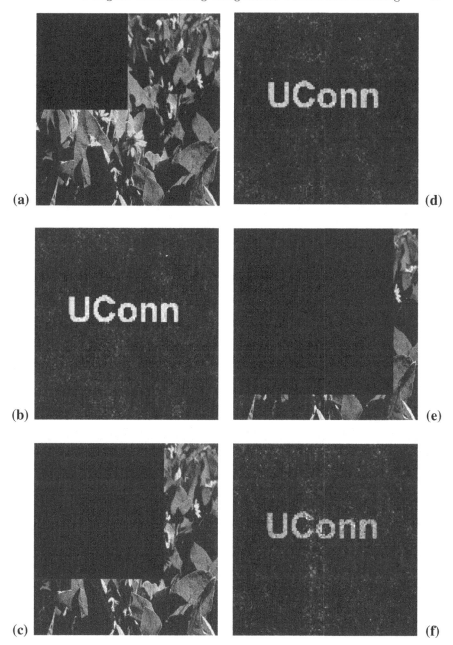

Fig. 3.5. The effect of occluding some of the watermarked image pixels: (**a**) 25% of the watermarked image pixels occluded; (**b**) recovered image with 25% occlusion; (**c**) 50% of the watermarked image pixels occluded; (**d**) recovered image with 50% occlusion; (**e**) 75% of the watermarked image pixels occluded; (**f**) recovered image with 75% occlusion.

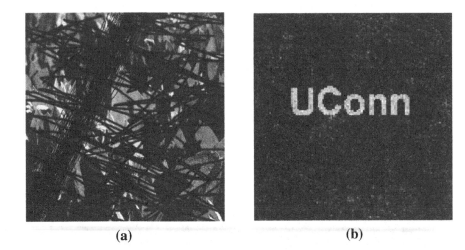

(a)   (b)

Fig. 3.6. The effect of the occlusion caused by scratches: (a) watermarked image with scratches covering 52% of the pixels; (b) recovered image.

Table 3.1. PSNR for the proposed system with some the pixels occluded.

	25% Uniform occlusion	50% Uniform occlusion	75% Uniform occlusion	52% Scratches
Watermarked image	12.5	9.8	8.6	11
Recovered image	21.5	20.7	18.9	20.3

(a)   (b)

Fig. 3.7. Output using only the real part of the watermarked image with $\alpha = 0.5$. (a) Watermarked image, PSNR = 24.3; (b) recovered hidden image PSNR = 18.

Table 3.2. PSNR for the proposed technique using only the real part of the watermarked image and with occlusion.

	25% Uniform occlusion	50% Uniform occlusion	75% Uniform occlusion	52% Scratches
Watermarked image	12.5	9.8	8.7	11.3
Recovered image	17.3	16.9	15.9	17

the next set of experiments, we test the system performance when using only the real part of the watermarked image and occluding some of the pixels at the same time. Table 3.2 shows the PSNR for both the watermarked image and the recovered image for different percentages of the occluded watermarked image occlusion. Figure 3.8 shows the resultant watermarked image and the recovered hidden image.

In the last set of experiments, we test the effect of applying Joint Photographic Experts Group (JPEG) compression on the real part of the watermarked image. The imaginary part of the watermarked image is not sent. The experiment is performed for JPEG compression of the watermarked image quality up to 75% and 50% from its original quality. Figure 3.9 shows the watermarked and the recovered hidden images using JPEG compression of the real part of the watermarked image. The resultant PSNRs are 17 and 15.2 for the watermarked image quality of 75% and 50% respectively.

In the following experiment, we test the information hiding system when the hidden image has fine details. Figure 3.10(a) shows the hidden image, and Figure 3.10(b) shows the watermarked image. The recovered image is shown in Figure 3.10(c), which has a PSNR of 21.3. Figure 3.10(d) shows the final output after simple Wiener filtering, where the PSNR of the recovered hidden image is improved to 22.6.

To further test the system performance under several kinds of distortions and attacks, we have performed additional simulations. In the following simulation, the watermarked image is distorted by fading. Figure 3.11(a) shows the distorted watermarked image, and Figure 3.11(b) shows the recovered hidden image, which has a PSNR of 20.9. Also, we test the system against attacks caused by image downsampling. Figure 3.11(c) shows the watermarked image after reducing the number of pixels to 25% of the original number; that is, averaging each four pixels into one pixel. Figure 3.11(d) shows the recovered image, which has a PSNR of 17.1. In the following experiment, we test the effect of filtering the watermarked image in the frequency domain. We used a symmetric Gaussian lowpass filter with a standard deviation equal to half the size of the watermarked image. Figure 3.11(e) shows the filtered watermarked image, and Figure 3.11(f) shows the recovered image, which has a PSNR of 20.97. In the last simulation, we added a colored noise to the watermarked image. Figure 3.11(g) shows the distorted watermarked image, and Figure 3.11(h) shows the recovered image, which has a PSNR of 20.3.

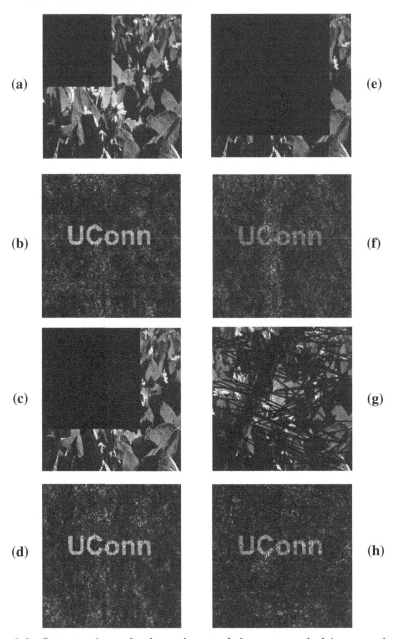

Fig. 3.8. Output using only the real part of the watermarked image and with some of the pixels occluded: (**a**) 25% of the watermarked image pixels are occluded; (**b**) recovered image with 25% occlusion; (**c**) 50% of the watermarked image pixels occluded; (**d**) recovered image with 50% occlusion; (**e**) 75% of the watermarked image pixels occluded; (**f**) Recovered image with 75% occlusion; (**g**) 52% of the watermarked image is scratched; (**h**) recovered image with 52% scratched.

Fig. 3.9. Watermarked and recovered images when applying JPEG compression to the real part of the watermarked image and without using the imaginary part. (**a**) 75% quality JPEG compressed watermarked image; (**b**) Corresponding recovered image for 75% quality; (**c**) 50% quality JPEG compressed watermarked image; (**d**) corresponding recovered image for 50% quality.

In the next experiment, we test the detection of the hidden image using a wrong code. Figure 3.12(a) shows the hidden image, and Figure 3.12(b) shows the recovered image using the wrong code. According to the simulation results for the images used, we can set a threshold of PSNR = 15 to decide whether we have a detection or not.

Fig. 3.10. (a) The hidden image. (b) The watermarked image. (c) The recovered image. (d) The recovered image after Wiener filtering.

3.5 Conclusion

In this chapter, we proposed a technique to hide images using double-phase encoding. Mathematical modeling of the technique indicates that the system can recover the hidden image with complex white Gaussian noise added. The variance of the additive noise depends on the host image. Also, we demonstrated the effects of distortions, such as occlusion, and using only the real part of the watermarked image. The PSNR metric is used to judge the quality of the recovered hidden image.

We conducted computer simulations to verify the ability to recover the embedded image and its ability to handle embedded-image removal trials. Computer simulations showed that the system is able to recover the embedded image, even under severe distortion. For example, the system recovered the embedded image when 75% of the pixels were occluded and the watermarked image is downsampled or lowpass filtered in the frequency domain. Moreover, simulations showed that the system may be used when the real part of the watermarked image is used to recover the hidden image.

Fig. 3.11. (**a**) The watermarked image with fading distortion. (**b**) The recovered image. (**c**) The watermarked with down sampling to 25% of the original number of pixels. (**d**) The recovered image. (**e**) The watermarked image after lowpass Gaussian filtering. (**f**) The recovered image. (**g**) The watermarked image with additive colored noise. (**h**) The recovered image.

Fig. 3.12. (a) The hidden image. (b) The recovered image using a wrong phase code.

Appendix A

Using Eq. (3.1), we can write:

$$\psi(x,y) = \sum_{\eta=0}^{N-1} \sum_{\xi=0}^{M-1} f(\eta,\xi) \exp[j2\pi p(\eta,\xi)] h(x-\eta, y-\xi). \qquad (3\text{A}.1)$$

Then,

$$E[\psi^*(x,y)\psi(x+\tau,y+\beta)] = \sum_{\eta=0}^{N-1} \sum_{\xi=0}^{M-1} \sum_{\lambda=0}^{N-1} \sum_{\gamma=0}^{M-1} f(\eta,\xi)^* f(\lambda,\gamma)$$
$$\times E\{\exp[j2\pi(p(\lambda,\gamma)-p(\eta,\xi))] h^*(x-\eta, y-\xi) h(x+\tau-\lambda, y+\beta-\gamma)\}$$
$$(3\text{A}.2)$$

However, since $p(x,y)$ and $b(v,w)$ are independent, then

$$E\{\exp[j2\pi(p(\lambda,\gamma)-p(\eta,\xi))] h^*(x-\eta, y-\xi) h(x+\tau-\lambda, y+\beta-\gamma)\}$$
$$= E_p\{\exp[j2\pi(p(\lambda,\gamma)-p(\eta,\xi))]\}$$
$$\times E_b\{h^*(x-\eta, y-\xi) h(x+\tau-\lambda, y+\beta-\gamma)\},$$
$$(3\text{A}.3)$$

where, E_p is the expectation with respect to $p(x,y)$, and E_b is the expectation with respect to $b(v,w)$. Since $p(x,y)$ is white noise uniformly distributed in [0,1], then it can be easily shown that

$$E_p\{\exp[j2\pi(p(\lambda,\gamma)-p(\eta,\xi))]\} = \delta(\eta-\lambda, \xi-\gamma), \qquad (3\text{A}.4)$$

where δ is the Kronecker delta function.

Using the two-dimensional discrete Fourier-transform definition,

$$h(x,y) = \frac{1}{N\times M} \sum_{v=0}^{N-1} \sum_{w=0}^{M-1} \exp[j2\pi b(v,w)] \exp[j2\pi(xv+yw)], \qquad (3\text{A}.5)$$

we can write

$$E_b\{h^*(x-\eta, y-\xi) h(x+\tau, y+\beta-\gamma)\}$$
$$= \frac{1}{N^2 M^2} \sum_{v=0}^{N-1} \sum_{w=0}^{M-1} \sum_{v'=0}^{N-1} \sum_{w'=0}^{M-1} E\{\exp[j2\pi(b(v',w')-b(v,w))]\} \qquad (3\text{A}.6)$$
$$\times \exp[j2\pi(v'(x+\tau-\lambda)-v(x-\eta))]$$
$$\times \exp[j2\pi(v'+\beta-\gamma)-v(y-\xi))].$$

Using Eqs. (3A.4) and (3A.6), we have

$$E_b\{h^*(x-\eta, y-\xi) h(x+\tau, y+\beta-\gamma)\} = \frac{1}{N^2 M^2} \sum_{v=0}^{N-1} \sum_{w=0}^{M-1} \exp[j2\pi(v\tau+w\beta)].$$
$$(3\text{A}.7)$$

According to the definition of the discrete delta function,

$$\sum_{v=0}^{N-1} \sum_{w=0}^{M-1} \exp[j2\pi(v\tau + w\beta)] = N \times M\delta(\tau, \beta). \qquad (3A.8)$$

From that, we get

$$E_b\{h^*(x - \eta, y - \xi)h(x + \tau, y + \beta - \gamma)\} = \frac{1}{N \times M}\delta(\tau, \beta). \qquad (3A.9)$$

Using Eqs. (3A.9) and (3A.4) in Eq. (3A.3), we get

$$E[\psi(x, y)\psi(x + \tau, y + \beta)] = \frac{1}{N \times M}\left[\sum_{\eta=0}^{N-1} \sum_{\xi=0}^{M-1} |f(\eta, \xi)|^2\right]\delta(\tau, \beta). \quad (3A.10)$$

Appendix B

In this appendix, we prove that the double-phase encoding is a unitary transform; in other words, it preserves the energy of the original image. Let the image to be encoded $f(x, y)$ use the same notation as before. The double-phase-encoded image is given by

$$\psi(x, y) = f(x, y)\exp[j2\pi p(x, y)] \otimes h(x, y). \qquad (3B.1)$$

For the transform to be unitary,

$$\langle \psi/\psi \rangle = \langle f/f \rangle, \qquad (3B.2)$$

where

$$\langle \psi/\psi \rangle = \sum_{x=0}^{N-1} \sum_{y=0}^{M-1} |\psi(x, y)|^2, \qquad (3B.3)$$

so

$$\langle \psi/\psi \rangle = \sum_{x=0}^{N-1} \sum_{y=0}^{M-1} (f(x, y)\exp[j2\pi p(x, y)] \otimes h(x, y)) \\ \times (f^*(x, y)\exp[-j2\pi p(x, y)] \otimes h^*(x, y)), \qquad (3B.4)$$

which simplifies to

$$\langle \psi/\psi \rangle = \sum_{x=0}^{N-1} \sum_{y=0}^{M-1} (f(x, y) \otimes h(x, y))(f^*(x, y) \otimes h^*(x, y)). \qquad (3B.5)$$

Using the fact that $H(v, w)$ is a phase-only filter,

$$|H(v, w)|^2 = 1. \qquad (3B.6)$$

Using Eq. (3B.6) and applying Parseval's theorem to Eq. (3B.5), we get

$$\langle \psi/\psi \rangle = \sum_{v=0}^{N-1} \sum_{w=0}^{M-1} f(v,w) f^*(v,w). \tag{3B.7}$$

Applying Parseval's theorem again, we get

$$\langle \psi/\psi \rangle = \sum_{x=0}^{N-1} \sum_{y=0}^{M-1} f(v,w) f^*(v,w) = \langle f/f \rangle, \tag{3B.8}$$

which proves that double-phase encoding is a unitary transform.

References

[1] N.F. Johnson, Z. Duric, and S. Jajodia, *Information Hiding: Steganography and Watermarking—Attacks and Countermeasures*, (Advances in Information Security, Volume 1), (Kluwer Academic, Dordrecht, 2001).

[2] W. Bender, D. Gruhl, N. Morimoto, and L. Lu, "Techniques for data hiding," *IBM Syst. J.* **35** (3,4), 313–336 (1996).

[3] J. Rosen and B. Javidi, "Hidden images in halftone pictures," *Appl. Opt.* **40**, 3346–3353 (2001).

[4] G.C. Langelaar, I. Setyawan, and R.L. Lagendijk, "Watermarking digital image and video data. A state-of-the-art overview," *IEEE Signal Process. Mag.* **17**(5), 20–46 (2000).

[5] C. Hosinger and M. Rabbani, "Data embedding using phase dispersion," *International Conference on Information Technology: Coding and Computing (ITCC2000)*, Las Vegas, NV, March 2000.

[6] P. Réfrégier and B. Javidi, "Optical image encryption using input plane and Fourier plane random encoding," *Opt. Lett.* **20**, 767–769 (1995).

[7] J.R. Fienup, "Phase retrieval algorithms: A comparison," *Appl. Opt.* **21**, 2758–2769 (1982).

[8] R.K. Wang, I.A. Watson, and C. Chatwin, "Random phase encoding for optical secutity," *Opt. Eng.* **35**, 2464–2460 (1996).

[9] B. Javidi, A. Sergent, G. Zhang, and L. Guibert, "Fault tolerance properties of a double phase encoding encryption technique," *Opt. Eng.* **36**, 992–998 (1997).

[10] F. Goudail, F. Bollaro, B. Javidi, and P. Réfrégier, "Influence of perturbation in a double phase-encoding system," *J. Opt. Soc. Am. A*, **15**, 2629–2638 (1998).

[11] S. Kishk and b. Javidi, "Information hiding technique using double phase encoding," *Appl. Opt.* **41**, 5470–5482 (2002).

[12] I.I. Gikhman and A.V. Skorokhod, *Introduction to the Theory of Random Process* (Dover, New York, 1969).

[13] B.V. Gnedenko, *The Theory of Probability*, (Chelsea, New York, 1962).

[14] B.V. Gnedenko, and A.N. Kolomogrov, *Limit Distributions for Sums of Independent Random Variables*, revised edition (Addison-Wesley, Reading, MA, 1968).

4

Steganography and Encryption Systems Based on Spatial Correlators with Meaningful Output Images

Joseph Rosen and Bahram Javidi

Summary. Three different optical security systems are surveyed in this chapter. Their common feature is the appearance of meaningful images on the system's output. In the first system, two phase-only transparencies are placed in a $4f$ correlator such that a known output image is received. In the second system, two phase-only transparencies are placed together in a joint-transform correlator for the same purposes. In both cases, the two phase masks are designed with an iterative optimization algorithm with constraints in the input and the output domains. In addition to simple verification, these security systems are capable of identifying the type of input mask according to the corresponding output image it generates. The third system is different from the two others in the sense that the system's input signal also is a meaningful image. This last system can offer various solutions for steganography, watermarking, and information coding. This chapter summarizes research first published in [1–3].

4.1 Introduction

Optical technologies have recently been employed in data security [4–7]. Compared with traditional computer and electrical systems, optical technologies offer primarily two types of benefits. (1) Optical systems have an inherent capability for parallel processing; that is, rapid transmission of information. (2) Information can be hidden in any of several dimensions, such as phase or spatial frequency; that is, optical systems have excellent capability for encoding information.

In several pioneering studies [4–6], the authors demonstrated different optical verification systems for information security applications based on optical correlations. These systems correlate two functions: one, the lock, is always inside the correlator, and the other, the key, is presented to the system by the user in the verification stage. Mostly, the systems determine whether the input is true or false by detecting the correlation peak in the output plane. The next generation of these security systems should offer a higher level of security and more sophisticated services than the simple verification offered by the existing systems. In this chapter, we survey three optical security systems [1–3] that are based on existing spatial correlators but have some additional benefits over those of the first generation. The common feature of all three systems is meaningful images on the output plane of the correlators.

The first property we intend to improve is the security level of the verification systems. It seems to us that the Achilles' heel of other systems is that the output of the optical system is a single narrow, intense spot of light, the correlation peak. This peak of light is detected by an intensity detector or a camera and converted to an electronic signal. If the signal is above some predefined threshold, the input mask is verified as the true input. We believe that this procedure has a weakness because unauthorized intruders may bypass the correlator and illuminate the camera from the outside with a sufficiently intense light spot to cause a false verification. In addition, the complete information of the key mask is given in the lock and vice versa. This is because the key function is equal to the complex conjugate of the Fourier transform of the lock function. That means that the reading of one phase mask by some phase-contrast technique permits a counterfeiting of the other mask. To overcome these drawbacks, we have suggested replacing the single spot with a collection of light points ordered in some predefined code or creating an image. This image is confidential and known only to the system designer. If and only if this image appears on the camera plane as a result of a correlation between two masks is the true input verified. Therefore, knowing one phase mask does not permit a person to know the distribution of the other. Even if a person in addition knows the expected image in the output, he cannot compute the other mask's values. He also needs to know the phase distribution of the output image to calculate the missing phase mask. However, the phase distribution of the output image can be measured only when the two masks exist inside the system performing the correlation process between their functions. As we shall see, the same system with the same filter can yield many images for different input masks. This property is an additional benefit of the proposed system. It can verify more than one kind of true input and identify the type of input. Let us compare the existing and the proposed systems with a real example. In a secured plant, for instance, the existing verification systems [4–6] can let someone enter or block that person from entering. Our system can do the same, but in addition it can identify the authorized person that asks to enter and distinguish him from other authorized persons. That is because each person gets a different key function, which yields a different image in the system's output when the key mask is introduced in the input.

To bypass the correlator illegally is impossible now unless the intruder knows the expected image and can project this image onto the output camera. One can argue that, because the correlator's yield is an image, this image should be automatically recognized. If a second optical correlator is added to recognize the yield of the first one, the output result of the second correlator is a correlation peak, which can be counterfeited in this stage. Our reply to this argument is that optical pattern recognition is not always the best option. If the image is binary with a simple shape or is some code such as a bar code and it appears alone on the output plane at more or less the same location, it can easily be recognized by a digital computer with appropriate software. Breaking into a digital pattern-recognition system seems harder than just illuminating the camera with an intense light spot. Moreover, in many cases the decision whether a presented identification card is legitimate or not is performed by a human rather than by an automatic machine. In these cases, a clear image in the output of the verification system is preferred over a peak of light. The human brain recognizes known images more easily than comparing an intensity of light to some threshold value. In conclusion, we claim that, although having the verification process in two stages adds complexity, it offers two new benefits: (1) improvement of the security level and (2) more information about the verified user or product.

The general concept described so far has been implemented in three different systems, which are described in this chapter. The first system [1], to be described in the next section, is an optical correlator in a $4f$ configuration. In this configuration, the key mask is displayed on the correlator's input plane, whereas the lock mask is displayed on the correlator's frequency plane. The joint transform correlator (JTC)-based security system [2] described in Section 4.3 overcomes inherent difficulties that exist in the $4f$ correlator. In Section 4.4, we survey a new development in our general paradigm called Concealogram. In this system, not only is the output a meaningful image but the input mask is as well [4].

4.2 Optical Security Systems Based on the $4f$ Correlator

An improved optical security system based on two phase-only computer-generated masks is described in this section. The two transparencies are placed together in a $4f$ correlator so that a known output image is received. In addition to simple verification, this security system is capable of identifying the type of input mask according to the corresponding output image it generates. The two phase masks are designed with an iterative optimization algorithm with constraints in the input and the output domains.

To make the concept clearer, let us precisely define the design problem of the proposed system. The system, shown in Figure 4.1, is an optical correlator in a $4f$ configuration with three domains: the input domain, in which the input mask h_1 is displayed; the Fourier domain, in which the filter mask H_2 is displayed; and the correlation domain, in which the camera should record the output predefined image. h_1 is employed as a kind of a key, whereas H_2 is used as a lock that always exists within the system. The predefined image is built up at the correlation plane P_3 only if the true key h_1 appears in the input. Otherwise, a scattered, meaningless light distribution is expected there. As in the systems of [4–6], and for the same reasons, both the masks h_1 and H_2 are chosen to be phase-only valued. That is because the phase

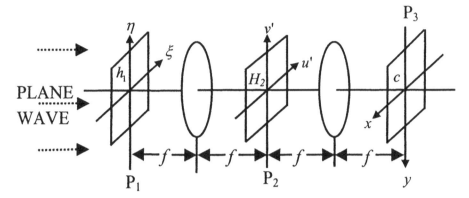

Fig. 4.1. The $4f$ correlator used for optical security verification.

distribution of phase-only transparencies, compared with the distribution of absorption masks, is hardly deciphered. The output image, obtained on the correlation plane, is constructed from an electromagnetic field projected onto this plane. Thus, the image is represented by the complex values of this field. However, the camera can record only the light intensity that is proportional to the square magnitude of the electromagnetic field. Therefore, the image's phase distribution actually creates a degree of freedom for the present problem, meaning that it can get any value between 0 and 2π. The problem is to find two phase masks located at two different planes of the correlator that together should yield on the output plane some function whose magnitude is equal to a predefined image. In other words, the problem is actually an optimization under constraints, in which one needs to find two transparency functions that yield the result closest to the desired image. In this study, we solve the optimization problem by a procedure similar to that suggested in [8]. This procedure is a generalization of the algorithm known by the name projection-onto-constraint sets (POCS) (sometimes "constraint" is replaced by "convex" [9]). Basically, in the POCS algorithm, a function is transformed back and forth between two domains. At each domain, the appropriate constraints are placed until the function converges, in the sense that the final error between the desired and the obtained images is minimal. Wang et al. [7] have proposed an algorithm similar to the POCS, called the phase-retrieval algorithm, for security applications. However, their algorithm produces the phase mask at the spatial-frequency plane (designated here as H_2) and not the input phase mask h_1, as in our case. Therefore, their algorithm is good only for producing a single pair of phase masks, one for the input plane and the other for the spatial-frequency plane. Creating many masks at the spatial-frequency plane for the same single input mask is useless because alignment problems do not permit use of the spatial-frequency plane as the input of the system. However, our algorithm can produce any desired number of input phase masks (many keys) for the same single phase-only filter (single lock) at the spatial-frequency plane. As a result, our method offers the additional service of identifying the type of input mask according to the corresponding output image it generates. The various output codes used to design the many input masks can give, in addition to simple verification, relevant information on the verified user or product. For example, if the input phase mask is part of a bill of paper money, as suggested in [4], we can design a verification system of bills that yields a series of codes, each of which would contain information on, for instance, the printing date and location of every bill. To the best of our knowledge, this additional service of coding information in the key function was not proposed in [7] or in other studies. The algorithm is explained in detail in Section 4.2.1, but before that we note that an additional application can be realized by the proposed system. The same setup and the same algorithm are suitable for encryption as well. Let us consider the image in the correlation plane P_3 as the information that we wish to encrypt. The same optimization algorithm yields two phase functions, h_1 and H_2. One of them, say h_1, is the encrypted data, whereas the other function, H_2, is employed as the decipher of this encrypted data. Placing h_1 in the input plane of the correlator, in which H_2 is positioned in its Fourier plane, is the only way to reconstruct the original image. In comparison with other optical encryption systems [10,11], the encryption process in our system is iterative and digital. The deciphering can be done either digitally or optically. However, the main advantage of this method comes from the nature of the encrypted data. Unlike in other methods [10,11], the encrypted data appear now as a phase-only function. This means that the amount of data in the encrypted function is half the general complex function with the same size; such phase functions are difficult to read with conventional detection devices.

4.2.1 Synthesizing the Key Mask in the $4f$ Correlator

With regard to the $4f$ correlator shown in Figure 4.1, the information is encoded into two phase-only computer-generated masks. One is located in the input plane, denoted by $h_1(\xi, \eta) = \exp[i\varphi(\xi, \eta)]$, and the other is in the spatial-frequency plane, denoted by $H_2(u, v) = \exp[i\Phi(u, v)]$. In this system the output at the correlation plane is given by

$$c(x, y) = \mathfrak{F}^{-1}\{\mathfrak{F}\{h_1(\xi, \eta)\}H_2(u, v)\}$$
$$= \mathfrak{F}^{-1}\{\mathfrak{F}\{h_1(\xi, \eta)\}\exp[i\Phi(u, v)]\} \qquad (4.1)$$

where $\mathfrak{F}$ and $\mathfrak{F}^{-1}$ denote the Fourier transform and its inverse transform, respectively. For notational simplicity, we assume that (u, v) are the spatial-frequency variables related to the spatial coordinates (u', v') by the relation $(u, v) = (u', v')/\lambda f$. The expected system's output is

$$c(x, y) = A(x, y)\exp[i\psi(x, y)], \qquad (4.2)$$

where $A(x, y)$ is the amplitude of the expected output image and $\psi(x, y)$ denotes the phase of $c(x, y)$. From Eq. (4.1) the input function is given by

$$h_1(\xi, \eta) = \mathfrak{F}^{-1}\left\{\frac{\mathfrak{F}\{c(x, y)\}}{H_2(u, v)}\right\} = \mathfrak{F}^{-1}\{\mathfrak{F}\{c(x, y)\}\exp[-i\Phi(u, v)]\}. \tag{4.3}$$

To design two phase-only masks that produce an output image with a given magnitude, we choose to use the generalized POCS algorithm. This iterative algorithm starts with a random function for the first $h_1(\xi, \eta)$. Then the function $h_1(\xi, \eta)$ is transformed by the correlation, defined in Eq. (4.1), into the output function $c(x, y)$ and then back through the inverse correlation defined by Eq. (4.3). At every iteration, in each of the two domains (x, y) and (ξ, η), the functions obtained are projected onto the constraint sets. In the (x, y) domain, the constraint set expresses the expectations to get the predefined image. In the (ξ, η) domain, the constraint set manifests the limitation on the input function to be a phase-only function. The algorithm continues to circulate between the two domains until the error between the actual and the desired output functions is no longer meaningfully reduced. As mentioned above, the constraint in the output plane should reflect the desire to get the image expressed by the positive function $A(x, y)$. Therefore, in the output plane, the projection P_1 on the constraint set is

$$P_1[c(x, y)] = \begin{cases} A(x, y)\exp[i\psi(x, y)] & \text{if } (x, y) \in S \\ 0 & \text{otherwise,} \end{cases} \tag{4.4}$$

where $A(x, y)$ is a real positive function representing the output image. S is an area support of the output image. In the input plane, we recall that $h_1(\xi, \eta)$ should be a phase-only function, and therefore the projection P_2 on the constraint set is

$$P_2[h_1(\xi, \eta)] = \begin{cases} \exp[i\varphi(\xi, \eta)] & \text{if } (\xi, \eta) \in W \\ 0 & \text{otherwise,} \end{cases} \tag{4.5}$$

where $\varphi(x, y)$ denotes the phase of $h_1(\xi, \eta)$; that is, $\exp[i\varphi(\xi, \eta)] = \frac{h_1(\xi, \eta)}{|h_1(\xi, \eta)|}$, and W is a window function that is necessary for reasons described in Section 4.2.2. The iteration process is shown schematically in Figure 4.2. Note that $H_2(u, v)$ is chosen only once before the beginning of the iterations, in a process that will be explained below. After $H_2(u, v)$ is defined, it becomes part of the correlator and is never changed during the circulating process. The convergence of the algorithm to the desired image in the nth iteration is evaluated by the average mean-square error e_n between the intensity of the correlation function before and after the projection as

$$e_n = \frac{1}{M}\iint \left| |P_1[c(x, y)]|^2 - \gamma_n|c_n(x, y)|^2\right|^2 dx\,dy, \tag{4.6}$$

where γ_n is a matching constant [12] determined to minimize e_n, and M is the total area of the output plane. When the reduction rate of this error function

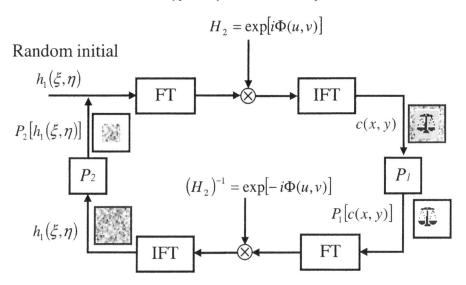

Fig. 4.2. Block diagram of the main POCS algorithm used to compute the phase-only $h_1(\xi, \eta)$.

becomes slower than some predefined value, the iterations are stopped. As discussed in [8], there are two conditions that guarantee that this error will never diverge. First, the correlator should be an energy-conserving operator. This property is easily achieved if $H_2(u, v)$ is a phase-only function, as indeed it is in the present case. The second condition to satisfy the nondiverging feature is realized if, among all the functions that belong to the constraint sets, the two projected functions in the nth iteration, $P_1[c_n(x, y)]$ and $P_2[h_{1,n}(\xi, \eta)]$, are the functions closest (by means of the mean-square metric) to the functions $c_n(x, y)$ and $h_{1,n}(\xi, \eta)$, respectively. It is easy to show that the second condition is also fulfilled in the present algorithm. Therefore, the POCS algorithm here can never diverge. Note that the nondiverging feature of the algorithm is an additional reason to favor phase-only functions in the spatial-frequency domain.

4.2.2 Simulation Results of the 4f Correlator

We wrote a computer simulation demonstrating our general concept discussed above. In our simulations, the algorithm was tested with two different binary images, as shown in Figure 4.3; Figure 4.3(a) is a duck smashing a computer, and Figure 4.3(b) is a scale. The images comprise 120×120 pixels, whereas the input and the output planes have 256×256 pixels each. In the input domain, the two correlated functions are made to cover only the central area of 128×128 pixels, designated as the window W. All the rest of the matrix outside this window is padded with zeros. This ensures that the computer

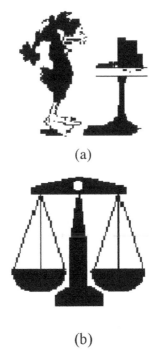

(a)

(b)

Fig. 4.3. Two expected output images of the correlator used in the computer simulation.

simulation based on a discrete Fourier transform truly simulates the analog optical system. The signal transformed by the discrete Fourier transform is considered periodic. Therefore, correlation between two functions that are extended beyond the central window W causes correlation between different cycles of the signals, a phenomenon that does not exist in the optical correlator. Padding the input plane with zeros outside the window is done on $h_1(\xi, \eta)$ at every iteration by the projection P_2, defined in Eq. (4.5). To generate $H_2(u, v)$ so that in the input domain $h_2(\xi, \eta)$ will also cover only the window area, a mini-POCS algorithm was introduced. This algorithm generates $H_2(u, v)$ with phase-only values, whereas its inverse Fourier transform $h_2(\xi, \eta)$ can get any complex value inside the window W and zero outside it. This mini-POCS algorithm is shown schematically in Figure 4.4. In this mini-POCS algorithm, the projection onto the constraint set in the Fourier domain is

$$P_1'[H_2(u, v)] = \exp[i\Phi(u, v)], \tag{4.7}$$

where, as defined above, $\exp[i\Phi(u, v)]$ is the phase of function $H_2(u, v)$. In the input domain, the projection on $h_2(\xi, \eta)$ is

$$P_2'[h_2(\xi, \eta)] = \begin{cases} h_2(\xi, \eta) & \text{if } (\xi, \eta) \in W \\ 0 & \text{otherwise,} \end{cases} \tag{4.8}$$

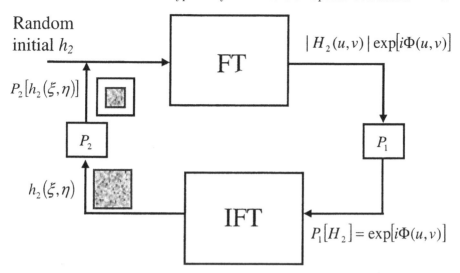

Fig. 4.4. Block diagram of the mini-POCS algorithm used to compute the phase-only mask $H_2(u, v)$.

where W is the window defined above. This time the average mean-square error function in the nth iteration is defined as

$$e'_n = \frac{1}{B} \int\!\!\int_B |h_{2,n}(\xi\eta)|^2 \, d\xi \, d\eta, \tag{4.9}$$

where B is the area surrounding the window W (i.e., $B \cup W = M$). In this simulation, the average error is less than 0.1% of the maximum value of $h_2(\xi, \eta)$ after only 30 iterations. Figure 4.5 shows the phase of $H_2(u, v)$ obtained with 30 iterations of the mini-POCS. $H_2(u, v)$ was calculated only once by the mini-POCS and then introduced into the correlator at the spatial-frequency plane. With the same $H_2(u, v)$, we calculated two different input functions $h_1(\xi, \eta)$ for the two images using the main POCS algorithm described in Section 4.2.1. Note that there is no limitation by H_2 on the number of different output patterns that can be created by the same single H_2 and many different h_1 functions. The only limitation is the number of patterns that can be drawn on a finite-sized matrix. For both images, the algorithm was terminated after one hundred iterations. The error plots for both experiments are shown in Figure 4.6. The final average errors are less than 2% of the average value of the image for both images. Figures 4.7(a) and 4.7(b) show the phase functions of the two masks h_1 for the expected output images shown in Figures 4.3(a) and 4.3(b), respectively. We examined the randomness of the resulting masks in comparison with the random masks with which the iterative process starts. It appears that the randomness is the same even after one hundred iterations. The phase values are still distributed uniformly, and the autocor-

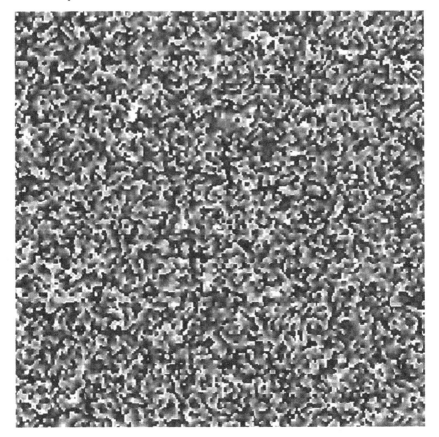

Fig. 4.5. Phase function of the mask $H_2(u,v)$.

relation width is still one pixel, which indicates that the phase values are mutually independent.

The resultant correlation functions $|c(x,y)|^2$ after one hundred iterations can be seen in Figure 4.8. The final question we consider here is whether the two phase functions $h_1(\xi,\eta)$ and $H_2(u,v)$ can be deciphered when we know the image function $A(x,y)$. Because each process of POCS starts with a random function as the first trial, the final solutions are always different from one experiment to another, although all of them yield the same desired image on the correlation plane. Therefore, even if some unauthorized intruder acquires the predefined image in the output plane, he or she would not be able to reproduce the masks $H_2(u,v)$ and $h_1(\xi,\eta)$ to get access to the system. This feature is demonstrated in the table shown in Figure 4.9. Five pairs of $H_2(u,v)$ and $h_1(\xi,\eta)$ were calculated for the same image of the scale by the mini- and the main POCS. This table shows the correlation intensity between any possible pair $h_{1,i}(\xi,\eta)$ and $\mathfrak{F}^{-1}\{H_{2,j}(u,v)\}$. Only the pairs, calculated together in the same process, yield the desired image, as seen along the diagonal of the table.

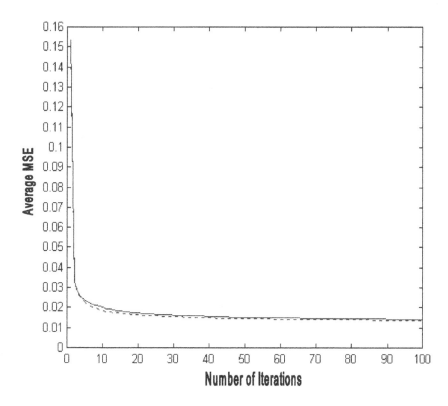

Fig. 4.6. Average mean-square error (MSE) versus the number of iterations for the POCS algorithm in the case of the scale (solid curve) and the duck (dashed curve).

All the rest of the cross correlations yield scattered meaningless distributions. The conclusion is that, even if the image is known, it is impossible to deduce the right $h_1(\xi, \eta)$ for an unknown $H_2(u, v)$.

In the next section, we describe a similar algorithm for a security system implemented in a JTC. The expected advantage from a JTC-based security system is the invariance of the system to in-plane shifts of both masks. Thus, the JTC can be a proper solution for the problem of misalignment sensitivity of the filter mask in the $4f$ correlator.

4.3 Optical Security Systems Based on JTC

In this section we review our security scheme implemented in the JTC [2]. The main advantage of this scheme over the $4f$ correlator described in the previ-

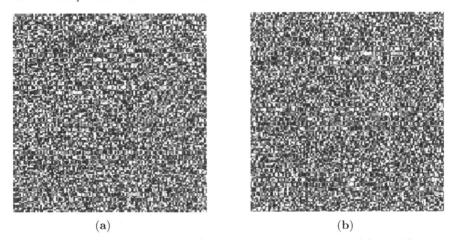

Fig. 4.7. Phase distributions of $h_1(\xi, \eta)$ for the output images of (**a**) the duck and (**b**) the scale.

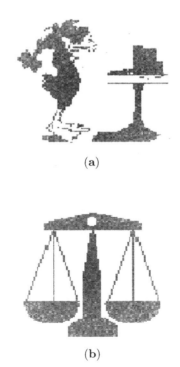

Fig. 4.8. Resultant images of $|c(x, y)|^2$ for (**a**) the duck and (**b**) the scale.

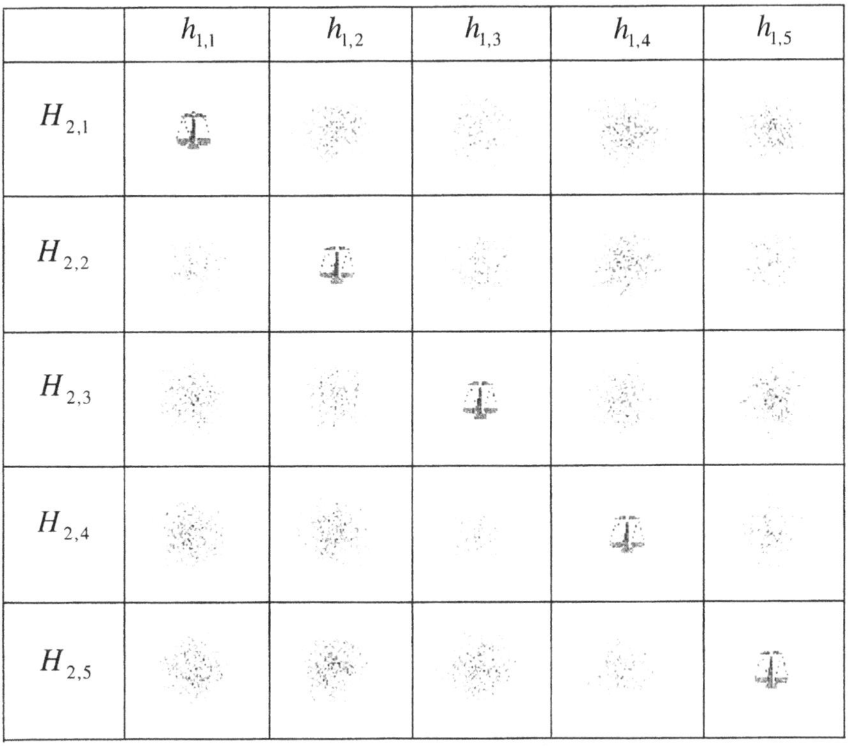

	$h_{1,1}$	$h_{1,2}$	$h_{1,3}$	$h_{1,4}$	$h_{1,5}$
$H_{2,1}$					
$H_{2,2}$					
$H_{2,3}$					
$H_{2,4}$					
$H_{2,5}$					

Fig. 4.9. Table of all cross correlations between functions h_1 and the inverse Fourier transforms of functions H_2. The second index of each function denotes the number of the process used to design the functions. Only the pairs that were designed together in the same process yield the image of the scale.

ous section is the less restrictive alignment requirements. The lateral distance between the two masks can be changed within a reasonable tolerance without changing the shape of the output image. Only the image location on the output plane is changed according to the relative distance between the two phase masks. On the other hand, in the $4f$ correlator, a slight mutual shift between the filter and the light distribution coming from the input considerably modifies the correlation results. The values of the phase mask used as the system's lock are determined once by a random-number generator. The system design problem is to find the second phase mask (the key) such that a correlation between the two masks yields an intensity pattern as close as possible to a predefined image. This design problem is actually equivalent to a nonlinear optimization problem. We choose to compute the mask's values by a modified version of the POCS algorithm, adjusted for the JTC [13]. This algorithm is a relatively rapid iterative process, but it usually achieves suboptimal solutions.

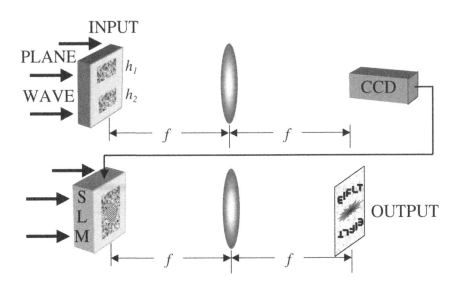

Fig. 4.10. The JTC used for optical security verification.

4.3.1 Synthesis of a Key Mask in the JTC

The proposed optical security system is based on the JTC configuration shown in Figure 4.10. The JTC input plane contains two phase functions, $h_1(x, y)$ and $h_2(x, y)$, located apart from each other. The output of the correlator contains three spatially separated diffraction orders,

$$o(x, y) = h_1(x, y) \otimes h_1(x, y) + h_2(x, y) \otimes h_2(x, y)$$
$$+ h_1(x, y) \otimes h_2(x - a, y - b) + h_2(x, y) \otimes h_1(x + a, y + b), \quad (4.10)$$

where $\otimes$ denotes the correlation operator, and a and b are the distances between h_1 and h_2 in the x and y directions, respectively. The useful terms are either the third or the fourth terms since both represent the cross correlation between the two input functions. We choose one of them, say the third term, as the output of the security system. This output distribution is complex-valued and is expressed by

$$c(x, y) = h_1(x, y) \otimes h_2(x, y) = A(x, y) \exp[i\varphi(x, y)], \quad (4.11)$$

where $A(x, y)$ is the amplitude of the expected output image and $\varphi(x, y)$ denotes the phase function of $c(x, y)$.

The computation of the phase-only mask h_1, which produces the desired output image from the cross-correlation intensity distribution with the other random phase-only mask h_2, is performed by the modified POCS algorithm. The POCS shown in Figure 4.11 is an iterative process, which in the present case transfers a function, by the JTC operator, from one domain to another. In

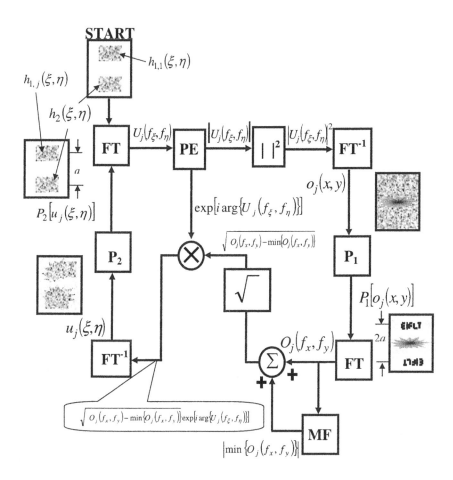

Fig. 4.11. Block diagram of the POCS process used to compute the phase-only mask $h_1(\xi, \eta)$; FT is the Fourier transform, PE is the phase extraction, MF is the minimum finder, and R_i is the ith projection.

every domain, the resulting function is projected onto a constraint set by two projection operators denoted by P_1 and P_2; both are defined in the following.

The convergence of the process is achieved when the function satisfies all the constraints in every domain simultaneously. This iterative algorithm starts with an initial random phase function denoted by $h_{1,1}$ and another random phase function h_2, which remains fixed thereafter. The overall input function of the JTC in the first iteration is

$$u_1(\xi, \eta) = h_{1,1}(\xi + a/2, \eta + b/2) + h_2(\xi - a/2, \eta - b/2). \qquad (4.12)$$

The support areas of the phase functions $h_1(\xi, \eta)$ and $h_2(\xi, \eta)$ are $[\bar{\xi}_1 \times \bar{\eta}_1]$ and $[\bar{\xi}_2 \times \bar{\eta}_2]$, respectively. The function $u_1(\xi, \eta)$ is Fourier-transformed into the spatial-frequency plane, where the phase factor is extracted (in the box PE) and memorized until the iterated function is transformed back to the (ξ, η) domain. This procedure is needed to keep as much information as possible on the projection of $u_j(\xi, \eta)$ in the jth iteration, such that the distance between $P_2[u_j(\xi, \eta)]$ and $u_{j+1}(\xi, \eta)$ is kept minimal. From the frequency plane, according to an ordinary JTC, the square magnitude of the spatial spectrum is inversely Fourier-transformed to yield the output function $o_j(x, y)$. The projected function $P_1[o_j(x, y)]$ is Fourier-transformed back to $O_j(f_x, f_y)$ in the frequency domain. The next necessary operation is to calculate the square root of the spectrum. $O_j(f_x, f_y)$ is a real function because $P_1[o_j(x, y)]$ is symmetric. However, the projection P_1 may cause $O_j(f_x, f_y)$ to have negative values. To avoid calculating the square root of negative values, the absolute value of the minimum of $O_j(f_x, f_y)$ is identified (in the box MF) and added to the function $O_j(f_x, f_y)$. The addition of uniform bias affects only the central value of the correlation plane and thus does not contradict the constraints in this plane, enforcing only on the first two diffraction orders, as seen in Eq. (4.13). After calculating the square root of a real positive spectrum, we multiply this result by the memorized phase factor of the spectrum in the forward step, as denoted in Figure 4.11 by $\exp[i \arg\{U_j(f_x, f_y)\}]$. This last product is inversely Fourier-transformed back to the (ξ, η) plane. Finally $u_j(\xi, \eta)$ is projected by P_2 and the next iteration starts.

As mentioned, at every iteration, in each of the two domains (x, y) and (ξ, η), the resulting functions are projected onto the constraint sets. In the (x, y) domain, the constraint set reflects the expectations to get the predefined image represented by the positive function $A(x, y)$. Therefore, in the output plane, the projection P_1 on the constraint set is

$$P_1[o_j(x, y)]$$
$$= \begin{cases} A(x - a, y - b) \exp[i\psi_j(x, y)] & |x - a| \leq W_x/2 \,\&\, |y - b| \leq W_y/2 \\ A(-x - a, -y - b) \exp[i\psi_j(x, y)] & |x + a| \leq W_x/2 \,\&\, |y + b| \leq W_y/2 \\ 0 & \text{otherwise}, \end{cases}$$

$$(4.13)$$

where W_x and W_y are the width and the height of the desired output image centered around $(x, y) = (a, b)$ and $(-a, -b)$. This constraint set guarantees that the spectrum $O_j(f_x, f_y)$ remains a real function. $\psi_j(x, y)$ is the phase function of $o_j(x, y)$, and it obeys the rule $\psi_j(x, y) = -\psi_j(-x, -y)$. In the input domain (ξ, η), the constraint set manifests the properties of the input functions as phase-only functions in a predefined area and zero elsewhere. Therefore, the projection P_2 is

$$
P_2[u_j(\xi, \eta)] = \begin{cases} \exp[i\theta_j(\xi, \eta)] & |\xi - a/2| \leq \bar{\xi}_1/2 \,\&\, |\eta - b/2| \leq \bar{\eta}_1/2 \\ h_2(\xi + a/2, \eta + b/2) & |\xi + a/2| \leq \bar{\xi}_2/2 \,\&\, |\eta + b/2| \leq \bar{\eta}_2/2 \\ 0 & \text{otherwise,} \end{cases}
$$

$$(4.14)$$

where $\exp[i\theta_j(\xi, \eta)]$ denotes the phase of $u_j(\xi, \eta)$. The algorithm continues to circulate between the two domains until the error between the actual and the desired output functions is no longer meaningfully reduced. Note that $h_2(\xi, \eta)$ is chosen in the initial step of the iterations, becomes a part of the correlator, and is never changed during the iteration process. Moreover, $h_2(\xi, \eta)$ is not in any way related to $h_1(\xi, \eta)$ or the output image and is not in any part of the system's memory. Therefore, $h_2(\xi, \eta)$ does not limit the quantity of key-mask output-image pairs that can be processed by the same key function $h_2(\xi, \eta)$. The present security system can be viewed as a generalization of the Fresnel computer-generated hologram. In this analogy, $h_2(\xi, \eta)$ plays the role of a generalized medium between $h_1(\xi, \eta)$ and the reconstructed output image $A(x, y)$ in a fashion similar to the quadratic phase factor representing the free-space medium in the Fresnel hologram reconstruction [14]. The medium function can be used as a lock to expose an image but does not contain any information on the image, and therefore its size does not limit the image capacity that can be utilized by the system.

The convergence of the algorithm to the desired image in the nth iteration is evaluated by the average mean-square error (MSE) between the intensity of the correlation function before and after the projection,

$$
e_j = \frac{1}{M} \iint \left| |P_1[o_j(x, y)]|^2 - |o_j(x, y)|^2 \right|^2 dx\, dy, \tag{4.15}
$$

where M is twice the area of the output images. Unlike the corresponding algorithm in the $4f$ correlator described in Section 4.2, the convergence of the process is not guaranteed. From our experience, the MSE here decreases to some saturation level and fluctuates slightly around this level. Apparently, the obtained output images resulting from the algorithm of Section 4.2 are less noisy than the results here. However, the comparison is not done in an equal condition and therefore it is hard to determine which algorithm leads to a lower MSE. We chose an arbitrary number of iterations, much larger than it takes to converge to the saturation level. During the iterations, we keep the key function h_1 that gives the minimum MSE in the memory, and this function is used as the final solution of the process.

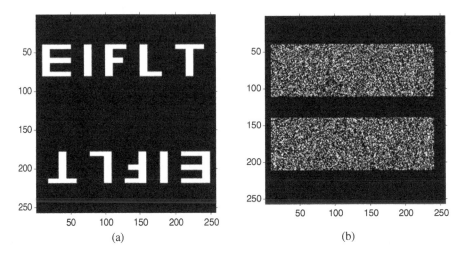

Fig. 4.12. (a) The expected output image used in the first computer simulation. (b) Phase functions of the mask h_1 (upper) and h_2 (lower) on the JTC input plane.

4.3.2 Simulation Results in the JTC

We performed a computer simulation demonstrating our proposed system discussed earlier. In the simulation, the algorithm was tested with two different binary output images. The first one is composed from the letters EIFLT, as shown in Figure 4.12(a), in a symmetrical position on the JTC's output plane. The JTC input and output planes have 256×256 pixels each. In the input domain, each of the two correlated functions covered only 70×230 pixels of phase-only values. All the rest of the matrix outside these two windows was padded with zeros, as shown in Figure 4.12(b). The lock phase mask was positioned in the lower part of the plane and was not being changed during the process. The algorithm was terminated after one hundred iterations, and the resultant correlation functions can be seen in Figure 4.13. The plot of the MSE, defined in Eq. (4.15), for this experiment is shown in Figure 4.14. The three orders of the correlation plane and the letter images in the first two diffraction orders can be clearly seen in Figure 4.13, demonstrating that our algorithm is effective in accomplishing its goal. Note that a low dynamic range of the joint power spectrum is not a problem here since both masks h_1 and h_2 are phase masks that yield an almost uniform joint power spectrum.

In the second experiment the output image was a picture of a bar code, shown in Figure 4.15(a). The reconstructed picture is shown in Figure 4.15(b). The error-function plot is shown in Figure 4.16. The behavior of the MSE along the iterations and the quality of the final result are similar to that of the first experiment.

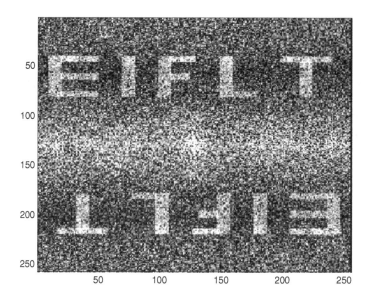

Fig. 4.13. Resultant image of the JTC output plane for the letters of Figure 4.12(a).

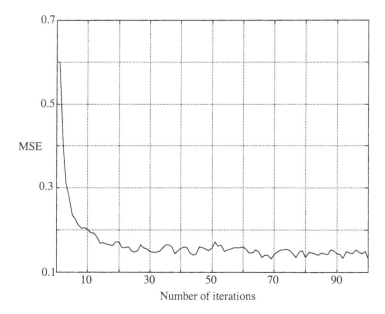

Fig. 4.14. MSE versus the number iterations of the POCS system for the experiment with the letters.

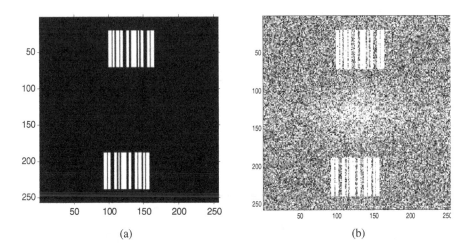

(a)   (b)

Fig. 4.15. (a) The expected output image used in the second computer simulation.
(b) The reconstructed image of the barcode shown in (a).

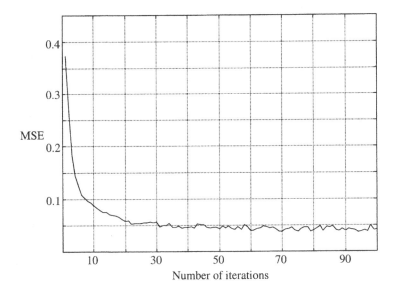

Fig. 4.16. MSE versus the number of iterations of the POCS system for the experiment with the barcode.

4.4 Concealogram: An Image within an Image

This section is devoted to a new development in our general concept [3]. Not only the output but also the input signal contains a meaningful image. Therefore, the input mask should incorporate two types of signals; on the one hand, the mask should be looked at as an ordinary meaningful image, but on the other hand it should include a secretive code that yields the other meaningful image on the system's output plane. Encoding the input image by a modified version of the halftone code satisfies these two requirements.

Halftone coding is a common method of representing continuous-tone images by binary values. In one of many techniques for halftone binarization [15], the various tone levels are translated to the area of binary dots. This method, termed binarization by a carrier [16], is related to the pulse-width modulation in communication theory. The locations of the dots inside their cells in the halftone picture usually do not represent any information. When the positions of the dots are not uniform from cell to cell, the nonuniformity is actually used to reduce the difference between the original gray-tone image and the resultant binary image as viewed by the detection system [15].

We describe here a method of encoding visual information in a halftone image using the locations of the dots inside their cells. The algorithm puts together two data files for two images, one that we want to print as an observable picture and the other that we want to conceal within the observable picture. The halftone image obtained is termed a Concealogram. In addition, we scramble the mathematical representation of the hidden image with a mathematical key. Once an image is encoded, only an authorized person who has the key can reveal the hidden image. The composite image can be printed on any printer. The print can then be read by a conventional optical scanner and processed by computer, or optical correlator, to access the hidden image. Like a Fourier hologram, the hidden image is concealed in a global manner. Every part of the visible image contains information on the entire hidden image, such that if one covers or destroys part of the Concealogram the entire hidden image can still be recovered from the rest. It is shown that the hidden image can be elicited even when most of the halftone picture is damaged or missing.

The scheme hides one image in another halftone image so that scanners of any kind can unlock and view the information. In general, this technique can be used as a pictured dot code. On one hand, it is a collection of dots used as a secret code, which can be deciphered only by a special key. On the other hand, this code is a picture in the sense that the code itself is a meaningful image, encoded independently of the hidden image. This feature is different from other known codes, such as the common bar code. Another application might be embedding steganographic information [17–19] in halftone pictures. The ordinary visible image is conventionally encoded by the dots' sizes, whereas the steganographic image is encoded by their positions. A possible application for this technique might be, for example, in identification cards [20]. A special halftone photograph of a person on an identification card can show the cardholder's picture, as usual. However, the same photograph can conceal confidential data such as an image of the person's signature, his or her fingerprint, or some other personal records. The cardholder in this case must be matched to both types of images and to all the rest of the data on the card. Thus, counterfeiting of identification cards by a person who resembles the authentic person, or switching of the photographs on their identification cards, without being discovered becomes much more difficult. The steganographic images are revealed by a special key in a particular processor that we discuss next.

Our proposed tool for revealing the hidden image is the well-known 2-D spatial correlator. The spatial filter of this correlator is the key function that enables the hidden image to appear on the output plane when the halftone figure is displayed on the correlator input plane. In other words, the hidden image is obtained as the correlation function between the halftone picture and a reference function. The reference function is related to the spatial-filter function by a 2-D Fourier transform. Using the correlator has the following advantages: (1) The image reconstruction from the halftone picture is relatively robust to noise. This is so because the hidden image can be memorized globally in all the halftone's dots. This means that every pixel in the output image is obtained as a weighted sum of the entire input picture's dots. Therefore, even if several pixels from the input halftone figure are distorted, the output result can still be recognized because of the contributions from the other, nondistorted pixels. (2) The spatial correlator has the property of the shift invariance, which means that, no matter where the halftone image appears at the input plane, the hidden output image is produced on the output plane. (3) The same deciphering system can be implemented as an optical, electrical, or hybrid system. This is so because spatial 2-D correlators can be implemented by the optical $4f$ correlator, by the hybrid JTC, or by a digital computer. The system that we show here is based on digital computing, although the use of optical correlators is also discussed. (4) When digital correlations are used, it is obvious to use the fast Fourier-transform algorithm as a tool for computing the correlations, both in the coding process and in reading the hidden images. Therefore, the computation time is relatively short compared with those of other, more general, linear space-variant processors [21].

4.4.1 Encoding of Images in a Halftone Picture

The coding process starts with the data of two images, the visible image $f(x, y)$ and the hidden image $a(\xi, \eta)$. They are defined in different coordinate systems because they are observed in two different planes. $f(x, y)$ is observed on the correlator's input plane, $a(\xi, \eta)$ on its output plane. Because they represent gray-tone images, both functions are real and positive. An additional function is determined once at the beginning of the process and is referred to the key function $H(u, v)$. $H(u, v)$ is the filter function displayed on the spatial-frequency plane, and its inverse Fourier transform is denoted $h(x, y)$. For reasons of algorithm stability explained below, $H(u, v)$ is a phase-only function of the form $H(u, v) = \exp[i\phi(u, v)]$, where $\phi(u, v)$ is a random function generated by a random-number generator of the computer and is uniformly distributed on the interval $-\pi$ to π. The computational problem is to find the halftone figure that, correlated with the predefined function $h^*(-x, -y)$, yields the hidden output image $a(\xi, \eta)$. The visible image $f(x, y)$ is used as the constraint on the input function. This means that, instead of a meaningless pattern of binary dots in the input, the halftone picture presents the image $f(x, y)$. The proposed algorithm is separated into two stages. In the first stage, we compute a phase function $\exp[i\theta(x, y)]$ of the complex function $g(x, y) = f(x, y) \exp[i\theta(x, y)]$. In other words, we are looking for a phase function $\exp[i\theta(x, y)]$ that, when it is multiplied by the image function $f(x, y)$ and passes through the correlator, results in a complex function with a magnitude that is equal to the hidden image $a(\xi, \eta)$. Therefore, one can get two independent images $f(x, y)$ in the input plane and $a(\xi, \eta)$ in the output. Both functions are the magnitude of the two complex functions. In the second stage, the complex gray-tone function $g(x, y)$ is binarized to a final halftone image. In other words, the phase function $\exp[i\theta(x, y)]$ is embedded in the binary pattern by modulating the dots' position, and the image $f(x, y)$ is encoded by modulating the dots' area. We next describe the first part of the algorithm; the second stage follows.

As we have mentioned, our goal for the first stage is to find the phase function $\exp[i\theta(x, y)]$ of the input function $g(x, y)$ such that a correlation between $g(x, y)$ and $h'(-x, -y)$ yields a complex function with the magnitude function $a(\xi, \eta)$. The phase of the output function is denoted $\exp[i\psi(\xi, \eta)]$, and the complex output function is denoted $c(\xi, \eta) = a(\xi, \eta) \exp[i\psi(\xi, \eta)]$. Therefore, the output correlation function is

$$
\begin{aligned}
c(\xi, \eta) &= \{f(x, y) \exp[i\theta(x, y)]\} \otimes h^*(-x, -y) \\
&= \mathcal{F}^{-1}\{\mathcal{F}\{g(x, y)\} \exp[i\phi(u, v)]\}, \quad (4.16)
\end{aligned}
$$

where $\otimes$ denotes correlation and we recall that the operators $\mathcal{F}$ and $\mathcal{F}^{-1}$ are the Fourier transform and the inverse Fourier transform, respectively. From Eq. (4.16), the input function is given by

$$
g(x, y) = \mathcal{F}^{-1}\left\{\frac{\mathcal{F}\{c(\xi, \eta)\}}{H(u, v)}\right\} = \mathcal{F}^{-1}\{\mathcal{F}\{c(\xi, \eta)\} \exp[-i\phi(u, v)]\}. \quad (4.17)
$$

To compute phase function $\exp[i\theta(x, y)]$, we choose again to utilize the POCS algorithm modified to operate with correlations. This iterative algorithm starts with an initial random function $\exp[i\theta_1(x, y)]$. Then the function $f(x, y) \exp[i\theta_1(x, y)]$ is transformed by the correlation described in Eq. (4.16). The function $c_1(\xi, \eta)$ is transformed backward by use of the inverse correlation defined by Eq. (4.17). At every iteration in each of the two domains (x, y) and (ξ, η), the functions obtained are projected onto the constraint sets. In both domains, the constraint sets express the expectation of getting the pre-defined images $a(\xi, \eta)$ at (ξ, η) and $f(x, y)$ at (x, y). The algorithm continues to iterate between the two domains until the error between the actual and the desired image functions is no longer meaningfully reduced.

The constraint on the output plane is defined by the requirement to obtain the hidden image $a(\xi, \eta)$. Therefore, in the output plane, projection P_1 onto the constraint set at the jth iteration is

$$P_1[c_j(\xi, \eta)] = \begin{cases} a(\xi, \eta) \exp[i\psi_j(\xi, \eta)] & (\xi, \eta) \in W \\ c_j(\xi, \eta) & \text{otherwise,} \end{cases} \tag{4.18}$$

where $\exp[i\psi_j(\xi, \eta)]$ is the phase function of $c_j(\xi, \eta)$ in the jth iteration. W is a window support of the hidden image. The window's area is smaller than, or equal to, the area of the output plane. Similarly, in the input plane, projection P_2 onto the constraint set at the jth iteration is

$$P_2[g_j(x, y)] = f(x, y) \exp[i\theta_j(x, y)], \tag{4.19}$$

where $\exp[i\theta_j(x, y)]$ is the phase function of $g_j(x, y)$ at the jth iteration. The iteration process is shown schematically in Figure 4.17. Note that $H(u, v)$ is chosen only once before the iterations. This $H(u, v)$ becomes part of the correlator, and it is never changed during the iteration process. Moreover, $H(u, v)$ is not in any way related to any of the encoded images and is not any kind of system memory. Therefore, as before in the two systems described in previous sections, the size of $H(u, v)$ does not limit the quantity of image pairs that can be revealed by the same key function.

The convergence of the algorithm to the desired images in the jth iteration is evaluated by two average mean-square errors between the two complex functions, before and after the projections in the two domains. Because the phase functions are not changed by the projections, the errors are the average mean square of the difference between magnitudes before and after the projections. The mean-square errors are

$$\begin{aligned} e_{c,j} &= \frac{1}{M_W^2} \int\int |P_1[c_j(\xi, \eta)] - c_j(\xi, \eta)|^2 \, d\xi \, d\eta \\ &= \frac{1}{M_W^2} \int\int |a(\xi, \eta) - |c_j(\xi, \eta)||^2 \, d\xi \, d\eta, \\ e_{g,j} &= \frac{1}{M^2} \int\int |P_2[g_j(x, y)] - g_j(x, y)|^2 \, dx \, dy \\ &= \frac{1}{M^2} \int\int |f(x, y) - |g_j(x, y)||^2 \, dx \, dy, \end{aligned} \tag{4.20}$$

where the size of the input planes is $M \times M$ and the size of the window support of the hidden image in the output plane is $M_W \times M_W$. When the reduction rate of these error functions falls below some predefined value, the iterations can be stopped.

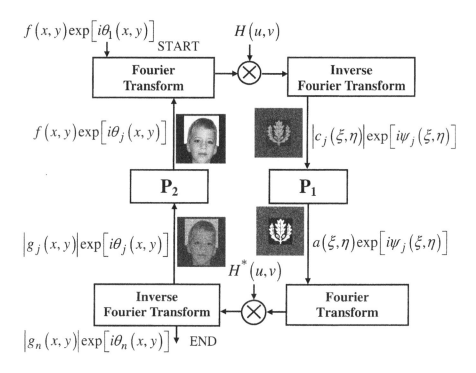

Fig. 4.17. Block diagram of the POCS algorithm used in the first stage of Concealogram production.

As discussed in [8], there are two conditions to guarantee that these errors will never diverge. First, the correlator should be an energy-conserving operator. This property is inherently achieved if $H(u,v)$ is a phase-only function, as is indeed so in the present case. The second condition is satisfied if, among all the functions that belong to the constraint sets, the two projected functions in the jth iteration, $P_1[c_j(\xi,\eta)]$ and $P_2[g_j(x,y)]$, are the functions closest (by means of the mean-square metric) to the functions $c_j(\xi,\eta)$ and $g_j(x,y)$, respectively. Because the phase distributions are the same before and after the projections in both domains, it is obvious that the second condition is also fulfilled. Therefore, the POCS algorithm here can never diverge, and at most the errors may stagnate at some values. Note that the nondiverging feature of the algorithm is the reason to favor phase-only functions as filters in the spatial-frequency domain. The optical realization of the correlator yields another reason to prefer phase-only filters. These filters theoretically do not absorb energy and thus promote maximum system efficiency.

The first stage of the algorithm is terminated in the nth iteration when the correlation between $g_n(x, y)$ and $h^*(-x, -y)$ yields a complex function whose magnitude, it is hoped, is close enough to the hidden image $a(\xi, \eta)$ by means of a relatively small mean-square error. Note that small error values are not guaranteed and depend on the nature of the given images $a(\xi, \eta)$ and $f(x, y)$. The algorithm is terminated before projection P_2, as indicated in Figure 4.17. This is so because, in the next stage, the function $g_n(x, y)$ is binarized, an operation that causes the output image to become only an approximation of the desired image. If we chose to terminate the algorithm after projection P_2, the error in image $a(\xi, \eta)$ would be increased because the magnitude of the correlation between $P_2[g_n(x, y)]$ and $h^*(-x, -y)$ is only an approximation of $a(\xi, \eta)$, and the binarization adds more error. The goal of the second stage in our process is to convert the complex function $g_n(x, y)$ into a binary function $b(x, y)$. By displaying $b(x, y)$ on the input plane, we should obtain the hidden image in the output of the correlator equipped with the same filter function $H(u, v)$. In the usual halftone binarization, only a single, positive, real gray-tone function is converted into a binary function. However, in the present case, there are two positive real functions to be encoded, phase $\theta_n(x, y)$ and magnitude $|g_n(x, y)|$, which is close enough to the visible image $f(x, y)$ if $e_{g,n}$ is indeed small. Following computer-generated hologram (CGH) techniques [22], we propose to encode magnitude $|g_n(x, y)|$ with the dot's area modulation and phase $\theta_n(x, y)$ with the dot's position modulation. Every pixel of the complex gray-tone function $g_n(x, y)$ is replaced by a binary submatrix of size $d \times d$. Inside each submatrix there is a dot represented by some binary value, say 1, on a background of the other binary value, say 0. The area of the (k, l)th dot is determined by the value of $g_n(x, y)$. The position of the (k, l)th dot inside the submatrix is determined by the value of $\theta_n(x_k, y_l)$. Without loss of generality, we choose the shape of the dot as a square. Each dot can be translated to two orthogonal axes, whereas each axis can store an independent phase function and thus a different hidden image. A schematic of one of the (k, l)th cells is shown in Figure 4.18. $b(x, y)$ is the final halftone binary picture, in which an approximation of the visible image $f(x, y)$ [i.e., $|g_n(x, y)|$] is encoded by the area of the dots. $\theta_n(x, y)$ is embedded into the halftone pattern by the position of the dots, and the hidden image $a(\xi, \eta)$ is exposed at the output plane of the correlator that is described next.

$b(x, y)$ is a 2-D grating, and its Fourier transform is an array of 2-D Fourier orders on the spatial-frequency plane separated by M pixels from one another. An example of a typical spatial spectrum of the grating $b(x, y)$ is depicted in Figure 4.19. Following the analysis of the detour-phase CGH [22], it is possible to show that an approximation of the complex function $G_n(u, v)$ (the Fourier transform of $g_n(x, y) = |g_n(x, y)| \exp[i\theta_n(x, y)]$) is obtained in the vicinity of

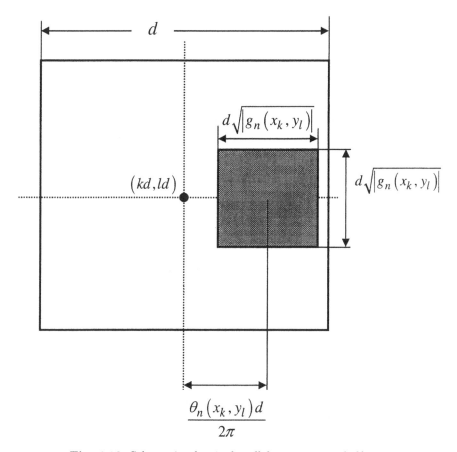

Fig. 4.18. Schematic of a single cell from an entire halftone.

the first Fourier-order component. Thus, the approximation is expressed as

$$B(u,v) \approx G_n(u,v) \qquad \text{if} \quad |u - M\Delta u| \le M\Delta u/2, \quad |v| \le M\Delta v/2, \quad (4.21)$$

where $\Delta u \times \Delta v$ is the size of the pixel in the spatial-frequency plane, and $B(u,v)$ is the Fourier transform of $b(x,y)$. The fact that the distribution about the first order is only an approximation of $G_n(u,v)$ introduces some error in the reconstructed image. This error is inversely dependent on the number of quantization levels used in the halftone picture. The number of quantization levels is naturally determined by cell size d. Future improvements in the phase coding may minimize this error in a fashion similar to the evolution of the CGH from the first detour-phase CGH [22] to the more recent and more accurate iterative CGHs [12]. Because the interesting distribution (that is, the approximation of $G_n(u,v)$) occupies only part of the spatial-frequency plane about the first-order component, we isolate this area of $M \times M$ pixels about point $(M,0)$. Next, the isolated area is multiplied by filter function

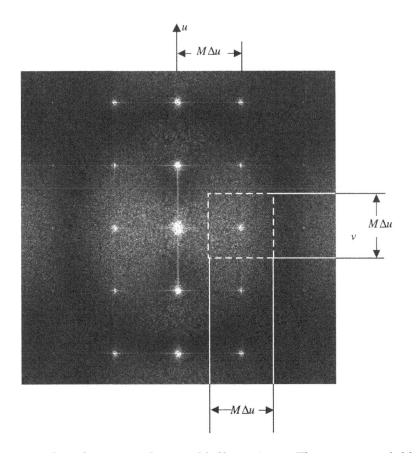

Fig. 4.19. Spatial spectrum of a typical halftone picture. The area surrounded by the white square is the region that is multiplied by the filter.

$H(u, v)$ and inversely Fourier transformed onto the output plane. Because the output distribution is approximately

$$c(\xi, \eta) \approx \mathcal{F}^{-1}\{G_n(u, v)H(u, v)\}, \tag{4.22}$$

the magnitude of output function $|c(\xi, \eta)|$ is approximately equal to the hidden image, $a(\xi, \eta)$.

The optical version of this correlator is shown in Figure 4.20. The halftone figure is displayed on plane $\mathcal{P}_1$ and illuminated by a plane wave. As a result, multiple diffraction orders are obtained on the back focal plane of lens L_1, each at a distance $\lambda f/d$ from its neighbors. The area of the first diffraction order of the size ($\lambda f/Md \times \lambda f/Md$) is multiplied by the phase-only filter

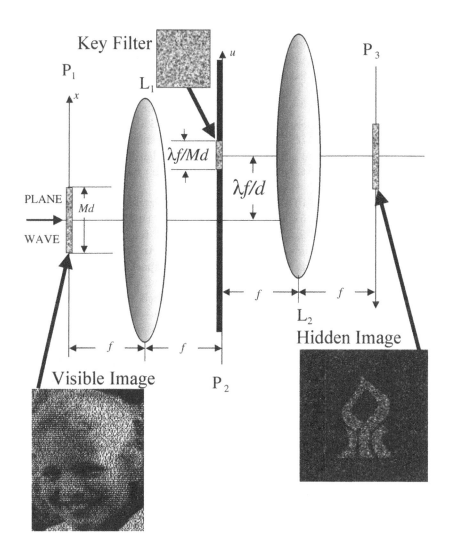

Fig. 4.20. Optical correlator that can be used to reveal the hidden image in the halftone picture.

mask $H(u, v)$, whereas the entire spectral area is blocked. The last product is Fourier-transformed again by lens L_2 onto plane $\mathcal{P}_3$, where the hidden image is assumed to come into sight. In this scheme, we assume that the input halftone picture is a transmission mask that modulates the plane wave. When the halftone picture is printed on regular opaque paper, it has to be recorded first by a digital camera. Then the recorded binary image can be displayed on a spatial light modulator and processed as shown in Figure 4.20.

4.4.2 Experimental Results with Concealogram

The proposed halftone-coding method was examined with a digital correlator. The first example is shown in Figure 4.21. Figure 4.21(a) is the visible halftone image, and Figure 4.21(b) shows the hidden image. Originally the boy's picture was a gray-tone image of the size of 512×512 pixels, and the hidden picture with the acronym "BGU" was a binary image of the size of 64×48 pixels. The size of each one of the three planes in the POCS algorithm was 512×512 pixels. The phase filter $H(u, v)$ distribution was generated by the random-number generator of the computer. The POCS algorithm was iterated on average as many as 50 times. Additional iterations have not meaningfully reduced the two errors $e_{c,j}$ and $e_{g,j}$.

After completing the POCS algorithm, we binarized the resultant complex functions $g_n(x, y)$ according to the above-mentioned rule. The size of each cell in these experiments is 9×9 pixels, and the gray-tone image is quantized with 19 levels of magnitude and 9 levels of phase. An enlarged region of the halftone figure is shown in Figure 4.22(a). For comparison, the same region, but without modulation of the dot position, is shown in Figure 4.22(b).

The robustness of the method was also examined. As mentioned above, this robustness is achieved because each pixel in the output is obtained as a weighted sum of many input pixels. The exact number of pixels that participate in this summation is equal to the size of $h(x, y)$. In the present study we did not take any action to narrow $h(x, y)$, as was done in the project described in Section 4.2, for instance. Thus we expect from our system a maximum degree of robustness to noise and distortions. In the first example of distortions illustrated in Figure 4.23(a), 16% to 40%, in 8% steps, of the pixel values of the halftone pictures were flipped randomly from their original values. The robust behavior was maintained for this type of noise, as is shown by the correlation results. The hidden images revealed from these covered halftone figures are shown in Figure 4.23(b). In another example, illustrated in Figure 4.23(c), the four images were covered in the vicinity of their centers with zero-valued squares of an area that varied from 22% to 55% in 11% steps. The hidden image can still be recognized, even when 55% of the area of the halftone picture is missing, as shown in Figure 4.23(d).

In the next example, six different images were hidden in a single colored halftone image. The original colored picture was separated into its three basic monochromatic red, green, and blue images. Each one of the basic pictures was

(a)

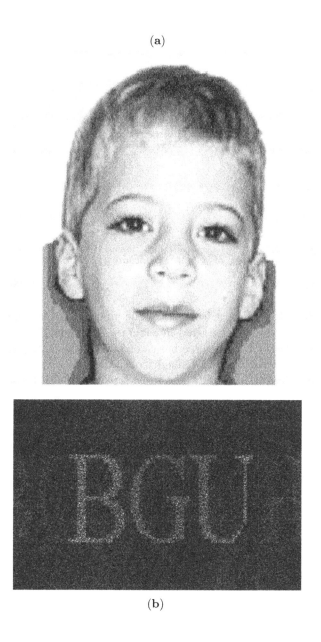

(b)

Fig. 4.21. (a) The resultant halftone image. (b) The hidden image revealed by the correlation between the image in (a) and the key function.

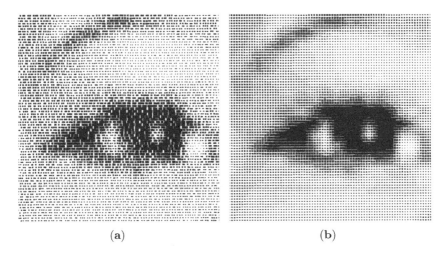

Fig. 4.22. Enlarged region of a halftone picture (a) with and (b) without dot position.

used to conceal two different icons, each one along one of the two Cartesian axes. The results are demonstrated in Figure 4.24. Figure 4.24(a) shows the three halftone images for the three basic colors after the coding process. The original and the reconstructed hidden icons, two in every basic color, are shown in Figure 4.24(b). Finally, the three separated halftone pictures were superposed to a single colored image shown in Figure 4.24(c). Using colored pictures and concealing images along two orthogonal axes enables six different images to be hidden in a single halftone picture.

4.5 Conclusions

We have described three different types of optical security systems based on computer-generated optical diffractive elements. In the first proposed system, one can design two phase-only transparencies for a $4f$ correlator in order to receive a chosen image. The resulting masks can be used for security and encryption systems, as the desired image will be received in the output plane only when the two specific phase masks are placed in the $4f$ correlator. Because computation of the two holograms starts from completely random functions, they cannot be reproduced, even if the output image is known. With the same phase mask in the spatial-frequency plane, the system can produce many images in the output by the introduction of different input masks. Therefore,

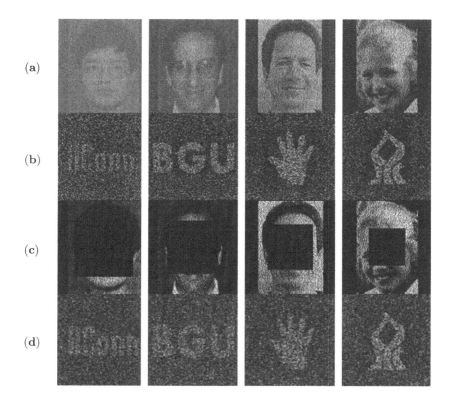

Fig. 4.23. (a) Set of halftone pictures in which various amounts of the pixel values have been randomly flipped from their original values. The number of flipped pixels is varied from 16% at the rightmost figure to 40% at the leftmost figure. (b) Correlation results between the set in (a) and the key function. (c) Set of halftone pictures covered by a zero-valued square with area values that vary from 22% of the picture area at the rightmost figure to 55% at the leftmost figure. (d) Correlation results between the set in (c) and the key function.

in addition to simple verification, the system can provide information on the identity of the authorized person.

The second proposed system is a modification of the first one. The algorithm is implemented on a JTC instead of the $4f$ correlator. According to this method, one can design two phase-only transparencies for a JTC to receive a chosen code or image. The resulting masks can be used for security systems such that the desired code is received on the output plane only when the specific phase masks are placed on the JTC input plane. As in the $4f$ correlator

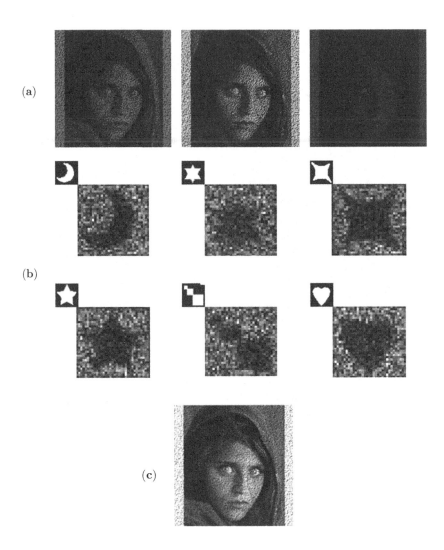

Fig. 4.24. (a) Three halftone pictures in the three basic colors: red, green, and blue. **(b)** The six recontructed icons, each two concealed in a different monochromatic halftone picture of (a). The small insets are the original icons. **(c)** The final colored halftone image obtained by superposition of the three basic images of (a).

case, since the computation of the holograms starts from completely random functions, they cannot be reproduced, even if the output image is known. With the same phase mask of the system's lock, the correlator can produce infinite output images by introducing different input key masks. Therefore, in addition to simple verification, the system can provide information on the identity of the authorized person. The implementation of the security system by a JTC avoids alignment difficulties between the various optical components.

The third system described in this chapter, the Concealogram, is an evolutionary stage in our general concept. This time both the input and the output signals of the system are meaningful images. The Concealogram is used as a method of concealing an arbitrary image in a different arbitrary halftone picture. A digital or optical correlator with a unique key filter can recover the hidden image. Every part of the hidden image is concealed globally in all the points of the Concealogram. This feature increases the robustness of the process to noise and distortions. The amount of the stored data can be significantly increased by shifting the halftone dots along the two orthogonal Cartesian axes. Additional expansion of the concealed data is achieved by use of a colored halftone image that can be considered as a composition of three monochromatic independent Concealograms. For future research, we believe that different optimization and coding algorithms can significantly reduce the noise and error of both input and output pictures. Also, a clever design of the correlator's filter function may extend the distortion-invariance properties of this system.

Acknowledgment

Different parts of the research described in this chapter have been done together with David Abookasis, Ortal Arazi, Haim Goldenfeld, Kathi Kreske, Youzhi Li, and Shlomi Moshkovits. The authors are grateful for their contributions.

References

[1]   Y. Li, K. Kreske, and J. Rosen, "Security and encryption optical systems based on a correlator with significant output images," *Appl. Opt.* **39**, 5295–5301 (2000).

[2]   D. Abookasis, O. Arazi, J. Rosen, and B. Javidi, "Security optical systems based on the joint transform correlator with significant output images," *Opt. Eng.* **40**, 1584–1589 (2001).

[3]   J. Rosen and B. Javidi, "Hidden images in halftone pictures," *Appl. Opt.* **40**, 3346–3353 (2001).

[4]   B. Javidi and J. L. Horner, "Optical pattern recognition for validation and security verification," *Opt. Eng.* **33**, 1752–1756 (1994).

[5]   B. Javidi, G.S. Zhang, and J. Li, "Experimental demonstration of the random phase encoding technique for image encryption and security verification," *Opt. Eng.* **35**, 2506–2512 (1996).

[6]   B. Javidi and E. Ahouzi, "Optical security system with Fourier plane encoding," *Appl. Opt.* **37**, 6247–6255 (1998).

[7]   R.K. Wang, I.A. Watson, and C. Chatwin, "Random phase encoding for optical security," *Opt. Eng.* **35**, 2464–2469 (1996).

[8]   J. Rosen, "Learning in correlators based on projections onto constraint sets," *Opt. Lett.* **18**, 1183–1185 (1993).

[9]   H. Stark, ed., Image Recovery Theory and Application, 1st ed. (Academic, New York, 1987).

[10] P. Réfrégier and B. Javidi, "Optical-image encryption based on input plane and Fourier plane random encoding," *Opt. Lett.* **20**, 767–769 (1995).

[11] B. Javidi, L. Bernard, and N. Towghi, "Noise performance of double-phase encryption compared to XOR encryption," *Opt. Eng.* **38**, 9–19 (1999).

[12] B.K. Jennison, J.P. Allebach, and D.W. Sweeney, "Iterative approaches to Computer-generated holography," *Opt. Eng.* **28**, 629–637 (1989).

[13] J. Rosen, and J. Shamir, "Application of the projection-onto-constraint-sets algorithm for optical pattern recognition," *Opt. Lett.* **16**, 752–754 (1991).

[14] O. Bryngdahl and F. Wyrowski, "Digital holography/Computer-generated holograms," *Prog. Opt.*, **28**, 1–86 (1990).

[15] O. Bryngdahl, T. Scheermesser and F. Wyrowski, "Digital halftoning: synthesis of binary images," *Prog. Opt.*, **33**, 389–463 (1994).

[16] D. Kermisch and P.G. Roetling, "Fourier spectrum of halftone images," *J. Oot. Soc. Am.* **65**, 716–723 (1975).

[17] B. Javidi, "Securing information with optical technologies," *Phys. Today*, March, 27–32 (1997).

[18] F. Hartung and M. Kutter, "Multimedia watermarking techniques," *Proc. IEEE* **87**, 1079–1107 (1999).

[19] F.A.P. Petitcolas, R.J. Anderson, and M.G. Kuhn, "Information hiding—A survey," *Proc. of IEEE* **87**, 1062–1077 (1999).

[20] R.L. van Renesse, ed., *Optical Document Security*, 2nd ed., p. 427, Artech House Boston, 1998.

[21] J.W. Goodman, *Introduction to Fourier Optics*, 2nd ed., p. 232, (McGraw-Hill, New York, 1996).

[22] A.W. Lohmann and D.P. Paris, "Binary Fraunhofer holograms generated by computer," *Appl. Opt.* **6**, 1739–1748 (1967).

5

Optoelectronic Information Encryption with Incoherent Light

Enrique Tajahuerce, Jesús Lancis, Pedro Andrés, Vicent Climent, and Bahram Javidi

5.1 Introduction

Optical processing techniques applied to security and encryption constitute an attractive alternative to electronic procedures because of their parallel operation, high space-bandwidth product, and large degree of freedom [1–4]. Besides, optical processors allow information to be hidden in any of several dimensions, such as phase, wavelength, spatial frequency, or polarization of the light, and they are especially useful for processing information already contained in the optical domain, such as images or holograms.

Several techniques exist to secure and store data by optical means [1,2,5–21]. One significant approach to optically encrypting information consists in filtering the input object with random phase masks by using optical-processor architectures working in the Fraunhofer or the Fresnel diffraction regions. In general, in these methods, the resulting encrypted data are fully complex, and the information must be recorded and stored holographically. This provides high security, as holograms are difficult to counterfeit, and also allows good quality in the reconstructed information. Nevertheless, information recorded in this way is difficult to transmit, and the original information must be reconstructed optically. To solve this problem, some methods based on digital holography have been proposed for securing information in the form of both 2-D or 3-D images [22–25]. These techniques avoid analog holographic recording and allow digital transmission of the information. In general, they involve an optical processor to encrypt, decrypt, or authenticate information and an interferometric system to record the digital holograms. It is important to note that these security optical configurations based on analog or digital recording are compelled, in principle, to work under coherent illumination, and thus they suffer from high sensitivity to misalignment and coherent artifact noise and are restricted to monochromatic signals.

The drawbacks above can be overcome by using optical processors designed to work under spatially incoherent and/or temporally incoherent illumination

[26–30]. These techniques offer several potential advantages over those performed with coherent light. Their multichannel nature makes them suitable
for processing information in a parallel way, and consequently for improving
the signal-to-noise ratio. Furthermore, these methods relax the requirements
of the light source (permitting the use of different nonlaser sources), reduce
the mechanical stability and tolerance requirements of the optical system, and
admit more general input objects.

In particular, the use of white-light point-source illumination (i.e., temporally incoherent but spatially coherent light) allows us to employ broadband
spectrum sources, such as gas-discharge lamps and light-emitting diodes, and
especially to deal with color input signals. Nevertheless, propagation of electromagnetic waves in free space is a physical phenomenon that explicitly depends on the wavelength of the light radiation. This well-known fact leads
to the chromatic dispersion of the field diffracted by a screen illuminated
with a broadband source, both in the Fraunhofer or in the Fresnel diffraction
regions. In this direction, the development of a broadband-dispersion compensation technique permits exploitation of the whole spectral content of the
incoming light [30]. The key to the compensation procedure lies in achieving a
wavelength-independent diffraction pattern (i.e., an incoherent superposition
of the monochromatic versions of a selected diffraction pattern in a single
plane and with the same scale for all the wavelengths of the incident light).

Some advantages of incoherent optical processing can also be achieved by
use of spatially incoherent monochromatic illumination. Spatially incoherent
processors are linear in irradiance instead of complex amplitude [26–29]. In
general, an incoherent processing system is an imaging system in which the
incoherent spatial impulse response has been tailored for a particular data-
processing task. Specifically, the irradiance distribution at the output plane
of the optical setup is given by a convolution integral between the incoherent
point-spread function (PSF) and the image-irradiance distribution predicted
by geometrical optics.

Several attempts have been made to extend conventional coherent optical
processing techniques to both white-light and spatially incoherent sources in
order to exploit the advantages of natural incoherent illumination. Nevertheless, the use of white light in the preceding incoherent systems introduces
an obstacle, the dependence of the scale of the incoherent PSF—the modulus
squared of the Fourier transform of the aperture amplitude transmittance—on
the wavelength of the incoming radiation. Again, broadband-dispersion compensation is required to obtain an undistorted PSF for all the monochromatic
channels. Recently, a totally incoherent optical processor has been demonstrated and applied to recognition of color signals under natural light [31–33].

In this work, we describe a method to extend optical encryption techniques to work under totally incoherent illumination. The optical encrypting
system is a totally incoherent dispersion-compensated processor constituted
by refractive and diffractive lenses. This configuration permits us to deal with
self-luminous color inputs and to perform, with a single filter, the same spatial-

filtering operation for all the spectral channels of the illumination simultaneously. We show that, with a random phase-only mask located at the aperture plane, the point-spread function (PSF) of our achromatic optical processor constitutes a white-light speckle pattern. The convolution in irradiance of this chromatically compensated PSF with the input signal allows us to encrypt information. The encrypted output is then recorded as an intensity image that can be easily stored and transmitted optically or electronically. Afterwards, decryption can be performed also with optical or digital methods.

The chapter is structured in the following parts. First, in Section 5.2, we briefly describe the principle of optical encryption with random phase codes by using an optical Fourier processor. In Section 5.3 we discuss the problems in achieving an encryption system with incoherent light. Section 5.4 deals with the design and development of a totally incoherent achromatic optical processor. In Section 5.5, we pay attention to the application of this incoherent optical processor to develop an encryption system with natural illumination by using a random phase mask. Finally, in Section 5.6 we present our conclusions.

5.2 Information Encryption by Phase-Mask Encoding

Security techniques by random phase encoding are based on using random phase masks in the input plane and/or the Fourier plane of an optical Fourier processor to encrypt the input information [1,2]. In this way, the optical processor converts the input image into a wide-sense stationary white noise. In this section, we review briefly the basis of this idea, although many other variants of this pioneering work have been proposed in the recent years [5–25].

Let $t(x,y)$ denote the image to be encrypted and $U(x,y)$ the encrypted image, where x and y denote the spatial coordinates, and the functions are sampled to have $N \times N$ pixels. Let $p(x,y)$ and $b(u,v)$ be two independent white sequences uniformly distributed in the interval $[0,1]$. Variables u and v denote the coordinates in the Fourier domain. A double-phase encryption of the input image $t(x,y)$ is obtained by two operations. First, $t(x,y)$ is multiplied by a phase-mask function $\exp[i2\pi p(x,y)]$. Second, the product $p(x,y)\exp[i2\pi p(x,y)]$ is convolved by a function $h(x,y)$, which is the impulse response of a phase-only transfer function $H(u,v)?\exp[i2\pi b(u,v)]$. In this way, the encrypted image $U(x,y)$ is given by the equation

$$U(x,y) = \{t(x,y)\exp[i2\pi p(x,y)]\} * h(x,y), \qquad (5.1)$$

where $*$ denotes the convolution operation.

Figure 5.1 illustrates the optical configuration used to encrypt an image. The input image, $t(x,y)$, on which the first random phase code, $\exp[i2\pi p(x,y)]$, is attached, is located in the input plane of the optical processor. This input plane is illuminated by a parallel coherent light beam. In the Fraunhofer plane of the optical processor, the Fourier transform of the input, the product of

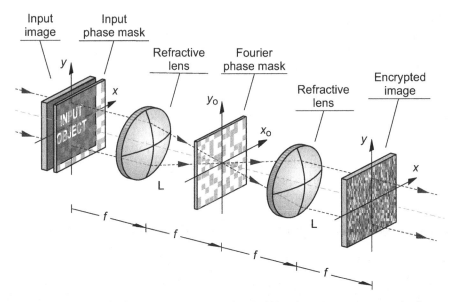

Fig. 5.1. Optical implementation of the double-phase-encrypting method.

the image, and the first random phase mask are multiplied by the second random phase code, $\exp[i2\pi b(u, v)]$. With another Fourier transform, we obtain the encrypted version of the input image in the output plane. This encrypted image can be decrypted only when the corresponding phase codes, referred to as keys, are used for the decryption. However, if $t(x, y)$ is real, the encoded image can be decrypted using only the second phase key.

A similar optical setup is used for decryption. The encrypted image $U(x, y)$ is located at the input plane of the optical processor, as is shown in Figure 5.2. This input is Fourier-transformed, and the result is multiplied by the complex conjugate of the Fourier phase mask used in the encryption method, $\exp[-i2\pi b(u, v)]$. At the output plane, the function $t(x, y)\exp[i2\pi p(x, y)]$ is obtained. The original image is then recovered in the space domain by using the complex conjugate of the first key, $\exp[-i2\pi p(x, y)]$. If the image is real and positive, the phase function over the image can be simply removed by using an intensity-sensitive device such as a CCD (charge-coupled device) camera.

Several practical implementations, by using different optical processors and some modifications of the original method, have been suggested to simplify the technique or to improve security. For example, random phase masks located at the Fresnel diffraction region instead of the Fraunhofer one have been proposed to encrypt the input information. Note that, in general, as the encrypted image is a complex function, a holographic method must be used to record its phase and amplitude. Some techniques use photorefractive crystals as recording media. Recently, digital holography has also been applied

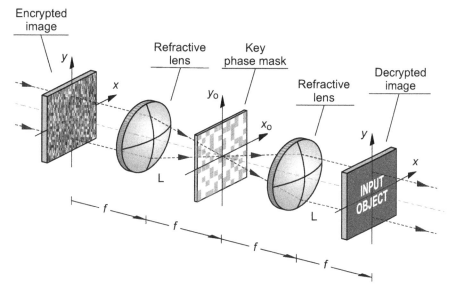

Fig. 5.2. Optical system for decryption.

to avoid conventional holographic recording and to allow electronic recording and transmission.

As was mentioned in Section 5.1, the extension of these optical systems to work under broadband point-source illumination permits exploitation of the color information of the input scenes. Furthermore, employment of spatially incoherent illumination allows one to use conventional light sources and to increase the mechanical stability of the optical system. However, these optical encryption techniques are restricted, in principle, to work under both temporal and spatially coherent illumination because of the strong chromatic dependence of light diffraction with the wavelength and the use of phase masks at the input plane. The design of chromatically compensated optical processors able to work with totally incoherent light will be the key to developing optical security systems with totally incoherent light.

5.3 Spatially Incoherent Optical Processor for Information Encryption

As is well-known, any spatially incoherent imaging system can be considered as a diffraction-based spatially incoherent processor. The cornerstone of the procedure consists in properly tailoring the pupil function of the imaging configuration. In this way, it is possible to achieve at the output plane the convolution operation between the input irradiance and the resulting incoherent PSF. Several applications based on this idea, working with monochromatic illumination, have been reported in the literature, in particular for implementing

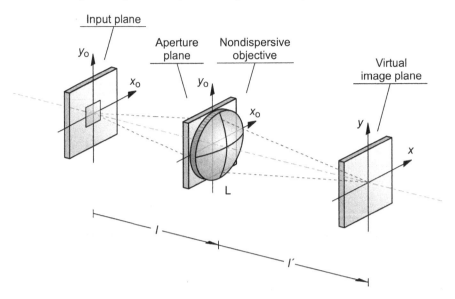

Fig. 5.3. Incoherent processor composed of a spatially incoherent imaging system.

optical correlations under incoherent light [34–39]. These optical processors exploit all the advantages of the use of incoherent light but also are restricted to work with real and nonnegative inputs and outputs. However, this last fact can be of interest if we want to record and transmit information electronically.

In incoherent imaging, the spectral bandwidth of the illuminating source is usually of no concern. Nevertheless, the scale of the PSF—the optical Fourier transform of the pupil transmittance—varies linearly with the wavelength. Thus, in optical processing operations, a broad-spectrum source produces severe chromatic errors. This fact restricts the spectral bandwidth that can be employed in an incoherent optical processor for encrypting applications, as we show next.

The irradiance distribution at the output plane of a conventional imaging system, such as that shown in Figure 5.3, is given by the convolution integral between the image-irradiance distribution predicted by geometrical optics, $I_O(x, y; \lambda)$, and the incoherent PSF for each wavelength λ. Thus, with both spatially and temporally incoherent illumination, the irradiance at the output plane, $I_F(x, y)$, is given by the incoherent superposition

$$I_F(x, y) = \int I_O\left(\frac{x}{M_o}, \frac{y}{M_o}; \lambda\right) * \left|\tilde{p}\left(\frac{x}{\beta(\lambda)}, \frac{y}{\beta(\lambda)}; \lambda\right)\right|^2 S(\lambda) d\lambda, \qquad (5.2)$$

where M_o is the lateral magnification between input and output planes and $S(\lambda)$ is the spectral-radiance power of the light source. In Eq. (5.2), $\tilde{p}(x, y; \lambda)$ denotes the Fourier transform of the aperture, $p(x, y; \lambda)$, for each wavelength, and $\beta(\lambda)$ is the wavelength-dependent scale factor of this transformation.

By locating a transparent rough surface with random height profile $h(x, y)$ at the aperture plane, $p(x, y; \lambda)$ can be written, for small scattering angles, as

$$p(x, y; \lambda) = A(x, y) \exp \left[j \frac{2\pi}{\lambda} (n - 1) h(x, y) \right],$$ (5.3)

where $A(x, y)$ denotes the circular aperture extension and n is the index of refraction of the transparent medium [40]. Function $\tilde{p}(x, y; \lambda)$ provides a zero-mean normally distributed random process provided that the transverse scale of the variations of $h(x, y)$ is much smaller than the size of $A(x, y)$, and the random phase function in Eq. (5.3) is uniformly distributed between 0 and 2π; that is, the root-mean-squar (rms) value of the optical-path difference in Eq. (5.3) associated to the height profile, σ_h, is such that [41]

$$\sigma_h \gg \lambda.$$ (5.4)

The speckle patterns represented by function $\tilde{p}(x, y; \lambda)$ depend on the wavelength in two ways. First, even with a nondispersive medium in the aperture, a dependence with the wavelength remains in the phase term of Eq. (5.3). Second, the scale factor of the Fourier transformation, $\beta(\lambda)$, changes with the wavelength. Consequently, the polychromatic speckle irradiance associated to the PSF of a conventional incoherent processor exhibits a radial structure. Thus the speckle irradiance is not a wide-sense stationary process. Furthermore, processing the input object with such a system will require recording the output irradiance with a multispectral sensor.

With respect to the first chromatic dependence, it has been shown that the speckle patterns corresponding to different wavelengths are almost equal if

$$\sigma_h \ll \lambda_0^2 / \Delta\lambda,$$ (5.5)

where $\Delta\lambda$ denotes the spectral bandwidth of the incident light and λ_0 is the mean wavelength [40]. By requiring this condition for the random filter, we guarantee a correlation between the speckles generated by different wavelengths. To fix the second problem, we propose to use an optical processor with a dispersion-compensated PSF.

5.4 All-Incoherent Dispersion-Compensated Optical Processor

The appropriate combination of refractive and diffractive lenses permits one to achieve achromatic Fraunhofer diffraction patterns of any transparency under white-light point-source illumination [42–47]. In particular, a scale-tunable achromatic Fourier transform has been demonstrated by means of a diffractive lens doublet [45]. Also, a very simple system that allows one to obtain the Fourier transformation of the input in the same axial position for all the wavelengths of the illumination source has been proposed [46]. The setups above

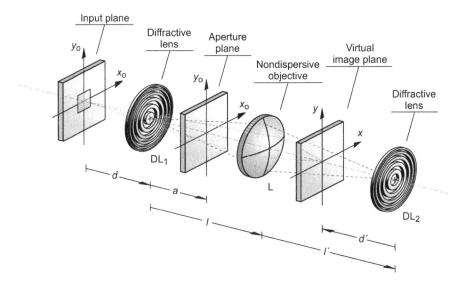

Fig. 5.4. Schematic diagram of the all-incoherent optical processor.

have made it possible to implement a variety of Fourier-based achromatic processing operations. In these setups, the chromatic behavior associated with diffractive lenses is fully exploited to achieve the chromatic compensation.

Based on the ideas above, a broadband spatially incoherent optical processor that performs the convolution of any color input scene with a reference pattern under natural (both spatial and temporally incoherent) illumination [31–33]. This novel all-incoherent optical processor, which simply consists of an hybrid lens triplet composed of two diffractive lenses and a nondispersive refractive objective lens located in between, is essentially a spatially incoherent wavelength-independent imaging system that is both linear and shift-invariant. In contrast to the conventional case, this optical design exhibits a dispersion-compensated point-spread function (PSF).

This hybrid (diffractive-refractive) imaging setup is shown in Figure 5.4. DL_1 and DL_2 are two diffractive lenses, with image focal lengths Z_0 and Z_0' for the reference wave number σ_0, and the nondispersive refractive objective L, with positive focal length f, is located in between. Axial distances s and s' denote arbitrary, but fixed, spacing between the optical elements. This optical architecture is a particular case of the general three-lens configuration studied by Morris [48].

Let the input object O, incoherently illuminated, be located at a distance d from DL_1. First, we are interested in achieving a final image O′, provided by the whole optical system, that is wavelength-independent, if we neglect the secondary spectrum of the lens L. Using elementary geometrical-optics concepts, we recognize that the position and the scale factor of O′ are the same for all the wavelengths when DL_1 and DL_2 are conjugated through L and the image of DL_1 has the same focal distance from DL_2 but the opposite sign. These two requirements are expressed in mathematical terms by

$$s' = -Ms \qquad (5.6)$$

and

$$Z_0' = -M^2 Z_0, \qquad (5.7)$$

where

$$M_0 = -\frac{f}{s - f} \qquad (5.8)$$

is the lateral magnification between these conjugated planes. Note that a diffractive study using Fresnel diffraction theory leads to the same results if we do not consider the finite size of the three lenses. In this way, the action of DL_1 is really cancelled by DL_2, and the image O' appears just at the conjugate plane of O through L; that is, at a distance d' from DL_2 such that

$$d' = -MM_0 d, \qquad (5.9)$$

where

$$M_0 = -\frac{f}{d + s - f} \qquad (5.10)$$

is the lateral magnification between the planes O and O'. In general, the final image is virtual and then an additional refractive nondispersive objective lens placed behind DL_2 is required to produce a real output.

Next, we consider the effect of a finite aperture in our setup. We assume that the aperture stop, with amplitude transmittance $p(x, y)$, is located between DL_1 and L. Now, we evaluate the monochromatic incoherent PSF, $h(x, y; \sigma)$, that is, the monochromatic irradiance distribution for the wave number σ at the image plane due to a single point radiator located at the center of the object plane (note that $\sigma = 1/\lambda$ instead of λ is used in this section for convenience). To this end, first we recognize that the amplitude distribution, for the wave number σ, over the aperture plane is

$$U(x, y; \sigma) = \exp\left[i\pi \frac{\sigma}{z}\left(x^2 + y^2\right)\right]^2 p(x, y). \qquad (5.11)$$

The symbol z stands for

$$z = a - \frac{1}{\frac{\sigma_0}{Z_0}\sigma - \frac{1}{d}}, \qquad (5.12)$$

where a denotes the distance between DL_1 and the aperture plane. Second, the incoherent PSF is obtained by propagating the function $U(x, y; \sigma)$ through the remaining part of the optical system by use of the Fresnel diffraction integral. The task above can be done in one step using the ABCD matrix corresponding to the propagation between the aperture plane and the output [49]. In this way, and aside from some irrelevant factors, we can write

$$
h(x, y; \sigma)
$$
$$
= \left| \int\!\!\!\int_{-\infty}^{\infty} p(x', y') \exp\left[i\pi\sigma \frac{A}{B}(x'^2 + y'^2) \right] \exp\left[-\frac{i2\pi\sigma}{B}(xx' + yy') \right] dx'\, dy' \right|^2 .
$$
(5.13)

Note that, in order to write Eq. (5.13), the propagation of interest must be modeled by the following collection of elements: a fictitious dispersive lens of focal length $-z$, which takes into account the spherical illumination of the aperture, a free-space propagation of length $(s - a)$, the objective L, a free-space propagation of length s', the diffractive lens DL$_2$, and, finally, a free-space propagation of length d'.

Taking into account the form for the ABCD matrices corresponding either to a free propagation or to the passage through a lens [49], it is straightforward to show that the overall matrix coefficients A and B are given by

$$
A = 0, \quad B(\sigma) = M_0 \left[\frac{ad\sigma_0}{Z_0\sigma} - (a + d) \right].
$$
(5.14)

It is easy to recognize that the condition $A = 0$ is equivalent to achieving a wavelength-independent image at the output plane. From Eqs. (5.13) and (5.14), we recognize that the function $h(x, y; \sigma)$ can be found from the modulus square of the Fourier transform of $p(x, y)$ for any σ. Mathematically,

$$
h(x, y; \sigma) = \left| \tilde{p}\left(\frac{x}{\beta(\sigma)}, \frac{y}{\beta(\sigma)} \right) \right|^2 ,
$$
(5.15)

where $\tilde{p}$ denotes the 2-D Fourier transform of p and the function $\beta(\sigma)$ stands for $B(\sigma)/\sigma$.

The functional dependence of the scale factor $\beta(\sigma)$ on σ indicates that the scaling of the incoherent PSF is wavelength-dependent. An achromatic correction for the scale can be achieved if the derivative of the function $\beta(\sigma)$ with respect to σ vanishes at a certain design wave number σ_0. In mathematical terms, we require

$$
\left. \frac{\partial\beta(\sigma)}{\partial\sigma} \right|_{\sigma_0} = 0.
$$
(5.16)

Equation (5.7) leads to the constraint

$$
a = \frac{dZ_0}{2d - Z_0},
$$
(5.17)

which fixes, in terms of d and Z_0, the axial location of the aperture plane. Finally, we find that the scaling of the dispersion-compensated PSF is

$$
\beta(\sigma) = \frac{M_0 d^2}{2d - Z_0} \frac{\sigma_0 - 2\sigma}{\sigma^2}.
$$
(5.18)

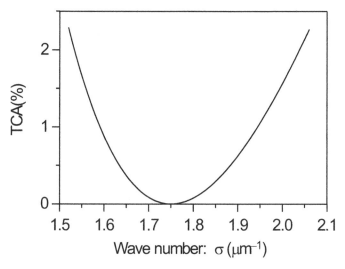

Fig. 5.5. Plot of the residual transversal chromatic aberration, TCA, associated with the incoherent PSF of the optical setup in Figure 5.4. We assume in this representation white-light illumination.

As a result of the achromatic correction, a low residual transversal chromatic aberration TCA still remains. A good indication of the value of TCA expressed as a percentage, could be given by

$$\mathrm{TCA}(\sigma) = 100\frac{\beta_0 - \beta(\sigma)}{\beta_0} = 100\frac{(\sigma - \sigma_0)^2}{\sigma^2}, \tag{5.19}$$

where $\beta_0 = \beta(\sigma_0)$. A plot of the function $\mathrm{TCA}(\sigma)$ is shown in Figure 5.5. In this plot, we select $\sigma_0 = 1.75\ \mu\mathrm{m}^{-1}$ and we assume that the effective end wave numbers of the broad-spectrum source are $\sigma_1 = \sigma_C = 1.52\ \mu\mathrm{m}^{-1}$ and $\sigma_2 = \sigma_F = 2.06\ \mu\mathrm{m}^{-1}$, which correspond, respectively, to the Fraunhofer lines labeled C and F of the visible spectrum. It appears that the optical setup in Figure 5.4 exhibits a dispersion-compensated PSF with a maximum chromatic error of less than 2.5% over the visible region.

Finally, we are concerned with the irradiance distribution at the output plane when an input object irradiance $I_0(x, y)$ is considered rather than a single radiator. Because the optical setup in Figure 5.4 is a spatially incoherent, wavelength-independent imaging system that is both linear and shift-invariant, the image-irradiance distribution, $I_F(x, y)$, can be written, in a first-order approximation, as

$$I_F(x, y) = I_0\left(\frac{x}{M_0}, \frac{y}{M_0}\right) * \left|\left(\tilde{p}\frac{x}{\beta_0}, \frac{y}{\beta_0}\right)\right|^2, \tag{5.20}$$

where the asterisk symbol denotes the convolution operation. Under the approximation above, a convolution integral between the spatially incoherent quasiwavelength-independent PSF and the image distribution predicted by geometrical optics is achieved. Now, it is possible to perform a wide variety of color optical operations with natural light by specifically tailoring the aperture transmittance $p(x, y)$.

5.5 Optical Security and Encryption with Totally Incoherent Light

In Section 5.3, we have described the effect of introducing a random phase mask at the aperture plane of a conventional incoherent processor. By locating the random phase mask at the aperture plane of the chromatically compensated incoherent processor reviewed in Section 5.4, the average size of the speckle, which is roughly equal to the size of the central lobe of the far-field diffraction pattern of the aperture extension $A(x, y)$, is the same for all the wavelengths of the incident light. Moreover, under an achromatic Fourier transformation, the only effect of the spectral bandwidth on the degree of polychromatic speckle correlation is on the speckle contrast, and this contrast is independent of the position in the speckle pattern [50]. Thus, if the random profile of the phase mask fulfills the above-mentioned requirements for σ_h, the average shape and size of the polychromatic speckle constituting the PSF is, in a first-order approximation, independent of the wavelength. The optical system is shown in Figure 5.6. Now, Eq. (5.2) can be written as

$$I_{\mathrm{E}}(x, y) = \left[\int I_{\mathrm{O}}\left(\frac{x}{M_0}, \frac{y}{M_0}; \lambda \right) S(\lambda) d\lambda \right] * \left| \tilde{p}\left(\frac{x}{\beta_0}, \frac{y}{\beta_0} \right) \right|^2, \qquad (5.21)$$

where the last term in Eq. (5.21), hereafter called $P_{\mathrm{E}}(x, y)$, is the totally incoherent achromatic PSF provided by the incoherent processor. This PSF is now a white random speckle pattern that constitutes a wide-sense stationary random process. The convolution of the incoherent image of the object with $P_{\mathrm{E}}(x, y)$ allows encryption of the information contained in the input into a random-like irradiance pattern.

Optical decryption can be performed, in principle, using again the optical system in Figure 5.6 with the same phase function at the aperture plane but rotated $180°$. By properly adapting the scale factors, a second convolution of $I_{\mathrm{E}}(x, y)$ with $P_{\mathrm{E}}(-x, -y)$ provides the autocorrelation of the PSF convolved with the input object irradiance. However, this method implies a low-pass filtering of the input. Hence, it is necessary to decrypt with an optical transfer function (OTF) inverse to that used in encryption. This implies, in general, a bipolar decryption PSF, $P_{\mathrm{D}}(x, y)$, generated as we show next. First, the incoherent encrypting PSF, $P_{\mathrm{E}}(x, y)$, associated with the key phase mask is measured with the optical system in Figure 5.6, and the encrypting OTF is evaluated by Fourier transformation. The inverse of the previous OTF constitutes the decrypting OTF. This function can be used directly as

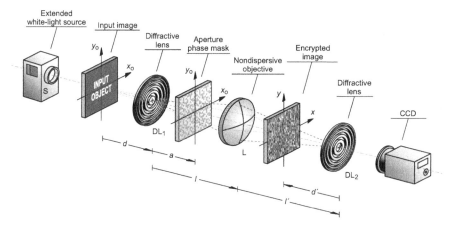

Fig. 5.6. Dispersion-compensated processor for optical encryption with incoherent light.

the key to decrypt digitally by inverse filtering. Optical decryption requires another inverse Fourier transformation to evaluate the required decrypting PSF, $P_D(x, y)$, which is decomposed into a positive and a negative component, $P_D(x, y) = P_D(x, y)P_D(x, y)$ [43]. Then, two decrypting aperture filters are generated to provide each of these PSFs optically. The final decrypted image is the difference between the optical convolutions of these PSFs and the encrypted image. If the encrypting OTF has zero values, the corresponding inverse is undetermined. In this case, a zero value is assigned to the decrypting OTF.

As a first experimental verification, an optical encryption of a two-dimensional image with totally incoherent light combined with digital decryption was carried out. Figure 5.7 shows a diagram of the whole process: first, recording of the encrypted image and the random PSF optically; and, second, decryption of the secured information digitally by computer by using the inverse OTF. The optical system in Figure 5.6 was implemented by using two diffractive lenses with focal length $Z_0 = 200$ mm and $Z_0' = 200$ mm for the reference wavelength $\lambda_0 = 535.5$ nm and two nondispersive objectives, L_1 and L_2, with focal distances $f = f' = 100$ mm. A transparency with information to be encrypted was located at the input plane. It was illuminated with a spatially incoherent light beam arising from a xenon white lamp.

In Figure 5.8(a), we show a gray-level picture of the irradiance distribution provided at the output plane by the whole optical system with no filter at the aperture. To encrypt this image, a random phase mask was located at the aperture plane. The phase mask was a diffuser of randomly varying thickness with a correlation length of 6 μm in both the x and y directions. The irradiance distribution corresponding to the PSF provided by this aperture mask is shown in Figure 5.8(b). It was registered by locating a commercial pinhole

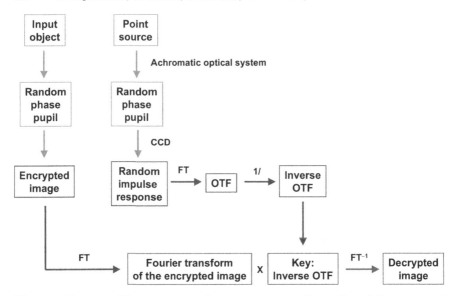

Fig. 5.7. Diagram of the encryption–decryption process. Recording of the encrypted image and the key (the random PSF) is performed optically. Decryption of the encrypted image with the key is performed digitally.

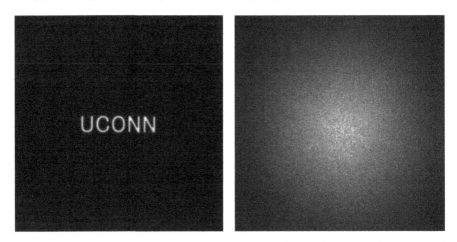

Fig. 5.8. Optical encryption experiment with totally incoherent light. Gray-level picture of: (a) the input object as imaged by the optical processor in Figure 5.6 without an aperture filter; and (b) the PSF associated with the system in Figure 5.6 with the random phase filter at the aperture plane.

Fig. 5.9. Gray-level picture of the irradiance corresponding to: (**a**) the encrypted image of the input in Figure 5.8(a) provided by the system in Figure 5.6, and (**b**) the result of the decryption.

with a diameter of 10 μm at the input plane instead of the input object. Note the lack of chromatic radial structure in the speckle pattern.

In Figure 5.9(a), we show the irradiance distribution of the encrypted image obtained when both the input object and the phase mask are located in the incoherent optical system. Decryption of the previous image was performed by optically measuring the OTF of the optical system, starting from the PSF in Figure 5.8(b), and digitally filtering the Fourier transform of the encrypted image with the corresponding inverse OTF. The resulting decrypted image is shown in Figure 5.9(b). The original input image is recovered, although with some noise.

Finally, Figure 5.10(a) shows the irradiance distribution corresponding to the PSF of a different phase mask located at the aperture plane of the optical system. The result of digitally filtering the Fourier transform of the encrypted image in Figure 5.9(a) with the inverse OTF of this erroneous PSF is shown in Figure 5.10(b). We can see clearly that the original information is not recovered in this case.

5.6 Conclusions

In summary, we have reviewed a method that allows us to encrypt information by optical means with totally incoherent light. To this end, first we have described an incoherent processor with a wavelength-compensated PSF. This processor performs, with a single filter, the same spatial-filtering operation for all the spectral channels of the illumination simultaneously. The encryption key is constituted by a phase-only mask, thus providing high security against

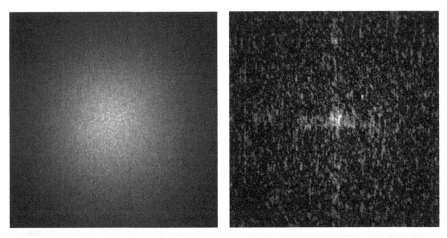

Fig. 5.10. Gray-level picture of the irradiance corresponding to: (**a**) teh erroneous PSF; and (**b**) the result of a wrong decryption of the image in Figure 5.9(a) with the incorrect PSF.

counterfeiting. Among others advantages of this method is an improvement of the tolerance to misalignment and the possibility of using color inputs and self-luminous displays. Output encrypted information is recorded as an intensity image that can be easily stored and transmitted optically or electronically. Decryption or authentication also can be performed optically or digitally. In the experiment described, a set of characters has been encrypted optically with a random phase function and decryption has been performed digitally.

Acknowledgments

E. Tajahuerce acknowledges financial support by the agreement between the Universitat Jaume I and the Fundació Caixa Castelló-Bancaixa (Ref. P1-1B2002-29), Spain.

References

[1]   B. Javidi and J.L. Horner, "Optical pattern recognition for validation and security verification," *Opt. Eng.* **33**, 1752–1756 (1994).

[2]   P. Réfrégier and B. Javidi, "Optical image encryption based on input plane and Fourier plane random encoding," *Opt. Lett.* **20**, 767–769 (1995).

[3]   H.-Y. Li, Y. Qiao, and D. Psaltis, "Optical network for real-time face recognition," *Appl. Opt.* **32**, 5026–5035 (1993).

[4]   C.L. Wilson, C.I. Watson, and E.G. Paek, "Combined optical and neural network fingerprint matching," *Proc. SPIE* **3073**, 373–382 (1997).

[5]   J.F. Heanue, M.C. Bashaw, and L. Hesselink, "Encrypted holographic data storage based on orthogonal-phase-code multiplexing," *Appl. Opt.* **34**, 6012–6015 (1995).

[6] R.K. Wang, I.A. Watson, and C.R. Chatwin, "Random phase encoding for optical security," *Opt. Eng.* **35**, 2464–2469 (1996).

[7] B. Javidi, A. Sergent, G. Zhang, and L. Guibert, "Fault tolerance properties of a double phase encoding encryption technique," *Opt. Eng.* **992**, 992–998 (1997).

[8] R.L. van Renesse, *Optical Document Security* (Artech House, Boston, 1998).

[9] N. Yoshikawa, M. Itoh, and T. Yatagai, "Binary computer-generated holograms for security applications from a synthetic double-exposure method by electron-beam lithography," *Opt. Lett.* **23**, 1483–1485 (1998).

[10] F. Goudail, F. Bollaro, B. Javidi, and Ph. Réfrégier, "Influence of a perturbation in a double phase-encoding system," *J. Opt. Soc. Am. A* **15**, 2629–2638 (1998).

[11] G. Unnikrishnan, J. Joseph, and K. Singh, "Optical encryption system that uses phase conjugation in a photorefractive crystal," *Appl. Opt.* **37**, 8181–8186 (1998).

[12] B. Javidi and E. Ahouzi, "Optical security system with Fourier plane encoding," *Appl. Opt.* **37**, 6247–6255 (1998).

[13] J.L. Horner and B. Javidi, eds., *Optical Engineering, Special Issue on Optical Security*, Vol. **38** (SPIE, Bellingham, WA, 1999).

[14] O. Matoba and B. Javidi, "Encrypted optical memory system using three-dimensional keys in the Fresnel domain," *Opt. Lett.* **24**, 762–764 (1999).

[15] P.C. Mogensen and J. Glückstad, "Phase-only optical encryption," *Opt. Lett.* **25**, 566–568 (2000).

[16] B. Javidi and T. Nomura, "Polarization encoding for optical security systems," *Opt. Eng.* **39**, 2439–2443 (2000).

[17] G. Unnikrishnan, J. Joseph, and K. Singh, "Optical encryption by double-random phase encoding in the fractional Fourier domain," *Opt. Lett.* **25**, 887–889 (2000).

[18] Z. Zalevsky, D. Mendlovic, U. Levy, and G. Shabtay, "A new optical random coding technique for security systems," *Opt. Commun.* **180**, 15–20 (2000).

[19] T. Nomura and B. Javidi, "Optical encryption using a joint transform correlator architecture," *Opt. Eng.* **39**, 2031–2035 (2000).

[20] B. Zhu, S. Liu, and Q. Ran, "Optical image encryption based on multifractional Fourier transforms," *Opt. Lett.* **25**, 1159–1161 (2000).

[21] O. Matoba and B. Javidi, "Encrypted optical storage with angular multiplexing," *Appl. Opt.* **38**, 7288–7293 (1999).

[22] B. Javidi and T. Nomura, "Securing information by means of digital holography," *Opt. Lett.* **25**, 29–30 (2000).

[23] E. Tajahuerce, O. Matoba, S.C. Verrall, and B. Javidi, "Optoelectronic information encryption using phase-shifting interferometry," *Appl. Opt.* **39**, 2313–2320 (2000).

[24] S. Lai and M.A. Neifeld, "Digital wavefront reconstruction and its application to image encryption," *Opt. Commun.* **178**, 283–289 (2000).

[25] E. Tajahuerce and B. Javidi, "Encrypting three-dimensional information with digital holography," *Appl. Opt.* **39**, 6595–6601 (2000).

[26] G.L. Rogers, *Noncoherent Optical Processing* (Wiley, New York, 1977).

[27] W.T. Rhodes and A.A. Sawchuck, "Incoherent optical processing," in S.H. Lee, ed., *Optical Information Processing: Fundamentals* (Springer-Verlag, Berlin, 1981), Chapter 3.

[28] H. Bartelt, S.K. Case, and R. Hauck, "Incoherent-optical processing," in H. Stark, ed., *Applications of Optical Fourier Transforms* (Academic, Orlando, 1982), Chapter 12.

[29] F.T.S. Yu, *White-Light Optical Signal Processing* (Wiley, New York, 1985).

[30] G.M. Morris and D.A. Zweig, "White-light Fourier transformations," in J.L. Horner, ed., *Optical Signal Processing* (Academic, San Diego, 1987), Section 1.2.

[31] P. Andres, V. Climent, J. Lancis, G. Mínguez-Vega, E. Tajahuerce, and A.W. Lohmann, "All-incoherent dispersion-compensated optical correlator," *Opt. Lett.* **24**, 1331–1333 (1999).

[32] A. Pe'er, D. Wang, A.W. Lohmann, and A. Friesem, *Opt. Lett.* **24**, 1469 (1999).

[33] G. Mínguez-Vega, J. Lancis, E. Tajahuerce, V. Climent, M. Fernandez-Alonso, A. Pons, and P. Andres, "Scale-Tunable Optical Correlation with Natural Light," *Appl. Opt.* **40**, 5910–5911 (2001).

[34] J.D. Armitage and A.W. Lohmann, "Character recognition by incoherent spatial filtering," *Appl. Opt.* **4**, 461–467 (1965).

[35] A.W. Lohmann, "Matched filter with self-luminous objects," *Appl. Opt.* **7**, 561–563 (1968).

[36] A.W. Lohmann and H.W. Werlich, "Incoherent matched filtering with Fourier holograms," *Appl. Opt.* **10**, 670–672 (1971).

[37] J. van der Gracht and J.N. Mait, "Incoherent pattern recognition with phase-only filters," *Opt. Lett.* **17**, 1703–1705 (1992).

[38] S. Gorodeisky and A.A. Friesem, "Phase filters for correlation with incoherent light," *Opt. Comm.* **100**, 421 (1993).

[39] J. Ding, M. Itoh, and T. Yatagai, "Optimal incoherent correlator for noisy gray-tone image recognition," *Opt. Lett.* **20**, 2411–2413 (1995).

[40] M. Françon, *Laser Speckle and Applications in Optics* (Academic, London, 1979).

[41] J.C. Dainty, *Laser Speckle and Related Phenomena* (Springer-Verlag, Berlin, 1975).

[42] R.H. Katyl, "Compensating optical systems. Part 3: Achromatic Fourier transformation," *Appl. Opt.* **11**, 1255–1260 (1972).

[43] G.M. Morris, "Diffraction theory for an achromatic Fourier transformation," *Appl. Opt.* **20**, 2017–2025 (1981).

[44] P. Andrés, J. Lancis, and W.D. Furlan, "White-light Fourier transformer with low chromatic aberration," *Appl. Opt.* **23**, 4682–4687, (1992).

[45] E. Tajahuerce, V. Climent, J. Lancis, M. Fernández-Alonso, and P. Andres, "Achromatic Fourier transforming properties of a separated diffractive lens doublet," *Appl. Opt.* **37**, 6164–6173 (1998).

[46] J. Lancis, E. Tajahuerce, P. Andres, G. Mínguez-Vega, M. Fernández-Alonso, and V. Climent, "Quasi-wavelength-independent broadband optical Fourier transformer," *Opt. Commum.* **171**, 153–160 (1999).

[47] J. Lancis, G. Mínguez-Vega, E. Tajahuerce, M. Fernández-Alonso, V. Climent, and P. Andres, "Wavelength-compensated Fourier and Fresnel transformers; a unified approach," *Opt. Lett.* **27**, 942–944 (2002).

[48] D. Faklis and G.M. Morris, "Broadband imaging with holographic lenses," *Opt. Eng.* **28**, 592–598 (1989).

[49] A.E. Siegman, *Lasers* (University Science Books, Mill Valley, CA, 1986).

[50] C. Brophy and G.M. Morris, "Speckle in achromatic-Fourier-transform systems," *J. Opt. Soc. Am.* **73**, 87 (1983).

6

Information Hiding: Steganography and Watermarking[1]

Lisa M. Marvel

Summary. Modern information-hiding techniques conceal the existence of communication. The primary facets of information hiding, also known as data hiding, are payload size, robustness to removal, and imperceptibility of the hidden data. In general, digital techniques can be divided into three classes: invisible watermarking, steganography, and embedded data. However, it is desirable for all classes to successfully conceal the hidden information from applicable detectors. Invisible watermarking, generally employed for copyright protection, traitor tracing, and authentication, forgoes a large payload amount for stringent robustness. Steganography, used for covert communication, pursues increased payload size while sacrificing robustness. Embedding information puts little emphasis on either robustness or covertness. In this chapter, we begin by introducing the motivation for information hiding. Following a brief survey of historic and digital examples, we will delineate the classes of information by describing objectives, applications, and terminology. Detection and attacks for these systems will then be outlined, and three representative techniques will be described in depth. Lastly, concluding remarks and suggestions for further reading will be presented.

6.1 Introduction

Our society has continually sought new and efficient ways to communicate. As more and more communication is conducted electronically, new needs, issues, and opportunities arise. At times when we communicate, we prefer that only the intended recipient have the ability to decipher the contents of a message. We want to keep the message *secret*. One solution is to use encryption to obscure the information content of the message. Encryption, once relegated to military and political information, is commonplace in electronic commerce and mail today.

---

[1] A portion of the material presented here appeared in part in the *IEEE Transactions on Image Processing*, **8**(8):1075–1083, August 1999, and at the *IEEE 3rd Annual Information Assurance Workshop*, West Point, NY, June 2002.

Instances also exist where we would prefer that the entire communication process be masked from any observer—that is, the fact that communication is taking place is secret. In this situation, we want the communication process to be *hidden*. Techniques that hide information can be used to hide or *cover* the existence of communication with other data. (These latter data are intuitively referred to as *cover data*.)

Whereas encryption masks the meaning of the message, information hiding masks the communication of the message. Information hiding is not meant to replace encryption but rather to enhance the secrecy—information can be encrypted and then covertly communicated by means of information hiding. We can think of information hiding as yet another tool to convey information and provide privacy. Consequently, well-designed information-hiding techniques do not rely on the secrecy of the hiding algorithm. Like cryptography, developers of information-hiding systems observe Kerckhoffs' principle [1], which states that the security of a system ought to lie only in the secrecy of the key and not the algorithm.

Privacy is not the only motivation for information hiding. By embedding one data item inside of another, the two become a single entity, thus eliminating the need to preserve a link between the distinct components or risk the chance of their separation. One application that would benefit from information hiding is the embedding of patient information within medical imagery. By doing so, a permanent association between these two information objects would be created. Additionally, information integrity can be provided using information hiding to embed authentication and tamper-detection information within the cover data. This is particularly advantageous in an age when the preservation and assurance of digital information is vital.

Digital media also introduce ownership and piracy issues. Digital copying is a lossless operation where a copy is the exact replica of the original data. Therefore, discerning rightful ownership is problematic. Governments have attempted to address these concerns by proposing a myriad of laws such as the U.S. Digital Millennium Copyright Act of 1998 (DMCA) [2] and the law or the *sui generis* right provided for in Chapter III of European Parliament and Council Directive 96/9/EC [3]. Both make the circumvention of effective technological measures designed to protect any rights related to copyright prohibited. Fortunately, those pursuing research or evaluating the effectiveness of copyright marking systems are exempt.

Furthermore, industry groups whose revenues are severely impacted by digital piracy also pursue scientific techniques such as information hiding to address copyright infringement and piracy. For instance, consider the Secure Digital Music Initiative (SDMI) [4]. This group, whose goal is to protect the playing, storing, and distributing of digital music, initiated a challenge by inviting the digital community to attempt to crack certain technologies (e.g., digital watermarks) they were considering for use in their systems to hinder unauthorized copying. The challenge made international news when researchers attempted to publish results on circumvention techniques at the

Fourth International Information Hiding Workshop [5]. According to a Web page [6] at the authors' institution, the authors decided not to present the paper after a lawsuit was threatened by the SDMI Foundation, among others. After the conference and much publicity, the authors finally published the paper a few months later [7].

In this chapter, we discuss information hiding in depth. We include historic examples and information hiding in the digital age. We then separate information hiding into major classes of invisible watermarking, steganography, and data embedding. The goals and applications of these classes will be expanded upon, followed by a discussion of attack and countermeasures. Subsequently, three representations will be exhibited in depth. We will conclude the chapter with some remarks on the science as well as suggestions for further reading.

6.2 Background

In this section, we examine the science of information hiding. We do this by first furnishing examples from ancient history and then providing some descriptions from the current era.

6.2.1 Historical Examples

Information hiding is not a new science. Some of the first documented examples can be found in the *Histories* of Herodotus, where the father of history relates several stories from the time of ancient Greece [8]. One is that of Histiaeus, who wished to inform his allies when to revolt against the enemy. To do so, he shaved the head of a trusted servant and then tattooed a message on his scalp. After allowing time for the slave's hair to grow back sufficiently to cover the tattoo, he was sent through enemy territory to the allies. To the observer, the slave appeared to be a harmless traveler. However, upon arrival, the slave reported to the leader of the allies and indicated that his head should be shaved, thereby revealing the message.

In another story from ancient times, one type of writing medium was a wooden tablet covered with wax. A person etched letters in the wax, and when he desired to remove the writing, the wax was melted to a smooth surface and the tablets reused. While exiled in Persia, Demeratus discovered that Greece was about to be invaded and wanted to convey a message of warning. The risk of exposure was great, so Demeratus concealed his message by writing directly on the wood and then covering it with wax. The seemingly blank tablets were then transported to Sparta, where the message was literally uncovered and his allies forewarned.

A less elegant method of hidden communication was adopted by Harpagus, a Median noble. He disguised a messenger as a hunter and hid a message in the body of an unskinned hare. The hunter carried the hare as if it were recently caught. Anyone encountering the messenger/hunter then would probably comment on his good fortune and be none the wiser. The message would then be delivered to the appropriate party without detection or interception.

The past century has yielded more advanced techniques. The use of invisible inks is one such method, where messages are written using substances that subsequently disappear. The hidden message is recovered using heat or other chemical reactions. Other methods employ routine correspondence, such as the application of pinpricks in the vicinity of particular letters to spell out a secret message. Advances in photography produced microfilm that was used to transmit messages via carrier pigeon. Improvements in film and lenses provided the ability to reduce the size of secret messages to a printed period. This technique, known as the microdot, was used in World War I.

6.2.2 Digital Developments

Much of today's communication occurs electronically using the Internet. Electronic mail and the World Wide Web are routinely used—perhaps more so than the telephone, facsimile, or traditional paper mail. With advancements in computational speed, broadband connectivity, mobile code (e.g., JAVA), and imaging, the Internet and Web content offer a wealth of potential candidates for hidden communication. In response, academics, hackers, and commercial researchers have created a plethora of information-hiding methods that take advantage of the various types of cover available. Many of these techniques employ multimedia signals, typically audio, video, or still imagery. But even computer software, network packet information, file systems, and electronic emissions can be used as cover media.

As an example, we present a scenario where public Web sites could be used to obscure the communication process. A user embeds a message within an image of an object using an information-hiding technique. That image is then placed on an auction Web site (e.g., eBay®, Amazon.com, etc.) offering the object for bid. A large number of users have access to the image and can view and download it. A user with the proper system and key can extract the hidden data from the downloaded image. Any observer will be unaware that the user that posted the image on the Web site and the user who downloaded the image communicated any information other than the image itself.

6.3 Objectives, Applications, and Terminology

Information hiding describes a varied body of research. It includes three classes: invisible watermarking, steganography, and the embedding of information where its existence may be known. Additionally, there are three primary parameters for any information-hiding technique. They are the hidden information size or *payload*, its imperceptibility, and its robustness to removal from the modified cover. The common pursuit in all classes is to provide imperceptibility of the hidden data in the cover. The measure of imperceptibility is determined by the detectors, and the form of the detector is, in most cases, a function of the type of cover employed. For instance, in the case of cover images, the detector is the human visual system as well as any potential computer analysis that an adversary may use. With audio cover, the detector is the human auditory system and computer analysis as well. If software is the cover, a compiler or operating system would be the primary detector, and so on.

Techniques that belong to the class of information embedding can be used for in-band captioning, bandwidth reuse, and forward/backward compatibility. The fact that the hidden information exists is of no consequence to the information-embedding objective. Two simple examples of embedded information systems are the exploitation of available bandwidth on home telephone lines and electrical wiring used for home networking. One can think of the telephone system or the electrical system as the detector in these examples. The embedded data are hidden in the cover, and the detectors are unaware of the existence of the hidden data and therefore operate as usual. A hiding failure would be to cause the disruption of the detector's operation (e.g., a telephone or electrical appliance).

Of the three classes of information hiding, we will focus our attention on invisible watermarking and steganography. Although many techniques and theory are shared between them, these two classes diverge when it comes to removal robustness and payload. To completely understand this divergence, we must investigate the objectives of watermarking and steganography.

6.3.1 Invisible Watermarking

Within the past decade or so, there has been a surge of research in the area of digital information hiding. A majority of the work in the area has been in the development of invisible digital watermarking. This thrust can be attributed to the desire for copyright protection and the deterrence of piracy. Visible watermarks are also used for copyright protection but are not of interest to us here because they are not hidden. (For brevity, we will use the term watermarking interchangeably with invisible watermarking.)

Invisible watermarking is used to embed an indelible mark on the cover data for the establishment of identity or ownership. It is primarily used for copyright protection and piracy deterrence. Because the cover is the fundamental commodity, imperceptibility of the mark is vital. Equally important is the watermark's resistance to removal as well as its successful recovery for the establishment of rightful ownership. The recovery should be reliable even after removal attempts by an adversary. A failure is determined when an adversary has the ability to render the watermark unrecoverable while still preserving desirable attributes of the cover.

Once a particular item of cover has been processed by the watermarking system, it is referred to as *marked*. As an example, if one were to apply a watermark to an image, the resultant would be called the *marked image*. In this chapter, we consistently referred to the detector as the observer that may discover the hidden information. However, it should be noted that at times in the popular literature, the watermark recovery process is also called detection. We refrain from doing so here for clarity.

Besides basic research, there are a few commercial interests pursuing digital watermarking technology. In the forefront is the Digimarc company, which has several patents in the watermarking area and has developed plug-in software for Adobe Photoshop (a prominent software package used by graphics professionals). Significant efforts are also under way at Philips Research Laboratories as well as Microsoft Research.

Depending on the features of the watermarking application, a *cover escrow scheme* may be used. Such methods require that the recipient possess the original cover in order to reveal the hidden information. To cite an example, let us consider the traitor-tracing scenario presented in [9]. Illegal copies of some cover are being distributed. The owner desires to detect the origin of the illegal copies—the traitor—so the owner marks each distributed copy of the cover with a unique tracking number. These marked copies are distributed to the potential traitors and the transactions along with the assigned tracking numbers are recorded. When an unauthorized copy is found, the owner uses the original cover to extract the tracking number and trace it back to the traitor.

Policy, a procedure by which some central authority (e.g., copyright office) monitors and enforces ownership rights, is a vital issue in copyright protection and determining rightful ownership. It is not adequate merely to show that an unauthorized party possesses a particular cover. Therefore, policy for many watermarking applications requires a trusted third party to maintain the only unaltered copy of the original cover data in a repository along with a time stamp. Hidden marking information as well as transaction details are also recorded by this central authority. When a copyright violation is suspected, this authority is employed. For further discussion on watermarking policy and ownership rights, the reader is referred to [10].

There is also a subclass of watermarking methods called *fragile watermarks*. The primary objective of these systems is authentication and tamper detection. They operate by marking the cover in some imperceptible way where the mark is designed such that it is robust to certain permissible modifications but, unlike typical watermarking, the mark is sensitive to all others. If an adversary attempts to modify the marked cover in a prohibited manner, the mark is impaired in some way. Consequently, the recipient will be unable to reliably recover the mark, leading the recipient to be reasonably sure that the item has been modified in some fashion, and in turn, to suspect the integrity or the origin of the object.

6.3.2 Steganography

Steganography literally means *covered writing*. Here the objective is to keep the *existence* of the hidden data secret. Failure is defined as a detector's proof of the existence of the hidden data [11]—not the decoding of its meaning. Steganography is used for such applications as covert communications and tamper detection. With steganography, robustness to removal is secondary to the objective of maximizing the amount of payload. Most steganography is two-party communication and little, if any, policy needs to be established.

We mentioned a cover escrow scheme used for watermarking, where the original cover is needed to extract the hidden data. However, for many steganographic applications, it is impractical to require the possession of the unaltered cover by the receiver. Of more pragmatic steganographic use are systems that operate in a *blind* manner. Blind information-hiding schemes permit the extraction of embedded data from the modified cover signal without any knowledge of the original cover information. Blind strategies are predominant among steganography of the present day.

A block diagram of a generic blind steganographic system is depicted in Figure 6.1. A message is embedded in the cover by the sender using the system encoder and key. The modified cover is produced. The resulting modified cover is transmitted over a channel to the recipient. During transmission, the modified cover can be monitored by unintended viewers, who should notice only the transmittal of the innocuous cover without discovering the existence of the hidden message. Additionally, as with any noisy communication channel, noise may be present and prevent the recovery of the hidden data. The recipient receives the modified cover and processes it by the system decoder using the the same key as the sender. The system represented in Figure 6.1 is a symmetric key system where both the sender and receiver have the same key. Public-key encryption could also be used in a blind steganography system.

Now, let us define a few basic steganography terms. The system that embeds hidden information is the *stegoencoder*. Conversely, the system that recovers the hidden information is the *stegodecoder*. The entire system consisting of both the encoder and decoder is defined as the *stegosystem*. The label for the modified cover, once data have embedded within it, is dependent on the type of cover and has *stego* as a prefix. For example, if the cover is an image, then the modified cover is the *stegoimage*. If the cover is audio, then the modified cover is the *stegoaudio*, and so on.

6.4 Detection and Attacks

Countermeasures for information-hiding methods are limited to watermarking and steganography. (Please note that successfully embedded information is benign to the detector and has some beneficial use. Therefore, no remedies or attacks are generally attempted for this class.)

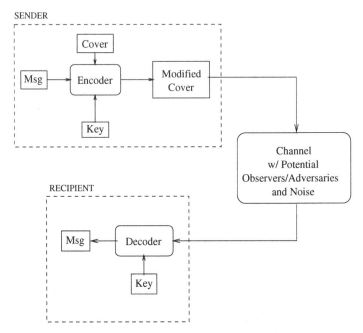

Fig. 6.1. Blind information hiding system.

Watermark attacks render the watermark useless by making it unreadable by the watermark–recovery system. Without reliable recovery, determination of the rightful owner is prevented. To counteract watermarking, the cover must be modified in such a way that it is still able to fulfill its function while destroying the watermark. There are many types of attacks, such as distortion, transformation, and conversion to analog and then back to the digital domain. There are also collusion attacks, where a number of watermarked items using the same cover are collected and used to obliterate the watermark. A simple form of the collusion attack is to average the collected watermarked copies. The attacker hopes that in doing so the watermark is disrupted enough that it cannot be reliably recovered. It should be noted that most attacks are cover specific.

Besides various techniques presented in the research literature, the following software tools are available as well. The image-watermarking attack tool StirMark was first developed in November 1997 at the University of Cambridge as a generic tool for simple robustness testing of image-watermarking algorithms [12, 13]. The software attempts to modify the marked image in such a way that the watermark and its recovery system are no longer synchronized and yet the image will retain much of its original appearance. Over the years, the project has been recognized as a benchmark for watermarking methods. Checkmark [14] is provided by the Computer Vision Group, University of Geneva, Switzerland. It provides tools to evaluate and rate watermarking technologies. There is also another benchmarking tool, Optimark, for still-image watermarking algorithms developed in the Artificial Intelligence and Information Analysis Laboratory at the Department of Informatica, Aristotle University of Thessaloniki, Greece [15].

As with home and car security systems, the mere proclamation that such a system is in place may act as a deterrent to illegal activities such as theft and piracy. Therefore, some Web sites where piracy is feared openly state that the information is watermarked. Such is the case with the State Hermitage Museum Web site in St. Petersburg, Russia. Images of their artwork and sculptures appear on the Internet with their image usage policy clearly noted. This policy states: "Images on this Web site have been invisibly watermarked; any attempt to remove the watermarks from these images is expressly forbidden."

As for attacks on fragile watermarking, an adversary could invoke a denial-of-service attack by altering the marked cover. This would prevent the recipient from authenticating the received data and therefore be denied its service. However, such activities will cause the recipient to become aware of the potential existence of the adversary; this suspicion could serve as valuable information in and of itself.

Attacks on steganography result in the detection of hidden information, not necessarily the decoding or deciphering of the information's meaning. Similarities with cryptography give name to the science of steganographic detection—*steganalysis*. Early researchers in digital information hiding engaged in developing a multitude of methods to hide information in various data. Comparably little effort was expended on schemes to thwart the communication of the hidden data. However, world events have led authorities to be suspicious that some may be using steganography for illegal ends [16], and researchers have taken heed by pursuing more steganographic detection.

In addition to the steganographic techniques appearing in the literature, there is a plethora of steganography software available on the Internet [17]. Steganography countermeasures are commonly designed to target a specific technique, such as hiding in the least significant bit (LSB) of a gray-scale image [18] or hiding in images that employ the JPEG format [19]. Others have targeted a specific type of cover, looking for anomalies or diversions from the classic models [20, 21]. Perhaps the most daunting aspect of detecting this type of steganographic communication is the shear number of candidate cover data that exist on the Web. Provos and Honeyman performed a study to detect hidden messages inside of JPEG-compressed images on the eBay® auction site [22] using their Stegodetect software. They were unable to detect any hidden messages in the 2 million images that they examined. This result raises the following question: Are people really using steganography (on eBay® or elsewhere), or were the detection techniques used in the experiment deficient in some way?

6.5 Three Representative Systems

We will illustrate our discussion of information hiding by describing three representative systems. Although differing in application, the systems make use of information hiding to accomplish the desired objectives. With the first system, we present an image-watermarking method that can be extended to other types of multimedia. Following this image-watermarking example, we will discuss a technique that uses a fragile watermark to convey authentication and tamper-detection information within mobile computer software. This system is designed to be bandwidth-efficient and requires negligible computational resources. Lastly, we demonstrate information hiding in a steganographic manner that employs still images as cover. We illustrate two approaches. The first is a simple substitution technique, and the second is a more complex technique that is resistant to commonplace image processing, such as compression.

6.5.1 Watermarking

We chose secure spread spectrum watermarking for multimedia by Cox et al. [23] as our representative watermarking method. In this paper, the authors give a worthy description of watermarking motivation, issues of varying attacks, and policy challenges. They designed the system to operate with digital images as cover data but propose that the system can be extended to other media forms such as audio and video. The system is an escrow system, in which the unmarked cover as well as the original watermark are needed for watermark recovery. As previously mentioned, although limiting, cover escrow systems are reasonable when a certifying authority is employed as part of the watermarking policy to determine rightful ownership.

The system uses the frequency domain to place the watermark. The rationale is that by doing so the watermarking data will be spread throughout the entire image. Additionally, the authors promote using the perceptually significant frequency information as sites for watermark placement. Although this is counterintuitive from an information-hiding standpoint, any attacks on the watermark, intentional or unintentional, will tend to minimize modification to the perceptually significant data in order to preserve the value of the cover. This will allow the watermark to remain intact, making it robust to attack. The specific location for watermark embedding is determined using a technique called perceptual masking. Masking locates perceptually significant regions in the spectrum that can support the small modification incurred by adding the watermark without impacting the image's perceptual fidelity.

Figure 6.2 embodies the basic watermarking process. The cover is transformed to the frequency domain using the Discrete Cosine Transform (DCT), and perceptually significant regions are then selected. The watermark is created using a random Gaussian noise distribution. The two are combined using an invertible computation. Then the inverse frequency transformation is ap-

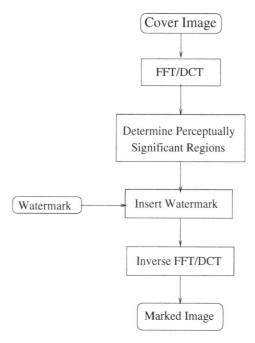

Fig. 6.2. Watermark insertion.

plied to convert the data to the spatial domain, and the result is the marked cover.

The watermark recovery process is illustrated in Figure 6.3. Here, the recovered image is transformed into the frequency-domain representation along with the escrowed original cover. The two are then subtracted to yield an estimate of the watermark. Next, this watermark is compared with the original watermark using a similarity measure. As an aside, the central authority is the only entity that is in possession of the unmarked original and the original watermark, both of which function as keys to the system, following Kerckhoffs' principle.

Expected attacks, deliberate or otherwise, consist of the following: lossy compression, geometric distortions such as rotation, translation, scaling, and cropping, other signal-processing modifications, and digital-to-analog conversion and its inverse (printing and scanning). Because the original unmarked cover is available during the watermark-recovery process, it can be used when attempting to reverse or counteract many of these attacks during a preprocessing stage. For instance, in the occurrence of geometric distortions, the recovered image can be registered against the original in an attempt to regain the state of the marked image prior to the attack. The reader is referred to the work of Cox et al. [23] for example images and experimental results for this technique.

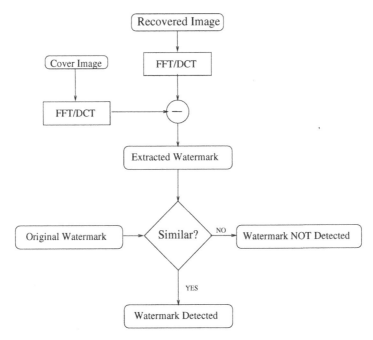

Fig. 6.3. Watermark recovery.

6.5.2 Fragile Watermarking

Next, we introduce a fragile watermarking method for mobile software. Although mobile code provides a highly flexible and beneficial form of computing, its use creates complex security considerations. The method described here is a bandwidth-efficient approach to the authentication of mobile Java codes using fragile watermarking. It is a desirable tool in applications where low bandwidth use is a requirement (e.g., wireless networks, low-power devices, and distributed computation). Analysis indicates that the system detects, with high probability, any degree of tampering within a reasonable amount of time, while avoiding increased bandwidth requirements.

The tool is called MOST (MObile Software Tamper detection) [24]. MOST enables a mobile software system to validate mobile Java codes with authentication data that are derived from the code itself without increasing bandwidth requirements, without the risk of separation of authentication data from code, and without relying on a third-party authentication agent. MOST embeds authentication data as a tamper detection mark (TDM), a cryptographic checksum, within the code as a way to address the issues of code integrity and authentication. This tool can be utilized within a Java environment to detect virtually any degree of tampering or alteration to a Java code. The Java platform was chosen due to its popularity as a mobile code and mobile agent language [25]. Authentication with the MOST tool is optional; a marked code

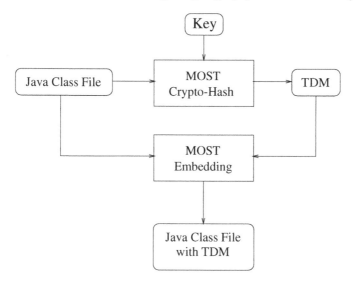

Fig. 6.4. Embed phase for the MOST tool.

is semantically equivalent to the original version of the code and can execute without any special preprocessing should authentication of the code not be desired or should the code execute on a system that is not MOST-aware.

MOST is designed to function either as a stand-alone authentication tool or as an authentication API for inclusion in a mobile code system. MOST operates in two phases, an embed phase and a validate phase. The embed phase typically takes place on the host that compiles the Java source code. We shall call this host the trusted host. A basic example using MOST is shown in Figure 6.4. The trusted host compiles a Java application, and then the TDM is created by computing a hash value of the Java class file and encrypting that hash value with the secret key. (Each TDM serves as a cryptographic checksum for the class files that compose the mobile code.) Then, the MOST system embeds the encrypted TDM within each of the application's class files. The program is then available for download.

Embedding the TDM in a Java class file is accomplished by permuting the order of the constant pool table. This table is similar in function to the symbol table of binary executable file formats (e.g., executable and linking format (ELF) [26]). The permutation algorithm used is that given in [27]. Manipulating the order of the constant pool table requires the entire class file to be updated to reflect the new table ordering, as there are many references to constant pool table objects throughout the entire Java class file. The embedded TDM is now part of the class file, represented by the ordering of the constant pool table.

After the TDM has been embedded within the code, the mobile code is ready for transmission to the local host. The new class files, or marked code, created during the embed phase are semantically equivalent to the original Java class files; the same computation is performed, and the runtime performance of the computation is not affected. These marked class files are able to execute on a regular Java Virtual Machine (JVM) with no special preprocessing.

Once the mobile code has arrived on the local host from the trusted host, the local host can validate the code with the MOST tool. During the validation phase, shown in Figure 6.5, the TDM of each class file is extracted and used to validate its respective class files. Extraction of the TDM from the marked class file involves finding the represented permutation of the constant pool table. This process is simply the inverse of the permutation algorithm used during the embed phase and yields the value of the encrypted TDM. The value is decrypted with the secret key to result in TDM'. This is compared with a locally generated mark, TDM''. Agreement of the local and extracted marks indicates successful code authentication and integrity. If the code has not been altered since marking and the proper keys have been used to create and validate, the validation result will return true. Any alteration to the code or incorrect key use will result in a failing validation phase.

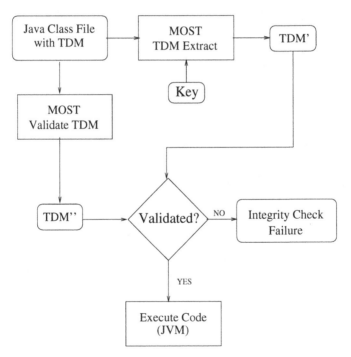

Fig. 6.5. Validate phase for the MOST tool.

Analysis indicates that the MOST tool detects, with high probability, any degree of tampering with a mobile Java code and can do so within a reasonable amount of time. This makes it desirable to designers and users of mobile code systems and software.

6.5.3 Steganography

LSB Steganography

We feel obliged to introduce one of the simplest forms of steganography. It involves the manipulation of the least significant bit (LSB) plane of the data (e.g., audio, imagery, video, etc.). Various techniques, such as direct replacement of the cover LSBs with message bits or an arithmetic combination between the two, are used. Several examples of LSB schemes can be found in [28–30]. LSB manipulation software has been written for a variety of image formats and can be found in [31]. These methods typically achieve both high payload (8:1 for gray-scale imagery) and low perceptibility. However, because the technique is well-known, LSB methods are vulnerable to detection and extraction by adversaries. In fact, there are several detection methods for LSB information hiding [18].

A More Effective Technique

We will use a method called spread spectrum image steganography (SSIS) to illustrate a more effective steganographic system [32, 33]. The SSIS system hides and recovers a message of substantial length within digital imagery while maintaining the original image size and dynamic range. The hidden message can be recovered using appropriate keys and operates blindly (without any knowledge of the original cover image). A message embedded by this method can be of any digital form. Applications for such a steganographic scheme include covert communication, tamper proofing, and authentication. We also postulate that this technique can easily be extended to other forms of multimedia cover data. SSIS is a more effective steganographic technique for natural images (those not manufactured using computer graphics) compared with the LSB methods because it uses noise statistically identical to the images' inherent noise to hide the message, thereby making it more difficult to detect [34].

SSIS was developed by applying a communication systems model. This model consists of a sender, receiver, and a channel through which the information is sent. In the steganographic case, the cover image functions as a channel through which the hidden information is conveyed. Because SSIS is a blind scheme and the extraction process is imperfect, errors in the recovered message exist. Therefore, we can consider this channel to be a noisy channel. Generally, channel estimation, error-control coding, and modulation techniques are employed to communicate over noisy channels. SSIS draws from these three areas to accomplish steganography. Because in this case the channel is the cover image, channel estimation is equivalent to image restoration in an attempt to restore the stegoimage to its original state.

Because wideband thermal noise, inherent in imagery of natural scenes captured by photoelectronic systems, can be modeled as additive white Gaussian noise (AWGN) [35], this type of noise is used in the SSIS system. In other types of coherent imaging, the noise can be modeled as speckle noise [35], which is produced by coherent radiation from the microwave to visible regions of the spectrum. We postulate that the concepts of SSIS can be extended to imagery with noise characteristics other than those modeled by AWGN. The additional noise that conceals the hidden message is a natural phenomenon and, therefore, if kept at typical levels, is not noticeable to the casual observer or computer analysis.

The major processes of the stegosystem encoder are portrayed in Figure 6.6. First, we begin with the message that the sender may optionally encrypt. The message is encoded via a low-rate error-correcting code, producing the encoded message, m. The sender enters key 1 into a wideband pseudorandom noise generator, producing a real-valued noise sequence, n. This noise sequence can then be scaled by a power factor. A modulation scheme is used to combine the encoded message with the noise sequence, thereby composing the embedded signal, s. This signal is then used as input into an interleaver using key 2. The resultant signal is added with the cover image, f, to produce the stegoimage, g. The result is appropriately quantized and clipped to preserve the initial dynamic range of the cover image. The final product is the stegoimage.

The interleaver in this scheme, which reorders the embedded signal before it is added to the cover image, serves a dual function. The first is to prevent a group or *burst* of errors. This allows the errors to occur almost independently

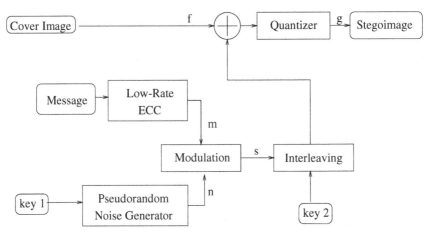

Fig. 6.6. SSIS encoder.

within a codeword, thus giving the error-correcting code an equal chance at correcting the errors in all codewords. Secondly, since the interleaver requires a key to stipulate the interleaving algorithm, this key can serve as another level of security to establish the proper order of the embedded signal before decoding.

To adjust the power of the embedded signal, a scale factor is applied to the stegosignal, as mentioned previously. The signal is then added to the cover image, which is subsequently quantized and clipped to result in the stegoimage. The scale factor is selected based on the detection of the stegosignal by the human visual system. Without the appropriate keys, the modulated signal is statistically indistinguishable from white Gaussian noise routinely present in the image.

The stegoimage is transmitted in some manner to the recipient, who maintains the same keys as the sender and uses the stegosystem decoder (shown in Figure 6.7) to extract the hidden information. The decoder uses image-restoration techniques to produce an estimate of the original cover image, $\hat{f}$, from the received stegoimage, $\hat{g}$. The difference between $\hat{g}$ and $\hat{f}$ is fed into a keyed deinterleaver to construct an estimate of the embedded signal, $\hat{s}$. With key 1, the noise sequence, n, is regenerated; the encoded message is then demodulated, and an estimate of the encoded message, $\hat{m}$, is constructed. Synchronization of these signals is trivial because the stegoimage is defined by the image format. The estimate of the message is then decoded via the low-rate error-control decoder, decrypted if necessary, and finally revealed to the recipient.

In Figure 6.8, we show a demonstration of SSIS. Figure 6.8(a) is the original cover image of size 256×256 pixels. Figure 6.8(b) is the stegoimage. To maximize payload, we presume the hidden message will be compressed and intolerant of errors, as is the case with Huffman and arithmetic coding. Con-

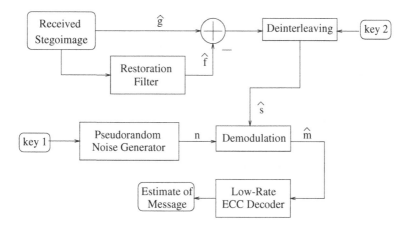

Fig. 6.7. SSIS decoder.

Fig. 6.8. SSIS example. **(a)** Cover image; **(b)** stegoimage

sequently, parameters necessary to accomplish this error-free recovery of the hidden data were used. This stegoimage resulted in a payload of 0.1667 bits per pixel, or 10 kilobytes. Notice the slightly increased noise in the sky area of the image, but overall the images look extremely similar, with little indication that hidden data exist in the stegoimage.

SSIS has been used as a case study in the first text on digital information hiding [11] and as an example of an additive Gaussian steganography system in the literature [36, 37].

It is not a coincidence that both the steganography and watermarking examples presented here are based on spread-spectrum methods [38]. Spread spectrum was designed for hidden radio communications prior to WWII, and much of our current wireless technology is based on spread-spectrum concepts.

6.6 Concluding Remarks and Further Reading

In this chapter, we have introduced information hiding and provided historic examples as well as recent developments. The main classes of information hiding were depicted along with corresponding detection and countermeasure schemes. Three systems were described in detail as examples of various information-hiding techniques.

Information hiding is an active research area. Current and future directions are concentrating on theoretical results that lend themselves to steganalysis. Of course, there is always room for more effective and efficient ways to hide information as well as identification of clever types of cover data.

There are several conferences devoted solely to the topic, such as the Information Hiding Workshop, the premier venue, and the SPIE International Conference on Security and Watermarking of Multimedia Contents. Both are well-attended by academic and industry researchers. There are also numerous conference sessions dedicated to the area based on the type of cover: IEEE International Conference on Image Processing; IEEE International Conference on Acoustics, Speech, and Signal Processing; and the Multimedia and Security Workshop (ACM Multimedia '99), to name but a few.

The reader is encouraged to examine several textbooks that have been recently published specifically on information hiding. Some concentrate exclusively on watermarking [39], while others provide an overview of existing information-hiding methods [11, 37, 40, 41].

References

[1] A. Kerckhoffs, "La cryptographie militaire," *J. Sci. Militaires* **9**:5-38 (1883).
[2] U.S. Copyright Office, Library of Congress. Digital Millennium Copyright Act of 1998 (DMCA). WWW page, 1998. `http://www.loc.gov/copyright/legislation/dmca.pdf`.
[3] European Parliament and Council Directive 96/9/EC. WWW page, 1996. `http://europa.eu.int/ISPO/infosoc/legreg/docs/969ec.html`.
[4] Secure Digital Music Initiative, May 2001. `http://www.sdmi.org`.
[5] I. Moskowitz, ed., *Information Hiding, Fourth International Workshop, Lecture Notes in Computer Science*, vol. **2137** (Springer-Verlag, Berlin, 2001).
[6] Status of the paper "Reading between the lines: Lessons from the SDMI challenge." WWW page, 2001. `http://www.cs.princeton.edu/sip/sdmi/`.
[7] S.A. Craver, M. Wu, B. Liu, A. Stubblefield, B. Swartzlander, D.S. Wallach, D. Dean, and E.W. Felten, "Reading between the lines: Lessons from the SDMI challenge," in *Proceedings of the 10th USENIX Security Symposium*, Washington, DC, August 2001.
[8] D. Kahn, *The Codebreakers—The Story of Secret Writing* (Scribner, New York, 1967).
[9] B. Pfitzmann, "Trials of traced traitors," In R. Anderson, ed., *Information Hiding, First International Workshop*, Lecture Notes in Computer Science, vol. **1174**, pages 49–64 (Springer-Verlag, Berlin, 1996).
[10] S. Carver, N. Memon, B. Yeo, and M. Yeung, "Resolving rightful ownerships with invisible watermarking techniques: Limitations, attacks, and implications." *IEEE J. Selected Areas Commun.* **16**:573–586 (1998).
[11] S. Katzenbeisser and F. Petitcolas, eds., *Information Hiding: Techniques for Steganography and Digital Watermarking.* (Artech House, Inc., Norwood, MA, 2000).
[12] F.A. Petitcolas, R.J. Anderson, and M.G. Kuhn, "Attacks on copyright marking systems," in D. Aucsmith, ed., *Information Hiding, Second International Workshop, Lecture Notes in Computer Science*, vol. **1525**, pages 219–239 (Springer-Verlag, Berlin, 1998).
[13] F.A. Petitcolas, "Watermarking schemes evaluation," *IEEE Signal Process.*, **17**(5):58–64 (2000).
[14] S. Pereira, S. Voloshynovskiy, M. Madueo, S. Marchand-Maillet, and T. Pun, "Second generation benchmarking and application oriented evaluation," in I. Moskowitz, ed., *Information Hiding, Fourth International Workshop*, Lecture Notes in Computer Science, vol. **2137** (Springer-Verlag, Berlin, 2001).

[15] V. Solachidis, A. Tefas, N. Nikolaidis, S. Tsekeridou, A. Nikolaidis, and I. Pitas, "A benchmarking protocol for watermarking methods," in *Proceedings of the IEEE International Conference on Image Processing*, Thessaloniki, Greece, October 2001.

[16] J. Kelley, "Terror groups hide behind web encryption," *USA Today*, June 2001. http://www.usatoday.com/life/cyber/tech/2001-02-05-binladen.htm.

[17] N.F. Johnson and S. Jajodia, "Exploring steganography: Seeing the unseen," *IEEE Computer*, pages 26–34, February 1998.

[18] J. Fridrich, M. Goljan, and R. Du, "Reliable detection of LSB steganography in grayscale and color images," in *Proceedings of the ACM Workshop on Multimedia and Security*, pages 27–30, Ottawa, Canada, October 2001.

[19] J. Fridrich, M. Goljan, and R. Du, "Steganalysis based on JPEG compatibility," in *Proceedings of the SPIE Multimedia Systems and Applications IV*, Denver, CO, August 2001.

[20] G.S. Lin and W.N. Lie, "A study on detecting image hiding by feature analysis," in *Proceedings of the 2001 IEEE International Symposium on Circuits and Systems*, pages 149–152, Sydney, Australia, May 2001.

[21] I. Avcibas, N. Memon, and B. Sankur, "Steganalysis using image quality metrics," *IEEE Trans. Image Process.*, **12**(2):221–229 (2003).

[22] N. Provos and P. Honeyman, "Detecting steganographic content on the internet," in *Proceedings of the Internet Society's Network and Distributed System Security Symposium (NDSS)*, San Diego, CA, February 2002.

[23] I.J. Cox, J.Kilian, F.T. Leighton, and T. Shamoon, "Secure spread spectrum watermarking for multimedia," *IEEE Trans. Image Process.*, **6**:1673–1687 (1997).

[24] M.J. Jochen, L.M. Marvel, and L.L. Pollock, "MOST: A tamper detection tool for mobile Java software," in *Proceedings of the 3rd Annual Information Assurance Workshop*. IEEE, June 2002.

[25] W. Amme, N. Dalton, and J. von Ronne, "SafeTSA: A type safe and referentially secure mobile-code representation based on static single assignment form," in *Proceedings of the ACM SIGPLAN 2001 Conference on Programming Language Design and Implementation (PLDI-01)* (ACM Press, City, 2001).

[26] Tools Interface Standard Committee. Tools interface standard (TIS) executable and linking format specification, 1993. http://developer.intel.com/vtune/tis.htm.

[27] D.E. Knuth, *The Art of Computer Programming: Seminumerical Algorithms*, volume 2 (Addison-Wesley, Reading, MA, 1981).

[28] R. Van Schyndel, A. Tirkel, and C. Osborne, "A digital watermark," in *Proceedings of the IEEE International Conference on Image Processing*, volume 2, pages 86–90, 1994.

[29] R.B. Wolfgang and E.J. Delp, "A watermark for digital images," in *Proceedings of the IEEE International Conference on Image Processing*, volume III, pages 219–222, Lausanne, Switzerland, September 1996.

[30] R. Machado, Stego, http://www.stego.com, 2000.

[31] Steganography information and archive. http://www.stegoarchive.com, 2003.

[32] L.M. Marvel, C.G. Boncelet, Jr., and C.T. Retter, "Spread spectrum image steganography," *IEEE Trans. Image Process.*, **8**(8):1075–1083 (1999).

[33] C.G. Boncelet, Jr., L.M. Marvel, and C.T. Retter, Spread spectrum image steganography (SSIS). U.S. Patent 6,557,103, 2003.

[34] J. Fridrich and M. Goljan, "Practical steganalysis of digital images—state of the art, in *Proceedings of SPIE Electronic Imaging—Photonics West*, San Jose, CA, January 2002.

[35] A.K. Jain, *Fundamentals of Digital Image Processing* (Prentice-Hall, Inc., Englewood Cliffs, NJ, 1989).

[36] W. Bender, W. Butera, D. Gruhl, R. Hwang, F.J. Paiz, and S. Progreb, "Applications for data hiding," *IBM Syst. J. MIT Media Lab.*, **39** (3 & 4) (2000).

[37] P. Wayner, *Disappearing Cryptography: Information Hiding—Steganography and Watermarking*, (Morgan Kaufmann Publishers, San Francisco, 2002).

[38] M.K. Simon, J.K. Omura, R.A. Scholtz, and B.K. Levitt, *Spread Spectrum Communications*, Volume I. (Computer Science Press, Rockville, MD, 1985).

[39] I.J. Cox, M.L. Miller, and J.A. Bloom, *Digital Watermarking: Principles & Practice* (Morgan Kaufmann, San Francisco, 2002).

[40] N.F. Johnson, Z. Duric, and S. Jajodia, *Information Hiding: Steganography and Watermarking—Attacks and Countermeasures* (Kluwer Academic Publishers, Boston, 2001).

[41] E. Cole, *Hiding in Plain Sight: Steganography and the Art of Covert Communication* (John Wiley and Sons, New York, 2003).

7

Watermarking Streaming Video: The Temporal Synchronization Problem

Edward J. Delp and Eugene T. Lin

Summary. This chapter examines the problems with temporal synchronization in video watermarking and describes a new method for efficient synchronization and resynchronization. In our method, efficient synchronization is achieved by designing temporal redundancy in the structure of the watermark. Temporal redundancy allows the watermark detector to establish and maintain synchronization without performing an extensive search in the watermark key space. Our method does not use synchronization templates that may be subject to attack and increase the visibility of the watermark. Another advantage of our technique is that the watermark structure is video-dependent, which enhances security.

The technique is implemented using a spatial-domain watermark on uncompressed video and a finite state machine watermark key generator. The implementation illustrates the effectiveness of using temporal redundancy for synchronization, and is shown to be resilient to desynchronization attacks such as frame dropping, frame insertion, local frame transposition, and frame averaging. Our method for synchronization is independent of the embedding method and can be used with a wide class of watermark embedding and detection techniques, including those for other time-varying signals such as digital audio.

7.1 Introduction

Many blind watermark-detection techniques (including the correlation-based detectors often used in spread-spectrum watermarks [1, 2]) require the detector to be synchronized with the watermarked signal before reliable watermark detection can occur. Synchronization is the process of identifying the correspondence between the spatial and temporal coordinates of the watermarked signal and those of an embedded watermark. Ideally, the watermark-detector will be given a watermarked signal such that the coordinates of the embedded watermark have not been changed since the embedding process. In this case, synchronization is trivial and the detector can proceed in the manner prescribed by the watermark detection technique [3]. However, if the coordinates of the embedded watermark have been changed (such as when the

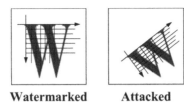

Watermarked   Attacked

Fig. 7.1. A synchronization attack.

watermarked signal is rescaled, rotated, translated, and cropped, as shown in Figure 7.1) the detector must identify the coordinates of the watermark prior to detection. Synchronization is crucial to successful watermark detection, and, if the detector cannot synchronize with its input signal, an embedded watermark may not be detectable even though it is present in the signal. Many of the techniques that are used to attack watermarked signals do not "remove" the watermark, as widely believed, but desynchronize the detector [4, 5].

Synchronization is a problem that cannot be ignored in most video watermarking applications, even in the absence of a malicious attacker. In applications such as secure digital television and broadcast monitoring, the watermark detector may be expected to detect the watermark starting from any arbitrary temporal location within the video signal (as opposed to starting detection from the "beginning" of the video signal), which is known as *initial synchronization.* In other applications, the video signal arriving at the watermark detector may have been damaged to the extent that the detector loses synchronization and must resynchronize before watermark detection can resume. One such application is streaming video [6–8], where parts of the video signal can be damaged or lost as it is transmitted over a network. The video signal may also be interrupted for an indeterminate time for reasons beyond the control of the user, such as network congestion. (We note that, in video streaming, the network is not under any constraint to preserve the perceptual quality of the video.) In these applications, it is essential that the watermark detector be able to resynchronize to the video, even after many frames of the video have been lost. Obviously, robust watermarking techniques should be robust against synchronization attacks by a hostile attacker.

In the worst case, establishing synchronization would involve an exhaustive search over the space of all possible geometric and temporal transformations to find the watermark. This is not practical for video watermarking applications that require real-time watermark detection. Methods for spatial synchronization in still images (see [9] for an overview) have been examined and typically involve efficient search techniques to deduce the geometric transformation or the use of embedding domains that are invariant to geometric transformations. However, some still-image watermark-synchronization techniques are too computationally expensive for real-time video applications, and those still-image techniques that are suitably efficient for real-time implementation do not consider temporal synchronization.

One oft-mentioned synchronization technique is the embedding of a pattern, known as a template, that a detector can examine to determine the orientation and scale of the watermark. (Templates have been suggested for video watermark synchronization [1, 10].) There are several disadvantages to template embedding. First, a template is designed to be easily detected; hence the template itself can be vulnerable to removal [11]. Second, the template must be robustly embedded into the video signal, which could introduce additional distortion in the watermarked video.

In this chapter, we consider the problem of temporal synchronization in video watermarking and temporal attacks. (These issues can also apply for time-varying signals other than video, such as audio.) Spatial synchronization issues, such as synchronization after rotation, scaling, translation, and cropping attacks, is not considered here. A method for synchronization in the presence of spatial shifts is described in [12]. The inclusion of the temporal coordinate dramatically increases the potential search space necessary for synchronization in video as compared with still images. More significantly, however, the temporal redundancy that is present in most video sequences can be exploited in attacks against watermarked video that are not possible for still images, such as frame cropping, insertion, transposition, and frame averaging (temporal collusion.)

We will propose a method (or protocol) in the design of video watermarks that allows efficient temporal synchronization by introducing redundancy in the structure of the embedded watermark. Our work exploits the use of frame-dependent watermarks [13, 14]. The detection mechanism takes advantage of the redundancy to reduce the search required for synchronization. The method does not involve the embedding of templates and generates a watermark that is dependent on the content of the video.

It is important to note that our new method is independent of the actual watermarking technique (embedder and detector). In fact, our method could be retrofitted into most existing video watermarking techniques.

7.2 Framework for Temporal Synchronization

When a watermark is embedded into a signal, a parameter that defines its structure is the embedding key K_E. The watermark can also be dependent on other factors. Some watermarking techniques embed side information into the watermark, known as the message or payload. Other watermarking techniques vary the watermark based on the unwatermarked signal itself, such as by using a visual model during embedding [3]. We assume the embedder uses K_E to generate a key schedule or uses a sequence of subkeys to watermark individual frames of the video. For notation, let $K(t)$ be the key used to generate the watermark for frame t. The goal of the detector is to determine $K(t)$ when frame

t is examined. If the detector can determine $K(t)$, temporal synchronization is (or has been) achieved. If the detector cannot determine $K(t)$, synchronization is lost. For simplicity, we focus on symmetric watermarking techniques in this discussion, where the embedding and corresponding detection keys are identical.

This framework, or model, is applicable to a large class of video watermarking techniques (including [1, ?, ?, ?, ?, ?]) and is not as restrictive as it may seem. For example, many watermarking techniques use the same watermark key for each frame, so the key schedule becomes $K(0) = K(1) = K(t) = K_E$. Other techniques use a completely different key to watermark each frame. For video watermarking techniques that use K_E as a seed for a pseudo-random number generator, the key schedule corresponds to the state of the pseudorandom-number generator at the start of each frame.

7.2.1 Temporal Redundancy

The amount of temporal redundancy in the key schedule can have a dramatic effect on the ease of synchronization and watermark security. In this context, temporal redundancy refers to the degree of randomness in the individual keys appearing in the watermark key schedule: A high degree of randomness corresponds to low temporal redundancy, and a low degree of randomness corresponds to high temporal redundancy. Special cases are used to illustrate the security and synchronization trade-offs (Figure 7.2.)

One type of key schedule is to use a single embedding key $K_E = K_0$ to watermark all video frames. This is known as a time-invariant key watermarks. The embedded watermark itself may be time-varying even though the same key is used to watermark all frames. Temporal synchronization in such water-

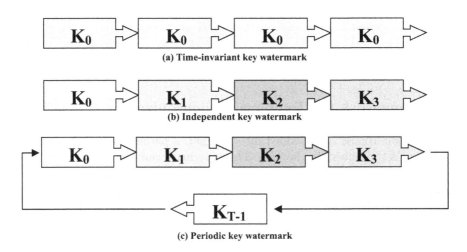

(a) Time-invariant key watermark

(b) Independent key watermark

(c) Periodic key watermark

Fig. 7.2. Special cases in watermark key schedules.

marks is a trivial problem because the detector is not required to search for the proper detection key. However, security may be a concern, particularly if the watermark is also time-invariant and susceptible to a temporal correlation attack to deduce the watermark (an example exploiting such a security vulnerability was demonstrated in [13]). These watermarks have high temporal redundancy in the key sequence (thus, synchronization is simple) but relatively low security.

Another type of key schedule is to use independent keys to generate the embedded watermark for each frame. Strictly speaking, the watermark keys $K(i)$ and $K(j)$, $i \neq j$, cannot be truly independent because all the keys in the key schedule are related to the embedding key K_E. This class of watermarks refers to those techniques whose keys do not repeat or repeat with an extremely long period. The watermarks embedded in successive frames of the video are uncorrelated, which is desirable from a security standpoint but undesirable in terms of robustness, as an attacker can perform temporal collusion (frame averaging) to remove the watermark. In addition, because attacks such as frame insertion and dropping remove the correspondence between the frame index t and the detection key $K(t)$, these attacks successfully desynchronize a naive detector. Recovering synchronization after it has been lost may involve a search over a large key space and could be nearly impossible to perform efficiently. These watermarks have little or no temporal redundancy in the key sequence (thus, synchronization is difficult) but relatively high security.

In a periodic key watermark, the keys form a repeating sequence (with relatively short period). Recovering synchronization from these watermarks is not difficult because the detector can restrict the search to a discrete, countable set of keys. For example, one method of synchronization recovery is to search for K0 if synchronization is lost. Periodic key watermarks do not possess the same degree of security as an independent key watermark, but resynchronization is practical even though not all the keys are identical.

The time-invariant key and periodic key watermarks indicate that efficient synchronization and resynchronization are possible if some degree of temporal redundancy is present in the watermark key schedule. Our technique focuses on the construction of a key schedule possessing temporal redundancy for efficient synchronization. The detector synchronizes by performing a limited search in the key space. This class of detectors is in contrast with detectors that embed side information for synchronization, such as templates. Time-invariant key watermarks, independent key watermarks, and periodic key watermarks are special cases in our technique.

7.3 A Protocol for Temporal Synchronization

7.3.1 Watermark Embedding Protocol

Our technique for generating the key schedule at the watermark embedder is illustrated in Figure 7.3, consisting of four principal components: watermark embedder, temporal redundancy control, feature extractor, and key generator.

Watermark embedder: The watermark embedder accepts as inputs a frame of the original (unwatermarked) video, an embedding key, and any other side information necessary to embed the watermark into the current frame. The embedding key is supplied from the temporal redundancy control component. Any watermark-embedding technique suitable under the assumptions described in Section 7.2 can be used.

Temporal redundancy control: This is the primary means of controlling the amount of temporal redundancy in the key schedule. Temporal redundancy is added to the key schedule by using a single key to watermark multiple frames of the video. Three parameters are used to define the key schedule: The master embedding key K_E is used as the initial key to generate the key schedule. The period, α, is the number of frames watermarked before the key generator is reset. The repeat, β, is the number of consecutive frames that are watermarked with the same key before a new key is generated ($1 \leq \beta \leq \alpha$). The functionality of the temporal redundancy control is shown below:

Watermark Embedding Procedure (WEP)

1. Set $a = 0$, $b = 0$, $K = K_E$, and reset the key generator state.
2. Read a frame of the input video.
3. Send the current key K to the watermark embedder to watermark the current frame.

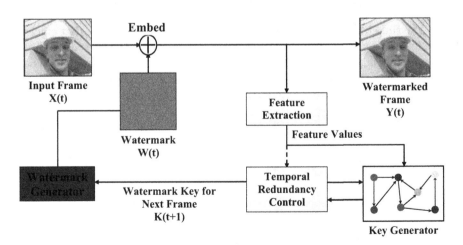

Fig. 7.3. Special cases in watermark key schedules.

time

Numbers indicate frame index

Fig. 7.4. Example key schedule ($\alpha = 8, \beta = 2$)

4. Increment a by 1. If $a < \alpha$, continue to step 5; otherwise, go to step 1.
5. Increment b by 1. If $b = \beta$, continue to step 6; otherwise, go to step 2.
6. Use the feature extractor and key generator (both described below) to create a new key K for subsequent frames, set $b = 0$, and go to step 2.

Figure 7.4 illustrates an example key schedule for the case of $\alpha = 8$ and $\beta = 2$, with frames depicted in the same shades indicating the same key is used to watermark those frames. Every two consecutive frames are watermarked with the same key. The watermark key is reset to K_E every eight frames, so frames 0, 1, 8, and 9 are watermarked with key K_E. The key used to watermark frames 2 and 3 is not necessarily identical to the key used to watermark frames 10 and 11. (The name "period" for α is somewhat a misnomer because the key schedule is not periodic in general.)

The period parameter α controls the degree of *global redundancy* in the key schedule, as it determines how frequently the watermark key generator is reset and K_E is used for embedding the watermark. The frames that are embedded with key K_E are known as *resynchronization frames* because the detector can search for K_E when it is lost. These frames also provide synchronization points when the detector performs initial synchronization. Increasing α decreases the frequency when K_E is embedded, thus reducing the temporal redundancy in the watermark. A key schedule generated by $\alpha = 1$ corresponds to embedding with key K_E for every frame, or a time-invariant key watermark.

The repeat parameter β controls the degree of *local redundancy* in the key schedule. A high repeat factor corresponds to increased redundancy and decreased key rate (number of distinct keys used for watermarking embedding per unit time.) Increased local redundancy is beneficial for resisting temporal attacks, such as frame-dropping, transposition, and averaging. As the experiments will show, a key schedule having minimal local redundancy (large α, $\beta = 1$) has behavior similar to an independent key watermark and can be desynchronized by frame-averaging and deletion attacks.

As mentioned in Section 7.2, the embedding key is just one parameter used to determine the embedded watermark, and frames watermarked by the same key do not necessarily imply that the embedded watermark signals in those frames are identical. One example is the technique in [13], which creates image-dependent watermark signals using a time-invariant key.

Feature extraction: Feature extraction allows the key schedule to be video dependent. It is invoked when a new watermark key is needed (step 6 of the WEP shown above) and is inactive during other times. The feature extractor examines the watermarked image and outputs a vector of features that is made available to the key generator.

For robust video watermarking, the features should change in value only when significant alterations are made to the watermarked frame. In our technique, it is possible to destroy temporal synchronization by performing spatial attacks on video frames because the key schedule is video-dependent. Ideally, the features are sufficiently robust so that such an attack is successful only when the attacked video no longer has any value in the application. The features should also resist (or be invariant to) spatial-synchronization attacks and be computationally efficient. In particular, if real-time embedding or detection is required by an application, the feature extraction process must also be in real time.

The features themselves should be kept secret and known only to the embedder and detector. For security, the feature values can involve K_E or an auxiliary key.

Key generator: The key generator is used to create new watermarking keys in the key schedule. The key generator, which is in general a state machine, is involved in steps 1 and 6 in the WEP. During step 1, the state of the key generator is reset to a predefined initial state. This is necessary so the watermark detector (which uses an identical key generator) will be able to recover the key sequence. In step 6, the key generator creates a new key using its current state information and input from the feature extractor. The state of the key generator can also change in step 6. During other steps of the WEP, the key generator is inactive.

The method of generating keys for future watermarks based on the current key and features also adds temporal redundancy in the key schedule. One way of viewing the operation of the key generator is that of a predictor, in which future watermark keys are predicted from current and past watermark keys. If the future keys are completely random (that is, there is no temporal redundancy), then the detector will not be able to determine the key schedule without an extensive search through the key space. However, if future keys are chosen in such a way that they can be deduced (or predicted) by the current and past keys and features, the detector will be able to identify the key schedule without performing a search.

One type of key generator is a finite state machine (FSM). One state in the FSM is identified as the initial state. Each state has an associated key transition function, or a description of how to generate the next key from the current key. Key transition functions can take as inputs the current key, the master key K_E, the feature vector supplied by the feature extractor, and other side information (for example, message payload). State transitions are (in general) dependent on the feature values; however, the key generators for

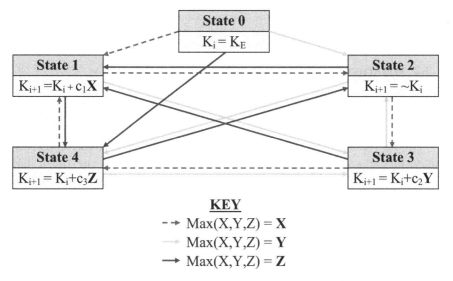

Fig. 7.5. An example key generator using a finite state machine.

time-invariant key and periodic key watermarks can be implemented as trivial FSMs that do not depend on features.

Figure 7.5 shows an example FSM key generator that uses three features (X, Y, and Z) from the feature extractor. The initial state is state zero. The key transition functions are shown with each state, and, with the exception of states 0 and 2, involve the current key value and the value of one of the features (c_1, c_2, and c_3 are constants). In state 2, the next key is generated by performing binary negation on the current key. The feature (X, Y, or Z) having the greatest value determines the state transitions.

Like the feature extractor, the structure of the key generator is a shared secret between the embedder and detector. The key generator can also be dependent on K_E or an auxiliary key, such as by using key-dependent key-transition functions or key-dependent state machines.

The flexibility of the key-generator and feature-extraction components provides a great deal of generality that can be exploited by the application or the degree of security needed. In this chapter, we chose a simple set of features and a trivial FSM key generator (Figure 7.5) to implement the protocol and evaluate its performance while varying the temporal-redundancy parameters (α, β). In future work, issues regarding the development of the feature extractor and key generator, such as the design criteria for maximizing robustness and security of the embedded watermark, will be studied.

Computational cost of the embedding protocol: The computational costs for watermark embedding can be critical in real-time implementations. By examination of the WEP, the worst-case computational cost is on the order of $(WE + FE + KG)$ per frame, where WE is the computational cost for watermark embedding, FE is the computational cost for feature extraction, and KG is the computational cost for key generation.

7.3.2 Watermark Detection Protocol

The watermark detection protocol uses the watermark detector and a queue to perform temporal synchronization. Figure 7.6 shows the watermark detection protocol.

The watermark detector in Figure 7.6 is the detector that corresponds to the technique used to embed the watermark into frames of the video. The feature extraction and key-generator components are identical to those components in the watermark-embedding protocol.

The memory is a queue that is used to perform the search for establishing and maintaining temporal synchronization. Each entry in the queue stores a watermark key and the state of the key generator, or a (key, state) tuple. The watermark detection procedure is as follows:

Watermark Detection Procedure (WDP)

1. The detector is initialized with an empty queue.
2. Read a frame of the input video.
3. Attempt to detect the watermark by using:
   a) The key K_E, and
   b) Every key that is stored in the queue.

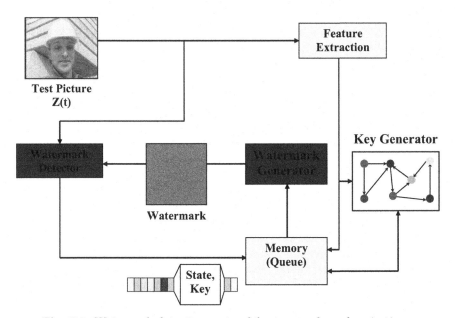

Fig. 7.6. Watermark detection protocol for temporal synchronization.

4. If no watermark is found during step 3, go to step 2 to try to detect a watermark in the next frame.
5. If the watermark detected did not correspond to key K_E, that key must already be present in the queue. Move that (key, state) tuple from its place to the head (front) of the queue.
6. Perform feature extraction on the watermarked image.
7. With the feature values obtained in step 6 and the state stored in the queue (use the initial state if K_E was detected) as inputs to the key generator, find (next key, next state) and insert (next key, next state) into the front of the queue. Because the queue is of finite size, this can cause the tuple at the tail of the queue to be lost.
8. Go to step 2 to read the watermark in the next frame.

In essence, the queue stores the most recently detected watermark keys and their corresponding next keys. A larger queue size (number of entries in the queue) allows more "history" in the key sequence to be kept and may provide increased robustness against frame-transposition attacks. A transposition attack involves interchanging frames of the video, so it is possible that the embedded watermark in a frame comes from a key observed in the past. However, the gain provided by a large queue is subject to diminishing returns, and large queues also increase the computational cost for processing one frame of the video.

The detector does not require knowledge of the embedder's temporal-redundancy parameters (α and β) as side information because the search performed during step 3 of the WDP allows the watermark to be detected without a priori knowledge of the correspondence between the key and the frame index. This is also beneficial for resilience against temporal attacks such as frame deletion, insertion, and transposition, which can be thought of as causing "jitters" in the key sequence. Also note that, unlike the watermark embedder, the feature-extraction and key-generator components may be active during every frame in the detector. The only frames that do not involve the use of the feature extractor and key generator are those in which no watermark can be found.

Computational cost of the detection protocol: The most significant portion of the computational cost is incurred during the watermark detections of step 3 of the WDP. The computational cost is on the order of $(((Q+1) \times WD) + FE + KG)$ per frame, where Q is the queue size, WD is the computational cost for watermark detection, FE is the computational cost for feature extraction, and KG is the computational cost for key generation. If FE and KG are computationally efficient, this reflects a linear increase in the computation costs for simple watermark detection. Furthermore, the multiple watermark detections for the key search may be implemented as parallel computations if necessary to fulfill real-time constraints.

7.4 Experimental Results

We have implemented both the watermark-embedding and detection protocols using a simple video-watermarking technique. The watermark is a zero-mean Gaussian pseudorandom signal that is added to the luminance pixel values of each frame. The watermark key is used as the seed to the random-number generator. The chrominance samples in each video frame were not watermarked. Detection is performed by using a filter to reduce the effect of the original image (as suggested in [12]) followed by correlation with the watermark signal. If the correlation exceeds a threshold value, the watermark is successfully detected in the frame.

The feature extractor partitions each frame into blocks (32×32 pixels in size) and randomly assigns each block to set X, Y, or Z. (This assignment occurs once, and the same block assignments are used for all frames of the video.) To obtain the feature value X for a particular frame, the average luminance of all blocks assigned to set X is found and then quantized (to make the feature value less sensitive to noise). A similar procedure is used for the features Y and Z. These features are then used as inputs to the FSM key generator shown in Figure 7.5. These features and the key generator were used because they are simple to implement and computationally efficient. It is realized that they may not be optimal for either security or robustness.

Three 352×288 uncompressed CIF videos (*Foreman*, *Akiyo*, and *Bus*) were watermarked with varying amounts of temporal redundancy and then attacked. Temporal redundancy was varied by adjusting the period (α) and repeat (β) parameters during embedding. After the attack, the percentage of watermarked frames successfully detected was measured as the performance measure. A percentage of 100% indicates that the attack was completely ineffective at desynchronizing the detector, while 0% indicates that the attack rendered the watermark undetectable. Ten trials were performed for each video and set of redundancy parameters, and the mean detection results of all trials and videos are shown in the results below. Unless otherwise noted, the queue size used was ten entries.

Figure 7.7 shows the performance of the technique after the video is attacked by dropping frames at random. Each frame in the video sequence has a fixed probability (0.05, 0.25, and 0.50) of being dropped (independent of the decision to drop other frames in the video). If there is no local redundancy in the key sequence ($\beta = 1$ frame), the performance of the technique is poor. However, if some local redundancy is introduced ($\beta = 5, 10$ frames), the detector is capable of maintaining synchronization and detecting a significant portion of the watermark even if 50% of the frames of the video had been lost. As expected, the rate of successful watermark detection decreases as the amount of temporal redundancy in the key schedule is decreased.

The technique is very effective against the insertion of unwatermarked frames, as shown in Figure 7.8. In this attack, frames of the original (unwatermarked) video are randomly inserted between frames of the watermarked

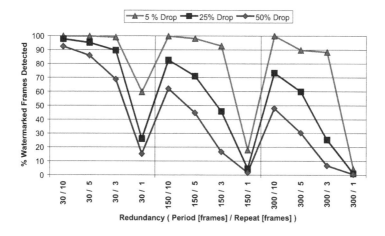

Fig. 7.7. Performance against frame drop attack

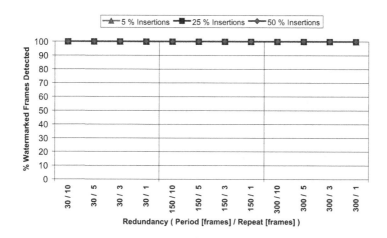

Fig. 7.8. Performance against frame insertion attack

video. The insertion of watermarked frames resembles a transposition attack, which is discussed next. Between each pair of frames (denoted Wa and Wb) in the watermarked video, the attacker inserts a single frame of the original video (denoted frame O) with fixed probability P. If a frame is inserted (so the video frame sequence appears as Wa-O-Wb), the attacker considers adding additional frames between frame O and frame Wb with an equal probability P for each additional frame. The results show that the attacker can insert an arbitrary number of unwatermarked video frames into the sequence and synchronization will be maintained. The contents of the inserted frames are irrelevant as long as they are unwatermarked (so they could be frames from

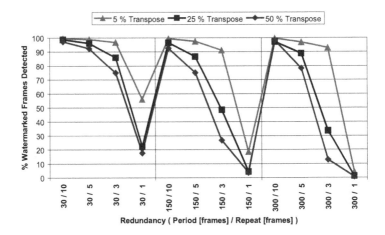

Fig. 7.9. Performance against frame-transposition attack (queue size = 10 entries).

the original video or some other arbitrary set of images). The performance is not at all surprising because the state of the queue is not changed if no watermark is detected in a given video frame.

Figure 7.9 and Figure 7.10 show the performance against a local frame-transposition attack. In this attack, the watermarked video is scanned from the first frame to the last, with each frame having a fixed probability ($P = 5\%, 25\%, 50\%$) of being interchanged with another (target) frame in the local neighborhood of the candidate frame. We chose the target frame by generating a Gaussian random number ($\sigma^2 = 5.0$), dropping fractions, and treating the

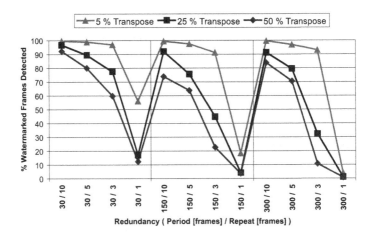

Fig. 7.10. Performance against frame-transposition attack (queue size = 4 entries).

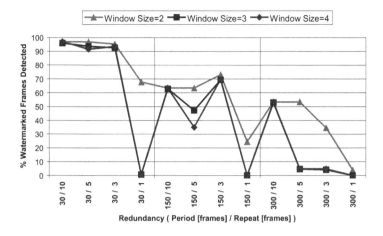

Fig. 7.11. Performance against frame averaging attack.

number as a relative time index where 0 is the current frame, -1 is one frame in the past, $+1$ is one frame in the future, and so on. Because transposing a frame with itself does not accomplish anything, we choose another random number if the relative time index is zero. The performance of the technique shows the same general trend as the frame-dropping attack. By comparing Figures 7.9 and 7.10, decreasing the queue size from ten entries to four entries does not affect the performance for low transposition rates ($P = 5\%$), but the performance can fall more sharply for larger transposition rates ($P = 50\%$).

The next attack examined was frame averaging. In this attack, a window of fixed size (2, 3, and 4 frames) was swept temporally across the video. At each step, a composite frame was constructed by computing, for each spatial location, the average pixel value over all the frames in the window. As Figure 7.11 shows, the frame-averaging process reduces the detection rate considerably. The primary culprit for the decrease in performance is the poor choice of features in our implementation. (Frame averaging can change pixel values greatly, particularly for video where a lot of motion occurs.) However, if sufficient local redundancy is present, the detector can still detect watermarks in 50% of the frames of the video despite the poor feature selection. As expected, a key schedule with little local redundancy is not very resilient against averaging attack.

Lastly, a combined attack was executed that consisted of the following attacks in order (each attack carried out with probability P): frame insertion, frame transposition, frame dropping, and frame averaging with a window size of 3. The performance under the combined attack is shown in Figure 7.12. After the attacks (and with the visual quality of the video being quite poor), the detector is still capable of detecting a portion of the watermarked frames in the video if sufficient temporal redundancy is present in the key schedule.

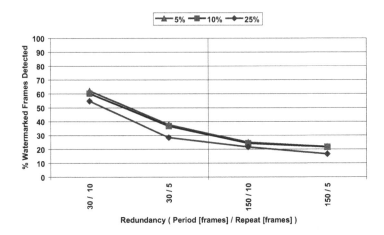

Fig. 7.12. Performance against combined attack (queue size = 4 entries).

7.5 Improvements and Further Work

The results presented in Section 7.4 show that if sufficient temporal redundancy is present in the key schedule, the watermark can maintain synchronization despite the presence of frame-dropping, insertion, transposition, and averaging attacks. However, the technique has a lot of room for improving watermark robustness and security.

In our implementation, one of the primary reasons why the detection performance was decreased after frame averaging is the feature selection used in the feature extractor. The features we used have the advantage of low computational cost but did not possess the desired amount of robustness. Sometimes, the detector was not able to create the same key schedule that was used to embed the watermark after frames were averaged because the feature values changed after the averaging attack. By finding and using more robust features, the performance against frame averaging can be improved.

Another method for improving our implementation is to adapt the temporal redundancy of the watermark with the temporal redundancy of the video instead of using the α and β constants. To illustrate the problem, consider the watermark key schedule shown in Figure 7.4 and a rapidly changing video. The watermark keys used for frames 4 and 5 are dependent on the feature values obtained in frame 3. However, if the feature values change rapidly (in particular, if the feature values change between frames 2 and 3), the detector must receive frame 3 to recover the key sequence used for frames 4 to 7. Because the feature values have changed, frame 2 does not carry any information about the key in frames 4–7. If frame 3 is dropped, then the detector will not be able to generate the key schedule for frames 4–7. (In effect, this represents a loss of temporal redundancy.) Instead of using α and β, the embedder can perform feature extraction in every frame for the purpose of deciding when to change the watermark key. The key is changed only after feature values are constant over a certain number of frames. In this manner, the watermark embedder can ensure that some temporal redundancy is available before the key is changed. This can significantly improve the performance against frame dropping and frame averaging.

Several improvements can be made to increase security. One of the foremost improvements is to make the embedded watermark in resynchronization frames strongly image-dependent (using the technique described in [13] or some other method). This prevents an attacker from using correlation to identify those frames. Another improvement is to embed resynchronization frames in nonuniformly spaced intervals (do not use α), which also complicates the process of identifying the resynchronization frames for the attacker. Lastly, the key generator needs to be examined more closely, and other key generators or key-transition functions should be considered. For example, one may consider the use of cryptographic hashes in key-transition functions $K_{i+1} = \text{hash}(K_E, K_i, \text{features})$. Such key-transition functions preserve the temporal redundancy of the key schedule but scatter the keys throughout the key space instead of localizing the keys near K_E. The hash function taking K_E as an argument makes it necessary for an attacker to have knowledge of K_E and not just K_i to deduce the key schedule.

In addition to improving the robustness and security, more temporal attack methods should be studied, including frame decimation and frame-rate conversion (temporal resampling), 3:2 pulldown and interlacing, and motion-compensated frame averaging. We are also interested in using the technique on compressed video.

We have investigated the use of more secure state machines, as well as the adaptive temporal redundancy described above, in [16]. In addition to the use of cryptographic hash functions for the key generator, we also investigated randomized state machines in [16].

7.6 Conclusions

This chapter describes a method for efficient temporal synchronization and resynchronization for video watermarking. With sufficient temporal redundancy in the key schedule, the embedded watermark can be resilient against temporal attack, such as frame dropping, insertion, transposition, and averaging. It was shown that a watermark with no temporal redundancy in the key schedule is not resilient against temporal attacks. This method also addresses initial synchronization and resynchronization. Synchronization was performed without the use of template embedding; however, the technique presented does not preclude the use of templates. Our new method is independent of the actual watermarking technique (embedder and detector) and could be retrofitted into most existing video-watermarking techniques.

Acknowledgment

This work was supported by a grant from Digimarc.

References

[1]   F. Hartung and B. Girod, "Watermarking of uncompressed and compressed video," *Signal Process.*, **66** (3):283–301 (1998).

[2]   I. Cox, J. Kilian, T. Leighton, and T. Shamoon, "Secure spread spectrum watermarking for multimedia," *IEEE Trans. Image Process.*, **6** (12):1673–1687 (1997).

[3]   R. Wolfgang, C. Podilchuk, and E. Delp, "Perceptual watermarks for digital images and video," *Proc. IEEE* **87** (7):1108–1112 (1999).

[4]   F. Petitcolas, R. Anderson, and M. Kuhn, "Attacks on copyright marking systems," *Proceedings of the Second International Workshop in Information Hiding*, April 14–17, 1998, pp. 218–238.

[5]   G. Braudaway and F. Mintzer, "Automatic recovery of invisible image watermarks from geometrically distorted images," *Proceedings of the SPIE Security and Watermarking of Multimedia Contents II*, vol. **3971**, San Jose, CA, January 24–26, 2000, pp. 74–81.

[6]   E.T. Lin, C.I. Podilchuk, T. Kalker, and E. J. Delp, "Streaming video and rate scalable compression: What are the challenges for watermarking?," *Proceedings of the SPIE Security and Watermarking of Multimedia Contents III*, San Jose, CA, January 22–25, 2001, pp. 116–127.

[7]   E. Lin, G. Cook, P. Salama, and E. Delp, "An overview of security issues in streaming video," *Proceedings of the International Conference on Information Technology: Coding and Computing*, Las Vegas, NV, April 2–4, 2001, pp. 345–348.

[8]   A. Eskicioglu and E. Delp, "An overview of multimedia content protection in consumer electronics devices," *Signal Process. Image Commun.*, **16**:681–699 (2001).

[9]   I. Cox, M. Miller, and J. Bloom, *Digital Watermarking*, (Morgan Kauffman Publishing, San Fransisco, 2002).

[10] F. Deguillaume, G. Csurka, J. O'Ruanaidh, and T. Pun, "Robust 3D DFT video watermarking," *Proceedings of the SPIE, Security and Watermarking of Multimedia Contents*, vol. **3657**, San Jose, CA, January 25–27, 1999, pp. 113–124.

[11] A. Herrigel, S. Voloshynovskiy, and Y. Rytsar, "The watermark template attack," *Proceedings of the SPIE, Security and Watermarking of Multimedia Contents III*, vol. **4314**, San Jose, CA, January 22–25, 2001, pp. 394–405.

[12] T. Kalker, G. Depovere, J. Haitsma, and M. Maes, "A video watermarking system for broadcast monitoring," *Proceedings of the SPIE, Security and Watermarking of Multimedia Contents*, vol. **3657**, San Jose, CA, January 25–27, 1999, pp. 103–112.

[13] M. Holliman, W. Macy, and M. Yeung, "Robust frame-dependent video watermarking," *Proceedings of the SPIE, Security and Watermarking of Multimedia Contents II*, vol. **3971**, San Jose, CA, January 24–26, 2000, pp. 186–197.

[14] B. Mobasseri and A. Evans, "Content-dependent video authentication by self-watermarking in color space," *Proceedings of the SPIE, Security and Watermarking of Multimedia Contents III*, vol. **4314**, San Jose, CA, January 22–25, 2001, pp. 35–44.

[15] G. Langelaar and R. Lagendijk, "Optimal differential energy watermarking of DCT encoded images and video," *IEEE Trans. Image Process.*, **10**(1):148–158 (2001).

[16] E. Lin and E. Delp, "Temporal Synchronization in Video Watermarking–Further Studies," *Proceedings of the SPIE, Security and Watermarking of Multimedia Contents V*, vol. **5020**, Santa Clara, CA, January, 2003.

Secure Display Using Encrypted Digital Holograms

Osamu Matoba and Bahram Javidi

8.1 Introduction

Recent advances in data communication based on optical-fiber technology enable us to send data at an ultrafast transfer rate that will be achieved commercially at 1 Tbit/s. With such ultrafast data transmission, three-dimensional television or movies are attractive applications. On the other hand, data security is one of the important issues in data networks. Optics can provide a high level of security in optical data storage and information processing [1–6].

In this chapter, we present secure three-dimensional display systems using digital information of the hologram. Digital holographic techniques that use an image sensor to record a hologram have become viable with the ongoing development of megapixel charge-coupled-device (CCD) or CMOS sensors that have sufficient dynamic range in each pixel [7–15]. It has been shown with numerical propagation that the fully complex field calculated from the digital holograms in the Fresnel domain can be used to reconstruct 3-D objects successfully. Storage of the hologram in a computer enables us to reduce the noise through image-processing techniques and numerically reconstruct the object with arbitrary views. The digital hologram is also in a convenient form for data transmission and object recognition. There are some papers that apply the digital holographic technique to data-security issues [9,10]. With digital reconstruction, however, it takes a long time to calculate an optical field, thus making it difficult to develop a real-time three-dimensional (3-D) display device.

We have been developing an optical-correlation-based reconstruction method using encrypted digital holograms [15]. The optical reconstruction using encrypted digital holographic techniques enables us to implement a real-time secure display system or a secure television because the transmission of successive encrypted images can be performed quickly and effectively via optical-fiber communications. In the following sections, we describe two systems for secure 3-D display based on encrypted digital holograms.

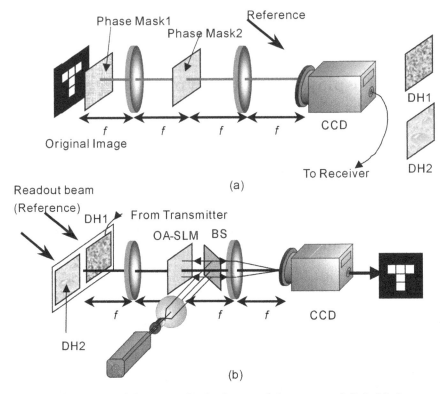

Fig. 8.1. Schematic of the secure display by use of the encrypted digital holograms. (a) The secure data transmission; and (b) optical retrieval system. DH1 and DH2 denote the digital holograms of the encrypted image and Fourier-transformed pattern of the phase mask in the Fourier plane used in the encryption process, respectively. OA-SLM denotes an optically-addressed spatial-light modulator.

8.2 Secure Three-Dimensional Display by Use of Digital Holograms

Figure 8.1 shows our concept of a secure data-communication system using encrypted digital holograms for a real-time secure display [15]. The system consists of two parts: a transmitter and a receiver. At the transmitter, as shown in Figure 8.1(a), original images are encrypted by a double-random phase-encryption technique [3] and then are recorded as digital holograms together with a plane reference wave in an image sensor such as a CCD camera. The digital hologram of the encrypted image (DH1) can be transmitted to a receiver via any conventional electronic or digital communication channel. The decryption key that has the phase information of the modulation mask used at the Fourier plane is also recorded as a key digital hologram (DH2). At the receiver, the original image can be retrieved by an optical-correlation-based reconstruction process by using both the digital holograms, as shown in Figure

8.1(b). Both the encryption and retrieval systems are operated by all-optical means; thus, a real-time secure display can be implemented using fast data transmission of successive encrypted images via optical-fiber communications.

8.2.1 System Operations

We describe the operation of the proposed system as shown in Figure 8.1. Let $f(x, y)$, $p(x, y)$, and $Q(v, \eta)$ denote the image to be encrypted and the random phase masks at the input and the Fourier planes, respectively. Spatial coordinates are taken as in Figure 8.1. In double-random phase encryption [3], the encrypted image, $e(x, y)$, can be written as

$$e(x, y) = f(x, y)p(x, y) \oplus q(x, y), \tag{8.1}$$

where $\oplus$ denotes convolution and $q(x, y)$ is the Fourier transform of $Q(v, \eta)$ We record a hologram (DH1), $I_s(x, y)$, in the CCD plane by making an interference pattern between the encrypted image and a plane reference wave, $r(x, y)$. We also record a key hologram (DH2), $I_k(x, y)$, which is used in the decryption process. The key hologram is recorded in the CCD by making an interference pattern between the plane reference wave $r(x, y)$ and a Fourier-transformed pattern of the phase mask used at the Fourier plane in the encryption process. Both the digital holograms can be described as

$$I_s(x, y) = |e(x, y) + r(x, y)|^2$$
$$= |e(x, y)|^2 + |r(x, y)|^2 + e^*(x, y)r(x, y) + e(x, y)r^*(x, y)' \tag{8.2}$$

and

$$I_k(x, y) = |q(x, y) + r(x, y)|^2$$
$$= |q(x, y)|^2 + |r(x, y)|^2 + q^*(x, y)r(x, y) + q(x, y)r^*(x, y)'. \tag{8.3}$$

Both digital holograms are transmitted to an authorized receiver via a communication network. At the receiver, the original image can be retrieved by using a joint transform correlator architecture as shown in Figure 8.1(b). Both holograms are placed side-by-side in the input plane. When the encrypted digital hologram (DH1) is shifted from the key hologram (DH2) at a distance of α along the x axis, a new input image, $s(x, y)$, is described as

$$s(x, y) = I_s(x + \alpha, y) + I_k(x, y). \tag{8.4}$$

The decryption is implemented optically by use of an optical-correlator-based system. In the optical reconstruction, this input image is displayed on a spatial-light modulator such as a liquid-crystal display and then is illuminated by the reference wave, $r(x, y)$, which is the same as that used in the recording. If the reference wave is tilted enough to spatially separate the diffracted light from the transmitted light, we can obtain, isolated, the following term:

$$t(x, y) = e(x + a, y) + q(x, y). \tag{8.5}$$

In the derivation of Eq. (8.5), we use $|r(x, y)|^2 = 1$. The diffracted light beam described in Eq. (8.5) is Fourier-transformed by a lens and then is recorded in an intensity-sensitive medium such as an optically addressed spatial-light modulator (OA-SLM). This process is the same as in conventional joint transform correlators used to obtain the correlation of the two objects at the input plane. The joint power spectrum at the Fourier plane is described as

$$
\begin{aligned}
|T(v, \eta)|^2 &= |E(v, \eta) \exp(-j\alpha v) + Q(v, \eta)|^2 \\
&= |E(v, \eta)|^2 + |Q(v, \eta)|^2 + E(v, \eta)Q^*(v, \eta) \exp(-j\alpha v) \\
&\quad + E^*(v, \eta)Q(v, \eta) \exp(j\alpha v) \\
&= |E(v, \eta)|^2 + |Q(v, \eta)|^2 + F(v, \eta) \oplus P(v, \eta) \exp(-j\alpha v) \\
&\quad + F^*(v, \eta) \oplus P^*(v, \eta) \exp(j\alpha v),
\end{aligned} \tag{8.6}
$$

where $T(v, \eta)$, $E(v, \eta)$, $F(v, \eta)$, and $P(v, \eta)$ are the Fourier transforms of $t(x, y)$, $e(x, y)$, $f(x, y)$, and $p(x, y)$, respectively. By illuminating the OA-SLM by another plane wave for readout and then Fourier-transforming of Eq. (8.6), we obtain the output signal as follows:

$$
\begin{aligned}
o(x, y) &= e(x, y) \otimes e(x, y) + q(x, y) \otimes q(x, y) \\
&\quad + f(x + \alpha)p(x + \alpha, y) + f^*(x - \alpha)p^*(x - \alpha, y).
\end{aligned} \tag{8.7}
$$

The third term on the right-hand side of Eq. (8.7) shows that the original image can be successfully retrieved by detection with the intensity-sensitive detectors. The recording and retrieval systems are operated all-optically; thus, we can construct a secure real-time display or television together with optical-fiber communications.

8.3 Numerical and Experimental Demonstrations

First, we present a numerical demonstration of the secure display system. Figure 8.2 shows the original binary image and its encrypted image. The original image is encrypted by using two random phase masks. Figure 8.2(b) shows that the encrypted image is a white-noise-like image. The key hologram is also calculated as a digital hologram. Note that the correct key is the digital hologram of a Fourier-transformed pattern of the phase mask in the Fourier plane used in the encryption process. Using the analysis from Eqs. (8.1) to (8.7) above, the reconstructed image is obtained as shown in Figure 8.3. Figures 8.3(a) and 8.3(b) show the reconstructed image using the correct and an incorrect decryption key hologram, respectively. In this figure, only a window centered with the reconstructed image is represented to extract the reconstruction results. The incorrect key is made by using a random phase distribution. We can see that the correct key can reconstruct the original image, but the white-noise-like image still remains with the incorrect key.

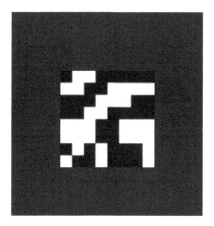 

Fig. 8.2. An example of numerical results. (**a**) Original image to be encrypted, (**b**) encrypted image.

We also experimentally demonstrate the optical reconstruction. Figure 8.4 shows an experimental setup. At this time, an OA-SLM with suitable resolution was not available to us to record the joint power spectrum as described in Eq. (8.6); therefore, we implement the optical reconstruction using a calculated joint power spectrum. The calculated joint power spectrum is displayed on a liquid-crystal spatial-light modulator (LC-SLM) with 1024×768 pixels. The collimated beam from a He–Ne laser at a wavelength of 632.8 nm illuminates the input joint power spectrum and then is Fourier-transformed by lens L with a focal length of 400 mm. The Fourier-transformed pattern of the joint power spectrum is recorded by a CCD camera. Figures 8.5(a) and 8.5(b) show the joint power spectrum patterns to be displayed on the LC-SLM by using

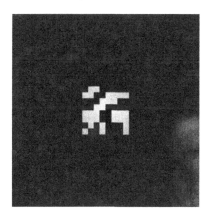 

Fig. 8.3. Numerical results of reconstructed images using an optical-correlation-based retrieval system when (**a**) the correct and (**b**) the incorrect key holograms are used.

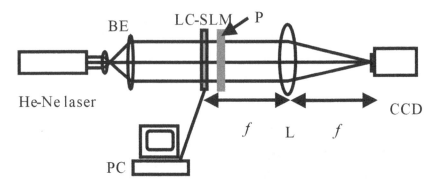

Fig. 8.4. Optical setup for the demonstration of the optical retrieval system. BE, LC-SLM, and P denote a beam expander, a liquid-crystal spatial-light modulator and a polarizer, respectively.

the correct phase key and the incorrect phase key, respectively. The reconstructed images are shown in Figure 8.6. Note that a window centered with the reconstructed image is represented to extract the reconstruction results. When the correct phase key is used, we obtain the original image as shown in Figure 8.6(a). We can see speckles in the reconstructed image. These speckles are caused by the limited space-bandwidth product of the optical system and limited amplitude levels by the LC-SLM. We will present the numerical analysis below. The reconstructed image is still a white-noise-like image, as shown in Figure 8.6(b), when the incorrect key is used. Thus, this system has been successfully implemented.

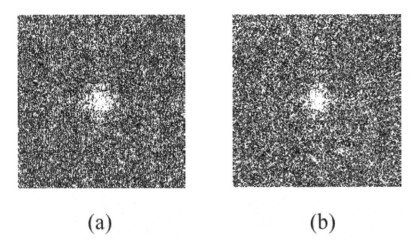

Fig. 8.5. The joint power spectrum when (**a**) the correct and (**b**) the incorrect digital holograms are used as keys.

Fig. 8.6. Experimental results of the reconstructed images using the optical-correlation-based retrieval system when (**a**) the correct and (**b**) the incorrect digital holograms are used as keys.

To implement a real-time operation of the secure display system, the OA-SLM is a key device. To evaluate the performance of an OA-SLM required for the real-time implementation, we investigate numerically the influence of the quantization level of the joint power spectrum in the reconstruction process. Here we use the following criteria to evaluate the error caused by the quantization of the joint power spectrum:

$$\text{Err} = \frac{\sum_i \{r_n(i) - r_Q(i)\}^2}{\sum_i \{r_n(i)\}^2}. \tag{8.8}$$

In Eq. (8.8), $r_n(i)$ and $r_Q(i)$ denote the intensity at the ith pixel of an ideal reconstructed image and a reconstructed image using a quantized joint power spectrum, respectively. The quantization level is changed from 1 bit to 14 bits. The ideal reconstructed image is obtained when the joint power spectrum is quantized into double precision. Figure 8.7 shows an example of the numerical results. Figure 8.7(a) shows the original binary image, and Figures 8.7(b)–(f) show the reconstructed images when the amplitude of the joint power spectrum is quantized into double precision, 10 bits, 8 bits, 6 bits, and 5 bits, respectively. We can see that 10 bits are required for successful reconstruction in the input binary images used in the calculation. Figure 8.8 shows the normized mean-squared error as described in Eq. (8.8) as a function of the number of bits in the quantization of the joint power spectrum. The error decreases as the number of bits increases. From this error curve, 12 bits is enough to give negligible error in the reconstructed image.

8.4 Three-Dimensional Display by Use of Phase Information of Digital Holograms

In this section, we present a reconstruction technique of a three-dimensional (3-D) object that uses only phase information calculated from phase-shifting

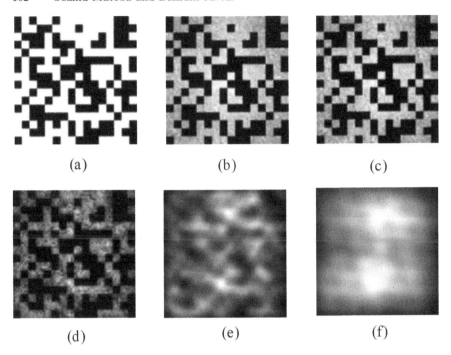

Fig. 8.7. Numerical results of (**a**) the input image; (**b**–**f**) the reconstructed images when the amplitude of the joint power spectrum is quantized into (**b**) double precision, (**c**) 10 bits, (**d**) 8 bits, (**e**) 6 bits, and (**f**) 5 bits.

digital holograms [16]. It is well-known that a phase-only hologram or phase information is used in kinoform [17,18] and diffractive optical elements for interconnection [19,20]. The method is suitable for real-time optical reconstruction of the 3-D object because the phase-only reconstruction allows us to use commercially available display devices such as liquid-crystal spatial-light modulators (SLMs). These devices can only be operated in either phase- or amplitude-modulation mode. Using only this phase information, we can reduce by half the storage requirements of the digital hologram. We also take advantage of minimal optical-power loss in the reconstruction process.

8.4.1 System Operations

An illustration of a 3-D object capture and display system is presented in Figure 8.9. A block diagram of the system is described in Figure 8.10. The system consists of two subsystems; one is a recording system of digital holograms of 3-D objects, and the other is an optical-reconstruction system using phase-only information of the digital hologram.

For the recording system, we use a Mach–Zehnder interferometer architecture as a phase-shifting interferometer. A laser beam is divided into two

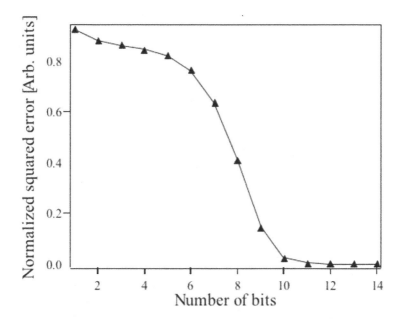

Fig. 8.8. Normalized squared error between the reconstructed images when the amplitude of the joint power spectrum is quantized into double precision and lower bits.

beams by beam splitter BS1. A 3-D object is illuminated by a collimated laser beam. The reflected and diffraction light propagate to a CCD camera that is located at a distance z along the optical axis. Since 3-D objects used in the experiments are small enough compared with the CCD size and the object is located far enough from the CCD, we consider the paraxial region where the scalar diffraction theory is valid. The light distribution at the CCD plane is described by using the Fresnel approximation as

$$\tilde{U}(x', y') = A(x', y') \exp[j\varphi(x', y')],$$
$$= \frac{1}{j\lambda} \iiint U(x, y, z) \frac{1}{z} \exp(jkz)$$
$$\times \exp\left\{\frac{jk}{2z}\left[(x' - x)^2 + (y' - y)^2\right]\right\} dx\, dy\, dz, \qquad (8.9)$$

where λ and k are the wavelength and the wave number of the illumination, respectively. In Eq. (8.9), $A(x', y')$ and $\varphi(x', y')$ are, respectively, the amplitude and the phase of the complex wave field at the CCD plane. The diffracted beam interferes in line with a reflected plane wave (reference beam) using beam splitter BS2. To reconstruct the complex field of the 3-D object as described in Eq. (8.9) from the digital holograms, we employ a phase-shifting digital holography technique [8]. This technique is effective in removing the

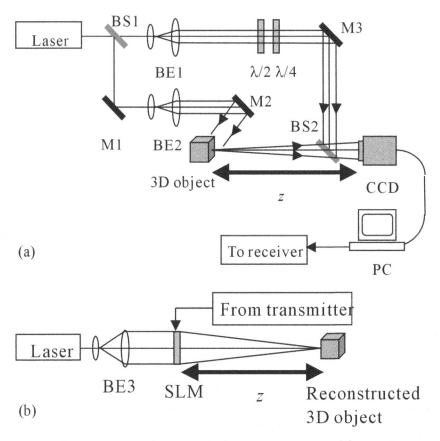

(a)

(b)

Fig. 8.9. A 3-D object recording and reconstruction system. (**a**) Recording system for digital holograms using a phase-shifting interferometer and (**b**) an optical-reconstruction system using phase-only information of digital holograms.

conjugate reconstruction even when an in-line hologram is employed. To implement phase-shifting digital holography, the reference beam passes through both a half-wave plate and a quarter-wave plate, as shown in Figure 8.9(a). Phase retardations in the reference beam of 0, $\pi/2$, π, and $3\pi/2$ can be achieved by controlling the directions of the fast and slow axes of the two plates. Using the resulting four digital interferograms, the complex field of the 3-D object in Eq. (8.9) can be calculated as described in [9].

It has already been shown that with computer reconstruction the fully complex field of the 3-D object calculated from the digital holograms can be successfully reconstructed. Using the complex field of the hologram, we can reconstruct the 3-D object numerically by the Fresnel–Kirchhoff integral,

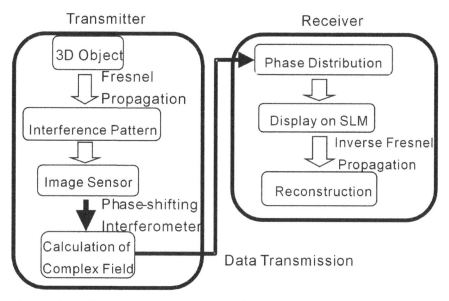

Fig. 8.10. A block diagram of the secure display system using encrypted digital holograms.

written by

$$R(x, y; z) = -\frac{1}{j\lambda} \iint \tilde{U}(x', y') \frac{1}{z} \exp(-jkz)$$
$$\times \exp\left\{ -\frac{jk}{2z}[(x - x')^2 + (y - y')^2] \right\} dx' \, dy'. \qquad (8.10)$$

In the proposed method that uses phase-only information, a reconstructed object is described as

$$P(x, y; z) = -\frac{a}{j\lambda} \iint \exp[j\varphi(x', y')] \frac{1}{z} \exp(-jkz)$$
$$\times \exp\left\{ -\frac{jk}{2z}[(x - x')^2 + (y - y')^2] \right\} dx' \, dy', \qquad (8.11)$$

where

$$a = \sqrt{\frac{\iint |R(x, y; z)|^2 dx \, dy}{\iint |P(x, y; z)|^2 dx \, dy}}. \qquad (8.12)$$

Note that the total powers of Eqs. (8.10) and (8.11) have been made equivalent by choosing a as indicated in Eq. (8.12). In the next subsection, we evaluate the quality of the reconstructed 3-D object using phase-only information.

8.4.2 Numerical Evaluation

To evaluate the error in the reconstructed 3-D object, we calculate a normalized root-mean-square (NRMS) criterion as,

$$
a = \frac{\sqrt{\displaystyle\iint \left[|R(x,y;z)|^2 - |P(x,y;z)|^2\right] dx\, dy}}{\sqrt{\displaystyle\iint \left[|R(x,y;z)|^2\right]^2 dx\, dy}}. \tag{8.13}
$$

The recording of digital holograms is obtained experimentally, but the reconstruction is calculated numerically. Digital holograms of 3-D objects are recorded optically as shown in Figure 8.10(a). An argon ion laser operated at a wavelength of 514.5 nm is used as a recording beams. The CCD array consists of 2028×2044 pixels. Each pixel is $9\mu m \times 9\mu m$ and has 1024 gray levels (10 bits). We use two 3-D objects in the experiments: one a die with dimensions 5 mm $\times$ 5 mm $\times$ 5 mm and the other a screw of similar size. The die and the screw are located at distances of 322 mm and 390 mm from the CCD, respectively.

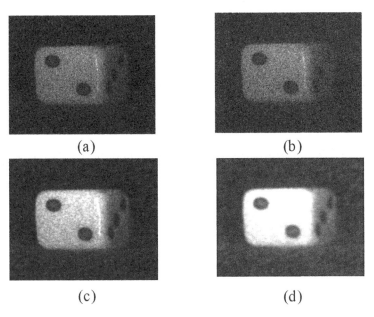

Fig. 8.11. Numerically reconstructed 3-D objects of a die from: (**a**) a fully complex field; (**b**) phase-only information; low-pass-filtered 3-D objects with a mean filter of (**c**) 11×11 pixels and (**d**) 21×21 pixels after reconstruction using phase-only information.

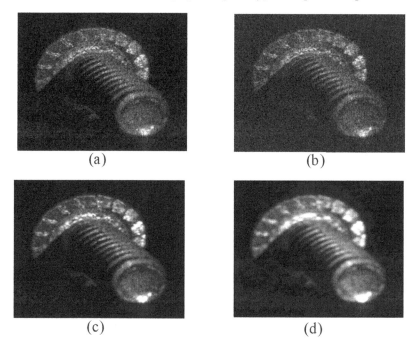

(a)   (b)

(c)   (d)

Fig. 8.12. Numerically reconstructed 3-D objects of a screw from: (**a**) a fully complex field; (**b**) phase-only information; low-pass-filtered 3-D objects with a mean filter of (**c**) 11×11 pixels and (**d**) 21×21 pixels.

Figures 8.11(a) and 8.11(b) show the numerically reconstructed 3-D die object using complex information and phase-only information, respectively. Figures 8.12(a) and 8.12(b) show the numerically reconstructed 3-D screw object using complex information and phase-only information, respectively. These figures show that the phase-only information can reconstruct the 3-D object successfully. We can also see that speckle noise is an influence on both 3-D reconstructed images. We evaluate the reconstructed 3-D object. The reconstructed 3-D objects are low-pass filtered by a mean filter to reduce the speckle noise in the reconstructed image. The low-pass-filtered images are presented in Figures 8.11(c), 8.11(d), 8.12(c), and 8.12(d). The size of the mean filter in Figures 8.11(c) and 8.12(c) is 11×11 pixels and in Figures 8.11(d) and 8.12(d) is 21×21 pixels. The intensity levels are scaled to enhance the contrast. We can see that speckle noise is removed to a large extent by low-pass filtering. The intensities of the die and the screw reconstructed from the phase-only holograms are compared with the intensities of the objects reconstructed from fully complex holograms for different mean-filtering neighborhoods. Figure 8.13 shows the NRMS difference as a function of the side length of the mean filter. We can see that the NRMS difference is reduced as the size of mean-filter increases for both 3-D objects. In the die 3-D object, the error is

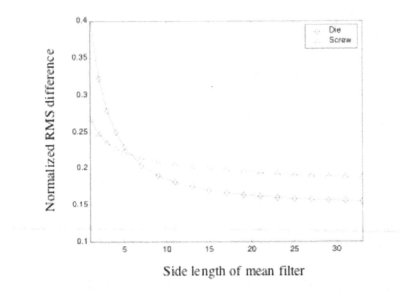

Side length of mean filter

Fig. 8.13. NRMS difference between reconstructed 3-D objects from a fully complex field and reconstructions from phase-only information as a function of the side length of the mean filter.

reduced from 40% to 15%. Since the screw has more detailed information than the die, the effect of the speckle noise in the screw reconstructions is stronger than in those of the die. The remaining error is due to the loss of amplitude information in the hologram. We expect that these errors can be decreased by manipulating the phase distribution by use of any error-reduction technique, such as simulated annealing or a genetic algorithm [21].

In practical systems, there is a limit on the number of phase levels that can be displayed in phase-only spatial-light modulators (P-SLMs). Die and screw reconstructions from phase-only information are compared with intensity images reconstructed from fully complex information for different numbers of quantization levels. For these calculations, we use a mean filter with 11×11 pixels. Figures 8.14 and 8.15 show the reconstructed 3-D objects without quantization reduction and with quantization reductions to 4 bits and 1 bit. We can discern the 3-D objects even when the phase level of the P-SLM is binary. Figure 8.16 shows the NRMS as a function of the number of phase levels of phase-only information. As few as 16 levels (4 bits) are sufficient to reconstruct the 3-D objects without significant loss of detail.

8.4.3 Optical Reconstruction

The optical setup to reconstruct 3-D objects is shown in Figure 8.9(b). A He–Ne laser operated at a wavelength of 632.8 nm is used. After the collimation

(a)   (b)   (c)

Fig. 8.14. Reconstructed 3-D object of the die (**a**) without quantization reduction and with quantization reductions to (**b**) 4 bits and (**c**) 1 bit.

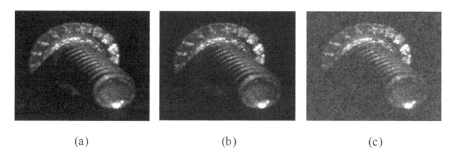

(a)   (b)   (c)

Fig. 8.15. Reconstructed 3-D object of the screw (**a**) without quantization reduction and with quantization reductions to (**b**) 4 bits and (**c**) 1 bit.

and the beamwidth expansion of the He–Ne laser beam, the beam illuminates a liquid-crystal SLM with 1024×768 pixels. The size of each pixel is approximately $18\ \mu\text{m} \times 18\ \mu\text{m}$. This SLM does not have enough pixels to display the complete 2044×2028 pixel phase-only hologram, so a 1024×768 pixel window is presented. By measuring phase retardation of the SLM in a Mach–Zehnder interferometer, the maximum amount of phase retardation possible with our SLM is 0.6π. The phase retardation is almost linearly proportional to the signal level from the computer, which has 256 levels at each pixel. Figures 8.17 and 8.18 show the reconstructed 3-D die and screw objects. Figures 8.17(a), 8.17(b), and 8.17(c) contain reconstructions with the CCD located at distances of 113 mm, 123 mm, and 133 mm from the SLM, respectively. Figures 8.18(a), 8.18(b), and 8.18(c) show reconstructions at distances of 135 mm, 155 mm, and 165 mm from the SLM, respectively. Figures 8.17(b) and 8.18(b) are the most in focus. From these figures, we can see that the die and the screw are reconstructed at different planes. The calculated positions of the in-focus reconstructions are 104 mm and 126 mm, respectively. These errors are caused by a small amount of phase retardation of the SLM, our use of a different readout wavelength, different pixel sizes, and a readout wave front not exactly the same as a plane wave.

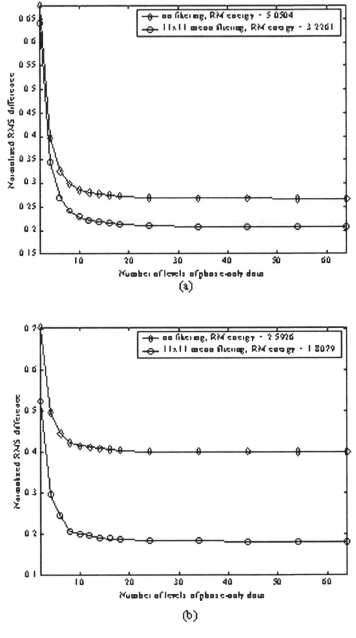

Fig. 8.16. NRMS difference as a function of the number of phase levels of phase-only information when 3-D objects are the (**a**) die and (**b**) screw, respectively.

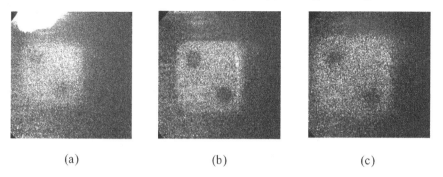

(a)   (b)   (c)

Fig. 8.17. Experimental results; (**a**), (**b**), and (**c**) are reconstructed images of the die, where the CCD is located at 113 mm, 123 mm, and 133 mm from the SLM, respectively.

(a)   (b)   (c)

Fig. 8.18. Experimental results; (**a**), (**b**), and (**c**) are reconstructed images of the screw, where the CCD is located at 135 mm, 155 mm, and 165 mm from the SLM, respectively.

8.5 Conclusions

We have presented a secure display system using encrypted digital holograms. Recording of the hologram in a computer by an image sensor enables us to send the holographic data by communication channels and enables us to reduce the noise through image-processing techniques. We have developed an optical-correlation-based reconstruction method of encrypted digital holograms. We have also developed an optical-reconstruction method that uses phase-only information of the digital holograms. We expect that the optical reconstruction using encrypted digital holographic techniques will enable us to implement a real-time secure display system or a secure television because the transmission of successive encrypted images can be performed quickly and effectively via optical-fiber communications.

References

[1] B. Javidi and J.L. Horner, *Opt. Eng.* **33**, 1752 (1994).

[2] B. Javidi, *Phys. Today* **50**, 27 (1997).

[3] P. Réfrégier and B. Javidi, *Opt. Lett.* **20**, 767 (1995).

[4] O. Matoba and B. Javidi, *Opt. Lett.* **24**, 762 (1999).

[5] O. Matoba and B. Javidi, *Appl. Opt.* **38**, 7288 (1999).

[6] O. Matoba and B. Javidi, *Appl. Opt.* **39**, 2975 (2000).

[7] U. Schnars and W. Juptner, *Appl. Opt.* **33**, 179 (1994).

[8] I. Yamaguchi and T. Zhang , *Opt. Lett.* **22**, 1268 (1997).

[9] B. Javidi and T. Nomura, *Opt. Lett.* 25, **28** (2000).

[10] B. Javidi and E. Tajahuerce, *Opt. Lett.* **25**, 610 (2000).

[11] S. Lai and M. Neifield, *Opt. Commun.* **178**, 283 (2000).

[12] M. Sutkowski and M. Kujawinska, *Opt. Lasers Eng.* **33**, 191 (2000).

[13] E. Tajahuerce, O. Matoba, and B. Javidi, *Appl. Opt.* **40**, 3877 (2001).

[14] Y. Frauel, E. Tajahuerce, M.A. Castro, and B. Javidi, *Appl. Opt.* **40**, 3887 (2001).

[15] O. Matoba and B. Javidi, *Opt. Lett.* **27**, 321 (2002).

[16] O. Matoba, T.J. Naughton, Y. Frauel, N. Bertaux, and B. Javidi, *Appl. Opt.* **41**, 6187 (2002).

[17] L.B. Lesem, P.M. Hirsch, and J.A. Jordan, Jr, *IBM J. Res. Dev.* **13**, 150 (1969).

[18] J.A. Jordan, Jr., P.M. Hirsch, L.B. Lesem, and D.L. Van Rooy, *Appl. Opt.* **9**, 1883 (1970).

[19] T.H. Barnes, T. Eiju, K. Matuda, H. Ichikawa, M.R. Taghizadeh, and J. Turunen, *Appl. Opt.* **31**, 5527 (1992).

[20] K.L. Tan, S.T. Warr, I.G. Manolis, T.D. Wilkinson, M. M. Redmond, W.A. Crossland, R.J. Mears, and B. Robertson, *J. Opt. Soc. Am. A* **18**, 205 (2001).

[21] N. Yoshikawa and T. Yatagai, *Appl. Opt.* **33**, 863 (1994).

9

Compression of Digital Holograms for Secure Three-Dimensional Image Storage and Transmission

Thomas J. Naughton and Bahram Javidi

Summary. We present the results of applying data compression techniques to encrypted three-dimensional objects. The objects are captured using phase-shift digital holography and encrypted using a random phase mark in the Fresnel domain. Lossy quantization is combined with lossless coding techniques to quantify compression rates. Our techniques are suitable for a range of secure three-dimensional object storage and transmission applications.

9.1 Introduction

Image encryption has received much attention as an application of optical science in recent years [1–17]. These applications take advantage of both the natural two-dimensional (2-D) imaging capabilities of optics and the parallelism achievable with optical processing. Of the encryption methods utilizing analog optics [1, 2, 5–17], most perform encryption with a random phase mask positioned in the input, Fresnel, or Fraunhofer domains, or a combination of domains. These invariably produce a complex-valued encrypted image. Digital holography [18–26], and phase-shift interferometry (PSI) [19, 21, 25] in particular, can be used to capture high-quality approximations of both the amplitude and phase of complex-valued optical wave fronts and has been proposed for 3-D object recognition and three-dimensional (3-D) display applications [27–33]. Recently, digital holography has been used in the encryption of 2-D real-valued images [12–14]. Of these, the techniques based on PSI make good use of detector resources in that they capture on-axis encrypted digital holograms [13, 14]. The PSI technique has also been extended to the encryption of 3-D objects [15].

The advantage of digital techniques over holographic encryption methods that use more traditional photorefractive media [9, 10] is that the resulting encrypted hologram can be easily stored electronically or transmitted over conventional communication channels. This motivates the study of how conventional compression techniques [34] could be applied to digital holograms.

Hologram compression differs from image compression principally because our holograms store 3-D information in complex-valued pixels and secondly because of the inherent speckle content that gives the holograms a white-noise appearance. It is not a straightforward procedure to remove the holographic speckle because it actually carries 3-D information. The noisy appearance of digital holograms causes lossless data-compression techniques to perform poorly on such inputs [31].

In this chapter, we apply quantization directly to the complex-valued holographic pixels as a means of achieving lossy data compression. Treatments of quantization in holograms can be found in the literature [35, 36], and compression of real-valued [37] and complex-valued [31, 32, 38] digital holograms has received some attention to date. Some studies have also been performed on the decrypted-domain effects of perturbations, including quantization, in the encrypted domain [39, 40]. This introduces a third reason why compression of digital holograms differs from compression of digital images: a change locally in a digital hologram will, in theory, affect the whole reconstructed object. Furthermore, when gauging the errors introduced by lossy compression, we are not directly interested in the defects in the hologram itself, only in how compression noise affects subsequent 3-D object reconstruction.

We examine the compression of encrypted digital holograms through lossless and lossy techniques [31, 41]. Holograms of 3-D objects were captured using a phase-shift interferometry setup [27, 29]. Encrypted versions of these holograms were obtained by perturbing the Fresnel diffraction of the 3-D objects with a random phase mask. For our experiments, this encryption step was performed electronically [42]. Each encrypted hologram contained 2048×2048 pixels. The compression technique we apply consists of lossy quantization of the real and imaginary components of each holographic pixel, combined with word-length reduction in the form of integer encoding, and followed by a lossless compression step. We use a reconstructed-object-plane root-mean-square (RMS) metric to quantify the quality of our decompressed and decrypted holograms.

The structure of the chapter is as follows. In Section 9.2, we outline how we perform encryption of the Fresnel propagation of 3-D objects using a random phase mask and how the complex wave front is subsequently captured using phase-shift digital holography. In Section 9.3, the decryption and reconstruction steps are explained. In Section 9.4, we examine the amenability of encrypted digital holograms to lossless compression using four well-known techniques, and in Section 9.5 we apply the lossy technique of quantization to the real and imaginary components of each holographic pixel. In this section too, we quantify quantization error by measuring deformation in the decrypted and reconstructed 3-D object intensities. We combine quantization with lossless compression in Section 9.6 to achieve far greater compression performance than using either technique alone, and we conclude in Section 9.7.

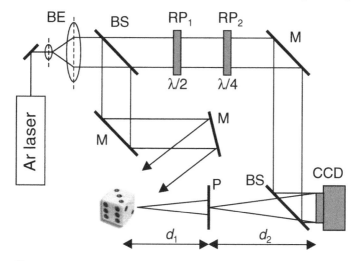

Fig. 9.1. Fresnel encryption of an illuminated 3-D object: L, coherent plane wave; P, random phase mask; CCD, camera.

9.2 Digital Hologram Encryption

Consider the lensless arrangement shown in Figure 9.1. A 3-D object is illuminated with a coherent plane wave, and light is scattered from the surface of the object. A random phase mask, placed in the path of the scattered light at a distance d_1 from the object, causes further diffraction, and the scattered light ultimately falls on a detector at a distance $(d_1 + d_2)$ from the object. Due to free-space propagation, the signal at the detector plane will have both its amplitude and phase modulated by the mask and will have a dynamic range suitable for capture by a CCD camera. In our implementations, the camera-plane complex signal is captured by PSI. The encryption key consists of the random phase mask, its position in 3-D space, the distance between the mask and the notional center of the object, and the wavelength of the illumination. The same key that was used for encryption is used for decryption. Such a key is an example of a so-called 3-D key [10].

Referring to Figure 9.1, the signal immediately after the phase mask $\Phi(x, y)$ has amplitude $A_P(x, y)$ and phase $\phi_P(x, y)$ and, under the Fresnel approximation [43], is given by the superposition integral

$$U_P(x, y) = A_P(x, y) \exp\left[i\phi_P(x, y)\right]$$

$$= -\exp\left[i\Phi(x, y)\right] \int\int_{-\infty}^{\infty} \int_{d_1-D}^{d_1+D} U_0(x', y', z) \frac{i}{\lambda z} \exp\left(i\frac{2\pi}{\lambda} z\right)$$

$$\times \exp\left\{i\frac{\pi}{\lambda z}\left[(x - x')^2 + (y - y')^2\right]\right\} dx' dy' dz, \qquad (9.1)$$

where λ is the wavelength of the illumination and D is the radius of the smallest sphere enclosing the 3-D object. $U_0(x, y, z)$ denotes the complex amplitude

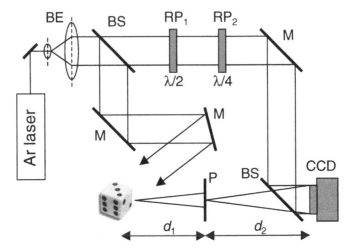

Fig. 9.2. Experimental setup for 3-D object encryption using phase-shift digital holography: BE, beam expander; BS, beam splitter; M, mirror; RP, retardation plate; P, phase mask.

signal describing the visible surface of the 3-D object, represented as a field of nonoccluding light sources in (x, y), each of which is a distance z along the optical axis from plane P. By free-space propagation, the light field in the camera plane is then given by

$$
\begin{aligned}
H_{\mathrm{E}}(x, y) &= A_{\mathrm{E}}(x, y) \exp\left[\mathrm{i}\phi_{\mathrm{E}}(x, y)\right] \\
&= \frac{-\mathrm{i}}{\lambda d_2} \exp\left(\mathrm{i}\frac{2\pi}{\lambda}d_2\right) \int \int_{-\infty}^{\infty} A_{\mathrm{P}}(x', y') \exp\left[\mathrm{i}\phi_{\mathrm{P}}(x', y')\right] \\
&\quad \times \exp\left\{\mathrm{i}\frac{\pi}{\lambda d_2}\left[(x - x')^2 + (y - y')^2\right]\right\} \mathrm{d}x'\mathrm{d}y'.
\end{aligned}
\tag{9.2}
$$

This encrypted wave front can be captured using PSI. In contrast to some phase-encryption schemes, the amplitude of the detected signal is also encrypted. If $\Phi(x, y)$ has a random distribution, then both $A_{\mathrm{H}}(x, y)$ and $\phi_{\mathrm{H}}(x, y)$ will look like random noise distributions. Once $H_{\mathrm{E}}(x, y)$ is in digital form, we compute the complex conjugate $H(x, y) = A_{\mathrm{E}}(x, y) \exp[-\mathrm{i}\phi_{\mathrm{E}}(x, y)]$. As such, we avoid the need for time-reversed Fresnel reconstruction expressions and can use the same phase mask (rather than its conjugate) in any optical reconstruction. Furthermore, in contrast to some Fourier-transform-based encryption schemes, the camera-plane signal has a dynamic range that is suitably scaled for capture by current CCD technology. (Fourier encryption schemes often require a second phase mask for this purpose.) A single random phase mask is therefore sufficient for this encryption scheme. We refer to $H(x, y)$ as an encrypted digital hologram of the 3-D object.

The encrypted complex-valued holograms can be captured using an optical setup (shown in Figure 9.2) based on a Mach–Zehnder interferometer architecture [27, 29]. A linearly polarized argon ion (514.5 nm) laser beam is divided into object and reference beams, both of which are spatially filtered and expanded. The first beam illuminates the 3-D object placed at an approximate distance $d_1 + d_2 = 350$ mm from a 10 bit 2028×2044 pixel Kodak Megaplus CCD camera. The reference beam passes through half-wave plate RP_1 and quarter-wave plate RP_2. This linearly polarized beam can be phase-modulated by rotating the two retardation plates. Through permutation of the fast and slow axes of the plates, we can achieve phase shifts of 0, $-\pi/2$, $-\pi$, and $-3\pi/2$. The reference beam combines with the light diffracted from the object and forms an interference pattern in the plane of the camera. At each of the four phase shifts, we record an interferogram. Using these four real-valued images, the complex camera-plane wave front can be approximated to good accuracy using PSI [27, 29].

In this system, the encryption key is $(\Phi, x, y, d_2, \lambda, e_x, e_y, d_1)$, consisting of the random phase mask, its position in 3-D space, the wavelength of the illumination, the dimensions of the detector elements (for a pixilated device), and the reconstruction distance, respectively. This key is also exactly the decryption key: a means of decrypting and reconstructing an arbitrary view of the 3-D object encoded in the hologram.

9.3 Decryption and Reconstruction

The decryption and reconstruction of the digital hologram is a two-stage process. Each stage could be performed optically or digitally. First, the hologram is propagated a distance d_2 to plane P. Secondly, the hologram is decrypted by multiplication with the phase mask and is reconstructed through further Fresnel propagation to focus in any chosen plane in the range $d_1 \pm D$. The decrypted and reconstructed object at a distance d_1 from plane P is obtained from

$$U(x, y; d_1) = \left| \frac{-i}{\lambda d_1} \exp\left(i\frac{2\pi}{\lambda} d_1\right) \exp\left[i\frac{\pi}{\lambda d_1}\left(x^2 + y^2\right)\right] \right.$$
$$\times \int_{-M}^{M} \int_{-N}^{N} \exp\left[i\Phi(x', y')\right] H_P(x', y') \exp\left[i\frac{\pi}{\lambda d_1}\left(x'^2 + y'^2\right)\right]$$
$$\left. \times \exp\left[-i\frac{2\pi}{\lambda d_1}\left(xx' + yy'\right)\right] dx' dy' \right|^2 , \qquad (9.3)$$

where the quadratic terms of the convolution exponent have been expanded, we assume that the phase mask has its origin centered on the optical axis, and $2N$ and $2M$ represent the height and width, respectively, of the larger of the two input signals (encrypted hologram and phase mask). $H_P(x, y)$ is the encrypted signal in the plane P, which can be expressed in convolution form

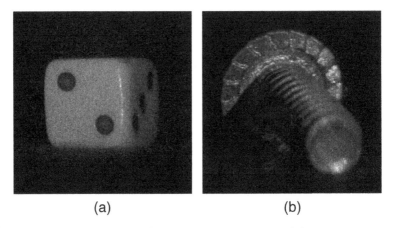

(a)   (b)

Fig. 9.3. Objects used in the study: (**a**) die, (**b**) screw.

as

$$H_{\mathrm{P}}(x,y) = \frac{-i}{\lambda d_2} \exp\left(i\frac{2\pi}{\lambda}d_2\right) \times H(x,y) \star \exp\left[\frac{i\pi}{\lambda d_2}\left(x^2 + y^2\right)\right], \quad (9.4)$$

where $\star$ denotes a convolution operation.

A decrypted digital hologram contains sufficient amplitude and phase information to reconstruct the complex field $U(x,y,z)$ in a plane in the object beam at any distance z from the camera. Furthermore, as with conventional holography [44], a windowed subset of the hologram can be used to reconstruct a particular view of the object. As the window explores the decrypted hologram in plane P, a different angle of view of the object can be reconstructed. Different views of the 3-D object can be obtained by multiplying Eq. (9.3) by an appropriate linear phase factor [27] within the angular range of the hologram. The range of viewing angles is determined by the ratio of the window size to the full CCD sensor dimensions. Our holograms have dimensions 2028×2044 pixels. A 1024×1024 pixel window has a maximum lateral shift of 9 mm over the face of the CCD sensor [29]. With an object positioned $d = 350$ mm from the camera, viewing angles in the range $\pm 0.74°$ are permitted.

Figure 9.3 shows two of the objects used for the experiments in this chapter. Both objects have approximate dimensions of 5 mm $\times$ 5 mm $\times$ 5 mm and were positioned 323 mm (for the die) and 390 mm (for the screw) from the camera. The intensity images in Figure 9.3 are reconstructed from digital holograms captured using a version of the apparatus shown in Figure 9.2 that did not contain a phase mask positioned in plane P [27, 29]. The objects were reconstructed at the correct $(d_1 + d_2)$ distance but again without the phase mask. These reconstructions serve as ground-truth data when quantifying lossy compression errors later in the chapter.

Fig. 9.4. Random phase mask used in the study.

In our experiments, we retrospectively encrypt holograms that were initially captured without a random phase mask [42]. By digitally simulating the encryption procedure we gain in terms of flexibility and security [42], while still accommodating the possibility for a real-time optical reconstruction [17, 32, 45]. The phase mask used in the experiments is shown in Figure 9.4. It consists of values chosen with uniform probability from the range $[0, 2\pi)$ using a pseudorandom number generator. The mask has dimensions 2048×2048 pixels, and in the encryption experiments our digital holograms were enlarged from 2028×2044 pixels to these dimensions by padding with zeros. For our experiments, the mask was positioned as shown in Figure 9.2 such that the ratio of the distances $d_1 : d_2$ was $35 : 65$.

In Figure 9.5, we show the amplitude and phase of the screw hologram before encryption and after encryption as described by Eq. (9.1) and Eq. (9.2). In Figure 9.6, we show the results of reconstructing an encrypted digital hologram according to Eq. (9.3) with and without the phase mask used in the encryption step.

9.4 Lossless Compression of Encrypted Digital Holograms

The digital holograms were treated as binary data streams and compressed using the lossless data-compression techniques of Huffman [46], Lempel and Ziv (LZ77) [47], Lempel, Ziv, and Welch (LZW) [48], and Burrows and Wheeler (BW) [49]. Huffman coding [46], an entropy-based technique, is one of the oldest and most widely used compression methods. It replaces each symbol in the input by a code word, assigning shorter code words to more frequent symbols.

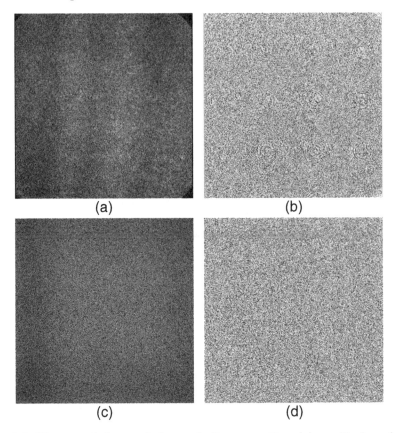

(a)   (b)

(c)   (d)

Fig. 9.5. The screw hologram before and after encryption: (**a**) amplitude and (**b**) phase of the original hologram, and (**c**) amplitude and (**d**) phase of the encrypted hologram.

The LZ77 algorithm [47] takes advantage of repeated substrings in the input data. In contrast to Huffman coding, a variable length string of input symbols is replaced by a fixed-size code word (a reference to the previous occurrence of that string). LZW [48] is a refinement of LZ77. It maintains a dictionary (or lookup table) of variable-sized code words and is less biased toward local redundancy. The more recent BW algorithm [49] transforms its input through a sorting operation into a format that can be compressed very effectively using standard techniques (in our particular implementation, Huffman coding).

Five digital holograms were used in the experiments [31, 41]. The holograms have dimensions 2028×2044 pixels and are originally in floating-point representation with 8 bytes of amplitude information and 8 bytes of phase information for each pixel. This amounts to a file size of 64,769 kB, where $1\,kB = 2^{10}$ bytes. The holograms were first compressed without any encryption. It has been shown that an intermediate hologram representation of two

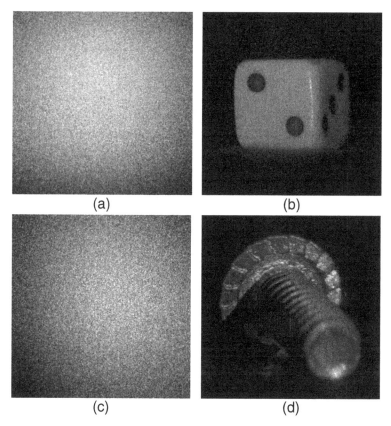

Fig. 9.6. Reconstruction of the die hologram (**a**) without the phase mask, and (**b**) with the phase mask, and reconstruction of the screw hologram (**c**) without the phase mask, and (**d**) with the phase mask.

separate binary data streams for real and imaginary components works well for lossless techniques [31]. Compressing separately the real and imaginary data streams achieves compression rates in the range 1.0–6.66, where compression rate r is calculated from

$$r = \frac{\text{uncompressed size}}{\text{compressed size}} \quad , \tag{9.5}$$

and where a rate of 1.0 was used when no compression was achieved or when the coded hologram was actually larger in size. By the term compression rate, we indicate the number of bits of uncompressed data that are effectively communicated with a single bit of compressed data.

Each bit of compressed data encodes, on average, 3.20 bits of uncompressed holographic data with LZW, 1.90 bits with Huffman, and 4.66 bits with BW. The full results for the three techniques are shown in Table 9.1. For this table, the real and imaginary streams were compressed separately and then the sizes

Table 9.1. Lossless compression of original (unencrypted) digital holograms; c.r., compression rate.

Hol. no.	Size (kB)	LZW (kB)	Huff. (kB)	BW (kB)	LZW c.r.	Huff. c.r.	BW c.r.
1	64,769	8931+ 8791	13,899+13,079	6389+ 6263	3.65	2.40	5.12
2	64,769	32,385+32,385	30,987+31,009	24,413+24,163	1.00	1.04	1.33
3	64,642	7747+ 8318	27,154+27,282	4512+ 5199	4.02	1.19	6.66
4	64,769	9047+ 8805	13,815+13,027	6512+ 6283	3.63	2.41	5.06
5	64,769	8839+ 8718	13,439+13,082	6336+ 6265	3.69	2.44	5.14
Averages:					3.20	1.90	4.66

Table 9.2. Lossless compression of encrypted digital holograms; c.r., compression rate.

Hol. no.	Size (kB)	LZW (kB)	Huff. (kB)	BW (kB)	LZW c.r.	Huff. c.r.	BW c.r.
1	65,536	32,768+32,768	31,264+31,265	31,935+31,934	1.00	1.05	1.03
2	65,536	32,768+32,768	31,259+31,260	31,918+31,918	1.00	1.05	1.03
3	65,536	32,768+32,768	31,257+31,258	31,912+31,911	1.00	1.05	1.03
4	65,536	32,768+32,768	31,258+31,257	31,912+31,913	1.00	1.05	1.03
5	65,536	32,768+32,768	31,256+31,257	31,912+31,913	1.00	1.05	1.03
Averages:					1.00	1.05	1.03

of the two compressed streams summed for the calculation of the compression rate. In Table 9.1, hologram nos. 1 and 2 correspond to the die and screw, respectively. As an example, the amplitude and phase of hologram no. 2 are shown in Figures 9.5(a) and 9.5 (b), respectively.

Each of the five holograms was encrypted with the phase mask shown in Figure 9.4. For our experiments, unencrypted holograms of the 3-D objects were captured optically and then encrypted according to Eq. (9.1) and (9.2) [42]. Prior to encryption, the holograms were padded with zeros to 2048×2048 pixels to make them of the same dimensions as the phase mask. The encrypted holograms also contained 2048×2048 pixels. The amplitude and phase of the encrypted version of hologram no. 2 are shown in Figures 9.5(c) and 9.5(d), respectively. The four lossless compression techniques were applied to each hologram (three are shown in Table 9.2). For Table 9.2, the real and imaginary streams were once again compressed separately. Each bit of compressed encrypted data encodes, on average, 1.05 bits of uncompressed encrypted holographic data with LZ77 or Huffman, 1.00 bits with LZW, and 1.03 bits with BW.

As might be expected, each encrypted hologram is compressed less effectively than its unencrypted counterpart. Although the original (unencrypted) digital holograms have a white-noise appearance [see Figures 9.5(a) and 9.5(b)], it is evident that they must be colored to some extent because the compressors have been able to exploit some redundancy or structure within the original hologram data (Table 9.1). This hypothesis is also supported by the fact that both real and imaginary streams were compressed at different rates. The structure in the original holograms is thought to be at the level of 2 to 3 pixel artifacts and due to the character of individual speckles.

Very little such redundancy or structure could be found in the encrypted hologram data (Table 9.2). The random phase mask, combined with Fresnel propagation, is very effective at removing apparent structure from the hologram data. Both real and imaginary streams appear equally random. For some encrypted holograms, with LZW in particular, the compressed sizes were even larger than the uncompressed. In these cases, the uncompressed encrypted file was used and a compression rate of 1.0 reported. These results illustrate the urgent need to explore lossy compression techniques suitable for encrypted digital holograms. One such lossy technique that has been successfully applied to 3-D digital holograms is quantization [31, 32, 38, 41].

9.5 Quantization of Encrypted Digital Holograms

The loss in reconstruction quality due to quantization in encrypted holograms was investigated. A combined rescale and quantization step was employed. The encrypted holograms were rescaled linearly to the square in the complex plane $[-1-i, 1+i]$ without changing their aspect ratio in the complex plane. The real and imaginary components of each holographic pixel were then quantized. Quantization levels were chosen to be symmetric about zero; as a result, b bits encode $(2^b - 1)$ levels. For example, two bits encode levels $\{-1, 0, 1\}$, three bits encode levels $\{-1, -2/3, -1/3, 0, 1/3, 2/3, 1\}$, and so on. The combined rescale and quantization operation is defined for individual pixels as

$$H'(x, y) = \text{round}\left[H(x, y) \times \sigma^{-1} \times \beta\right] \times \beta^{-1} \qquad (9.6)$$

and was applied to each pixel (x, y) in the encrypted hologram H, where

$$\sigma = \max\left\{ \left|\min\left[\text{Im}(H)\right]\right|, \left|\max\left[\text{Im}(H)\right]\right|, \right.$$
$$\left. \left|\min\left[\text{Re}(H)\right]\right|, \left|\max\left[\text{Re}(H)\right]\right| \right\}, \qquad (9.7)$$

and where $\beta = 2^{(b-1)} - 1$. Here, b represents the number of bits per real and imaginary value, $\max(\cdot)$ returns the maximum scalar in its argument(s), and

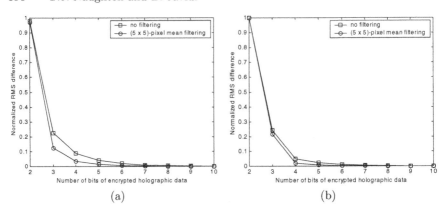

Fig. 9.7. NRMS intensity difference in decrypted and reconstructed 3-D object images plotted against quantization level, showing the effect of mean filtering to remove noise: (**a**) die, and (**b**) screw.

round(α) is defined as $\lfloor \alpha + 0.5 \rfloor$. After quantization, each real and imaginary value will be in the range $[-1, 1]$.

The procedure for quantifying reconstruction loss due to quantization was as follows. An encrypted digital hologram $H(x, y)$ was quantized as $H'(x, y)$ according to Eq. (9.6). A view of the encoded 3-D object $U'(x, y)$ was decrypted and reconstructed according to Eq. (9.3). The quality of the reconstruction $U'(x, y)$ was calculated by a comparison with the reconstruction $U_0(x, y)$ from an unencrypted digital hologram of the same object. The unencrypted digital hologram was captured with the same optical setup as $H(x, y)$ but without the presence of a phase mask in plane P. The two reconstructions were compared in terms of the normalized RMS (NRMS) difference of their intensities, defined as

$$
D = \left[\sum_{m=0}^{N_x-1} \sum_{n=0}^{N_y-1} \left\{ |U_0(m, n)|^2 - |U'(m, n)|^2 \right\}^2 \right.
$$
$$
\left. \times \left(\sum_{m=0}^{N_x-1} \sum_{n=0}^{N_y-1} \left\{ |U_0(m, n)|^2 \right\}^2 \right)^{-1} \right]^{1/2}, \tag{9.8}
$$

where (m, n) are discrete spatial coordinates in the reconstruction plane and N_y and N_x are the height and width of the reconstructions, respectively. In order to lessen the effects of speckle noise, we examine only intensity in the reconstruction plane and apply a low-pass filtering operation prior to calculating the NRMS.

Figure 9.7 shows plots of the NRMS difference against the number of bits per (real and imaginary) data value in the holograms of both objects, with and without mean filtering over a neighborhood of 5×5 pixels. Figure 9.8 shows

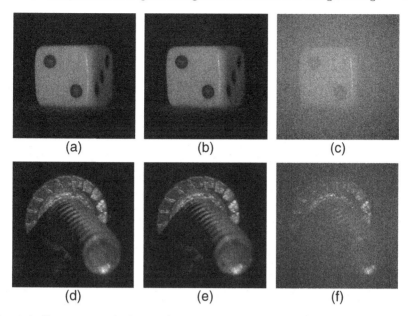

(a)   (b)   (c)

(d)   (e)   (f)

Fig. 9.8. Reconstructed objects (with 5×5 pixel mean filtering) from encrypted digital holograms with various numbers of bits of information in each real and imaginary value: die object (**a**) 4 bits, (**b**) 3 bits, (**c**) 2 bits; and screw object (**d**) 4 bits, (**e**) 3 bits, (**f**) 2 bits.

reconstructed object intensities for both objects for selected quantization resolutions. Note that quantization at 4 bits (with 5×5 pixel mean filtering) reveals little visible loss in reconstruction quality and (from Figure 9.7) small NRMS errors of 0.05 and 0.02 for the die and screw, respectively. As was recently suggested [31], it should be possible to successfully apply lossless compression techniques to the quantized encrypted values, even though lossless compression completely failed on the full-resolution encrypted data (Table 9.2). With the added step of word-length reduction, quantization at 4 bits (a reduction from 8 bytes for each real and imaginary value) corresponds to a compression rate of 16 [31]. In the next section, we employ the more sophisticated lossless techniques introduced earlier in the chapter.

9.6 Combining Quantization with Lossless Data Compression

By quantizing the hologram data, we reduce the number of symbols (different real and imaginary values) required to describe that hologram. This lossy step introduces structured defects into the hologram but, strictly speaking, does not compress it. By reducing the number of possible values in each pixel, however, we make it possible to reduce the number of bits required to describe

Table 9.3. Lossless compression applied to encrypted and quantized die hologram data; c.r., compression rate

Bits	Size (kB)	LZ77 (kB)	LZW (kB)	Huff. (kB)	BW (kB)	LZ77 c.r.	LZW c.r.	Huff. c.r.	BW c.r.
2	65,536	49	42	1028	32	1337	1560	64	2048
3	65,536	992	867	1264	931	66	76	52	70
4	65,536	2009	1844	1876	1967	33	36	35	33
5	65,536	2981	2837	2888	2890	22	23	23	23
6	65,536	3886	3876	3806	3801	17	17	17	17
7	65,536	4646	4964	4668	4700	14	13	14	14
8	65,536	5358	6091	5505	5566	12	11	12	12

Table 9.4. Lossless compression applied to encrypted and quantized screw hologram data; c.r., compression rate

Bits	Size (kB)	LZ77 (kB)	LZW (kB)	Huff. (kB)	BW (kB)	LZ77 c.r.	LZW c.r.	Huff. c.r.	BW c.r.
2	65,536	47	42	1027	32	1394	1560	64	2048
3	65,536	1138	1006	1317	1097	58	65	50	60
4	65,536	2120	1963	1991	2084	31	33	33	31
5	65,536	3097	2969	3021	2985	21	22	22	22
6	65,536	4003	4018	3923	3901	16	16	17	17
7	65,536	4732	5124	4784	4795	14	13	14	14
8	65,536	5460	6236	5613	5659	12	11	12	12

it. The term "quantization compression" therefore always implies the use of a separate (usually lossless) compression stage after quantization.

We perform two lossless compression steps on the quantized encrypted hologram data. In the first, we uniformly reduce the word length in each real and imaginary data stream by a factor of 8 with an integer-encoding routine that converts each quantized 8 byte floating-point real into a signed 8 bit integer. In the second, the real and imaginary streams were concatenated together and processed by one of the lossless techniques (LZ77, LZW, Huffman, BW) outlined earlier.

Tables 9.3 and 9.4 show the results of this three-step compression process for the die and screw, respectively. Compared with Table 9.2, there are dramatic increases in compression rate for both holograms, for all quantizations, and across all lossless compression algorithms. For example, for both the die and screw, with 2 bit quantization, compression rates of 2048 are achievable with BW. As shown in Figure 9.8, however, 2 bits is not sufficient to reconstruct the object with any great fidelity. With 3 bit quantization, compression rates in the range 65–76 are possible with LZW for reasonable decryption and reconstruction quality (according to Figure 9.8). With 4 bit quantization, compression rates in the range 31–36 can be expected with very low NRMS errors of 0.035 and 0.021 for the die and screw, respectively (from Figure 9.7, for 5×5 pixel mean filtering).

In order to quantify the gains made through lossless algorithms independent of quantization, we compare the lossless algorithms to the simple bit-packing technique [38]. Bit packing is the most general form of lossless compression through word-length reduction. It is also the most basic step that can be applied after data have been quantized. It therefore serves as a lower bound on the compression rate due to quantization. Bit packing makes up for its lack of sophistication with its ease of implementation and its potential for very fast software and hardware implementations. It is most suitable in time-critical applications where quantization has been applied but where the data are still too noisy for a run-length or Huffman technique to perform well. It has already been successfully implemented in a digital hologram networking application [38].

Bit packing uniformly reduces the length of each real and imaginary value's description but does not exploit any redundancy or repetition in the data. In this technique, data reduction is performed by extracting the appropriate b bits from each real and imaginary value. The data values are converted into integers in the range $[-2^{(b-1)} + 1, 2^{(b-1)} - 1]$. The parity bit and low-order $(b - 1)$ bits from each quantized value are accumulated in a bit buffer and packed into bytes. A digital hologram of $M = N \times N$ pixels at b bits resolution requires exactly $\lceil (2M \times b)/8 \rceil$ packed bytes. This packed size is used in the calculation of the compression rate.

The plots in Figure 9.9 graphically illustrate the improvement over quantization plus bit packing that can be achieved by additionally employing one of the more sophisticated lossless techniques. For example, for 3 bit quantization, compression rates in the range 65–76 can be expected with LZW. This is a significant increase over the compression rate of 21.3 that is achievable using quantization and bit packing alone. The plots in Figure 9.9 are also remarkable in that the lossless compression algorithms perform equally well on quantized data up to 4 bits. In such cases, one's choice of lossless algorithm would be determined by one's application requirements. For a secure 3-D digital hologram networking application, for example, one's concerns could include network latency constraints, reliability of the underlying network, memory available to the compressor or decompressor, whether the data are streaming or not (i.e., must one start (de)compressing the beginning of the data stream before receiving all of the data), and so on.

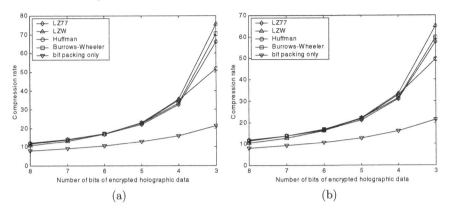

Fig. 9.9. Compression rate plotted against quantization combined with various lossless data compression techniques: (**a**) die, and (**b**) screw.

Comparing the curve representing no lossless compression (bit packing only) to the other curves in Figure 9.9, shows that lossless techniques perform significantly better as the number of different symbols in the data is reduced (as the quantization becomes more severe). For Huffman, this can be explained by the fact that as the number of quantization levels drops, the set of different symbols gets smaller. For LZ77 and LZW, this can be explained in terms of a resulting increase in the frequency of repeated symbol sequences.

Finally, the designer of secure 3-D digital hologram applications should have some way of directly relating the combined performance of quantization and lossless compression to NRMS error in the decrypted and reconstructed objects. This is provided, in the case of BW lossless compression, by the plot in Figure 9.10. This plot combined the results from Table 9.3 with the NRMS data from Figure 9.7(a). For illustration purposes, the points on the curve corresponding to 3 bit and 7 bit quantization are highlighted.

9.7 Conclusions

This chapter outlines an optical encryption technique, based on phase-shift interferometry, that is suitable for secure 3-D object storing and transmission applications. This technique makes the best use of both the massive parallelism inherent in optical systems and the flexibility offered by digital electronics. With this technique, phase-only perturbation of the Fresnel diffraction from a 3-D object will encrypt both the amplitude and phase of the resulting hologram. A single phase mask is therefore sufficient for this encryption scheme. Decryption and reconstruction of a particular view of the 3-D object can be performed optically or electronically. Reconstructions without the appropriate phase mask result in an unintelligible wave front. Multiple keys at

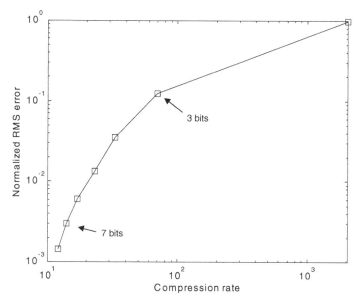

Fig. 9.10. NRMS intensity difference plotted against compression rate (combined quantization and BW lossless compression), with 5×5 pixel mean filtering for the die hologram.

different locations could be used to further increase the level of encryption. The encrypted hologram data are in a form suitable for digital electronic storage, transmission, or manipulation.

Lossless and lossy compression techniques were applied to the digital hologram data. Lossless techniques, such as LZ77, LZW, Huffman, and BW, perform very poorly on digital hologram data due to its white-noise characteristics. We find that the encrypted holograms are compressed even less effectively. Quantization has been applied to good effect on the encrypted hologram data, and reductions to as few as 3 bits in each real and imaginary value have resulted in good-quality decompressed and decrypted 3-D object reconstructions. Not only does quantization perform significant compression itself (measured through the use of a basic bit-packing algorithm) but also reduces the number of symbols (for Huffman) and introduces structure into the bit stream (for LZ77 and LZW) to allow them to perform more effectively. Quantization at 3 bits corresponds to a compression rate of approximately 21 (with bit packing). When some of the more sophisticated lossless algorithms are applied to 3 bit data, compression rates of up to 76 are possible for encrypted digital holograms.

Acknowledgments.

The authors wish to thank Enrique Tajahuerce and Yann Frauel for use of their hologram data. The first author wishes to acknowledge support from Enterprise Ireland.

References

[1]   B. Javidi and J.L. Horner, *Opt. Eng.* **33**, 1752 (1994).
[2]   P. Réfrégier and B. Javidi, *Opt. Lett.* **20**, 767 (1995).
[3]   S. Fukushima, T. Kurokawa, and Y. Sakai, *IEEE Photon. Technol. Lett.* **3**, 1133 (1991).
[4]   M. Madjarova, M. Kakuta, M. Yamaguchi, and N. Ohyama, *Opt. Lett.* **22**, 1624 (1997).
[5]   J.F. Heanue, M.C. Bashaw, and L. Hesselink, *Appl. Opt.* **34**, 6012 (1995).
[6]   R.K. Wang, I.A. Watson, and C.R. Chatwin, *Opt. Eng.* **35**, 2464 (1996).
[7]   L.G. Neto and Y. Sheng, *Opt. Eng.* **35**, 2459 (1996).
[8]   B. Javidi and E. Ahouzi, *Appl. Opt.* **37**, 6247 (1998).
[9]   G. Unnikrishnan, J. Joseph, and K. Singh, *Appl. Opt.* **37**, 8181 (1998).
[10] O. Matoba and B. Javidi, *Opt. Lett.* **24**, 762 (1999).
[11] P.C.Mogensen and J. Glückstad, *Opt. Lett.* **25**, 566 (2000).
[12] B. Javidi and T. Nomura, *Opt. Lett.* **25**, 28 (2000).
[13] S. Lai and M.A. Neifeld, *Opt. Commun.* **178**, 283 (2000).
[14] E. Tajahuerce, O. Matoba, S.C. Verrall, and B. Javidi, *Appl. Opt.* **39**, 2313 (2000).
[15] E. Tajahuerce and B. Javidi, *Appl. Opt.* **39**, 6595 (2000).
[16] E. Tajahuerce, J. Lancis, B. Javidi, and P. Andrés, *Opt. Lett.* **26**, 678 (2001).
[17] O. Matoba and B. Javidi, *Opt. Lett.* **27**, 321 (2002).
[18] J.W. Goodman and R.W. Lawrence, *Appl. Phys. Lett.* **11**, 77 (1967).
[19] J.H. Bruning, D.R. Herriott, J.E. Gallagher, D.P. Rosenfeld, A.D. White, and D.J. Brangaccio, *Appl. Opt.* **13**, 2693 (1974).
[20] T.-C. Poon and A. Korpel, *Opt. Lett.* **4**, 317 (1979).
[21] J. Schwider, B. Burow, K.E. Elsner, J. Grzanna, and R. Spolaczyk, *Appl. Opt.* **22**, 3421 (1983).
[22] L. Onural and P.D. Scott, *Opt. Eng.* **26**, 1124 (1987).
[23] U. Schnars and W.P.O. Jüptner, *Appl. Opt.* **33**, 179 (1994).
[24] U. Schnars, *J. Opt. Soc. Am. A* **11**, 2011 (1994).
[25] I. Yamaguchi, T. Zhang, *Opt. Lett.* **22**, 1268 (1997).
[26] G. Pedrini, P. Frning, H. Fessler, and H.J. Tiziani, *Appl. Opt.* **37**, 6262 (1998).
[27] B. Javidi and E. Tajahuerce, *Opt. Lett.* **25**, 610 (2000).
[28] E. Tajahuerce, O. Matoba, and B. Javidi, *Appl. Opt.* **40**, 3877 (2001).
[29] Y. Frauel, E. Tajahuerce, M.-A. Castro, and B. Javidi, *Appl. Opt.* **40**, 3887 (2001).
[30] Y. Frauel and B. Javidi, *Opt. Lett.* **26**, 1478 (2001).
[31] T.J. Naughton, Y. Frauel, B. Javidi, and E. Tajahuerce, *Appl. Opt.* **41**, 4124 (2002).
[32] O. Matoba, T.J. Naughton, Y. Frauel, N. Bertaux, and B. Javidi, *Appl. Opt.* **41**, 6187 (2002).

[33] B. Javidi and F. Okano, *Three-Dimensional Television, Video, and Display Technologies* (Springer, Berlin 2002).

[34] M. Rabbani, *Selected Papers on Image Coding and Compression*, SPIE Milestone Series **MS48** (SPIE Press, Bellingham, WA, 1992).

[35] J.W. Goodman and A.M. Silvestri, *IBM J. Res. Develop.* **14**, 478 (1970).

[36] W.J. Dallas and A.W. Lohmann, *Appl. Opt.* **11**, 192 (1972).

[37] T. Nomura, A. Okazaki, M. Kameda, Y. Morimoto, and B. Javidi, "Digital holographic data reconstruction with data compression." in, *Algorithms and Systems for Optical Information Processing V, San Diego, July 2001*, Proceedings of the SPIE **4471** (SPIE Press, Bellingham, WA, 2001).

[38] T.J. Naughton, J.B. McDonald, and B. Javidi, *Appl. Opt.* (submitted August 2002).

[39] B. Javidi, A. Sergent, G. Zhang, and L. Guibert, *Opt. Eng.* **36**, 992 (1997).

[40] F. Goudail, F. Bollaro, B. Javidi, and P. Réfrégier, *J. Opt. Soc. Am. A* **15**, 2629 (1998).

[41] T.J. Naughton and B. Javidi, "Compression of encrypted digital holograms of three-dimensional objects." In preparation.

[42] T.J. Naughton and B. Javidi, "Optical encryption of three-dimensional objects." In preparation.

[43] J.W. Goodman, *Introduction to Fourier Optics* (McGraw-Hill, New York, 1996).

[44] H.J. Caulfield, *Handbook of Optical Holography* (Academic Press, New York, 1979).

[45] M. Sutkowski and M. Kujawinska, *Opt. Lasers Eng.* **33**, 191 (2000).

[46] D.A. Huffman, *Proc. IRE* **40**, 1098 (1952).

[47] J. Ziv and A. Lempel, *IEEE Trans.* **IT-23**, 337 (1977).

[48] T.A. Welch, *IEEE Computer* **17**, 8 (1984).

[49] M. Burrows and D.J. Wheeler, *Digital SRC Rep.* **124** (1994).

Optical Image Encryption Using Optimized Keys

Takanori Nomura and Bahram Javidi

Secure optical storage based on a configuration of a joint transform correlator (JTC) using a photorefractive material is presented. A key code designed by an optimized algorithm so that its Fourier transform has a uniform amplitude distribution and a uniformly random phase distribution is introduced. Original two-dimensional data and the key code are placed side-by-side at the input plane. Both of them are stored in a photorefractive material as a joint power spectrum. The retrieval of the original data can be achieved with the same key code. We can record multiple two-dimensional data in the same crystal by angular multiplexing and/or key-code multiplexing.

10.1 Introduction

Optical information-processing methods have been proposed for various applications. Recently, optical information technologies for security and encryption systems [1–8] and holographic memory systems [9–12] have been studied.

For secure storage, the reference beam, object beam, or both of them can be encoded optically. However, a holographic configuration of optical systems is required. It is difficult to make a practical system because fully complex data are needed for phase encoding. In this chapter, we propose a key code represented by real-valued data. Owing to adoption of real-valued data for a key code, we can use a configuration of a JTC because two-dimensional data and a key code can be placed side-by-side at the input plane using an amplitude-only spatial-light modulator. Furthermore, this optical setup is more compact than an ordinary holographic system because the object and reference beams share a single $4f$ system. Multiple secure data recording by angular multiplexing and/or key-code multiplexing is possible in the proposed system. Optics have many degrees of freedom (wavelength, polarization, 3-D topology, etc.). Some of these parameters are used in the chapter for securing data. We believe that optical technologies provide a more complex environment and are more resistant to attacks compared with digital electronic systems.

In Section 10.2, we describe the principle of the proposed data-storage system using a JTC configuration. In Section 10.3, the method to design the key code is presented. In Section 10.4, we show computer simulations for characteristics of the key code and then we confirm the performance of the proposed system. In Section 10.5, optical experimental results of angular multiplexing and key-code multiplexing are shown to confirm the proposed system.

10.2 Secure Optical Data Storage Based on a Joint Transform Correlator Architecture

The proposed secure optical data storage based on a JTC architecture is shown in Figure 10.1. Referring to Figure 10.1, we describe the proposed secure data-storage system. Let $f(x)$ and $h(x)$ denote the image to be recorded and the key code. We use one-dimensional notation for simplicity. In the write-in process, $f(x)$ and $h(x)$ are placed side-by-side at the input plane shown in Figure 10.1(a). For simplicity, we have ignored the spatial shifts in the input images. The input plane is the front focal plane of a Fourier transform lens. We can place an input phase mask on the input image for double-random phase encoding [2, 4, 7, 8] to obtain optical encryption based on input-plane and Fourier-plane random encoding. In this section, we have ignored the input phase mask to avoid complexity. The joint power spectrum $E(u)$ is given by

$$E(u) = |F(u)|^2 + |H(u)|^2 + F(u)H^*(u) + F^*(u)H(u), \qquad (10.1)$$

where $F(u)$ and $H(u)$ denote the Fourier transforms of $f(x)$ and $h(x)$, respectively, and the notation * denotes a complex conjugate. In a readout process, the joint power spectrum is illuminated by a Fourier transform of the key code. Namely, we have

$$\begin{aligned} D(u) &= H(u)E(u) \\ &= H(u)|F(u)|^2 + H(u)|H(u)|^2 + F(u)|H(u)|^2 + F^*(u)H^2(u) \end{aligned} \quad (10.2)$$

If $|H(u)|^2$ is equal to unity, we obtain

$$D'(u) = H(u)|F(u)|^2 + H(u) + F(u) + F^*(u)H^2(u). \qquad (10.3)$$

With the inverse Fourier transform of $D'(u)$ we finally get $d'(x)$ as

$$d'(x) = h(x)*f(x)\star f(x) + h(x) + f(x) + f(x)\star h(x)*h(x), \qquad (10.4)$$

where the notations * and $\star$ denote convolution and correlation operations, respectively. It is clear that we can obtain the image to be recorded in the third term on the right-hand side of Eq. (10.4). To realize the proposed optical data-storage system, we have to design the key code $h(x)$ so that its

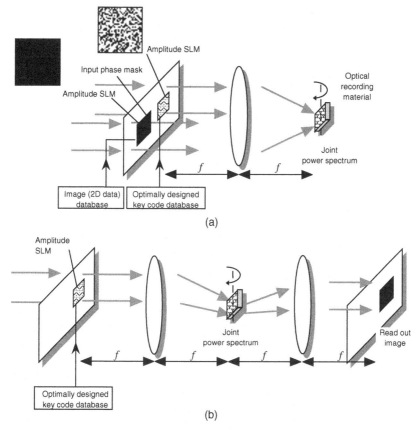

Fig. 10.1. Scheme of secure optical data storage based on a joint transform correlator architecture: (**a**) write-in system; (**b**) readout system.

Fourier transform has only phase distribution. Our approach is different from so-called computer-generated holograms, including conventional binary detour phase [13–15] and digital halftoning computer-generated holograms [16]. In our approach, the designed wave front is reconstructed as the zero-order diffraction beam in the proposed methods. For secure recording, the phase distributions of the Fourier transform of the key code are supposed to be random. Multiple images can be recorded by angular multiplexing. Furthermore, it is necessary that multiple independent key codes exist for multiple image recording by key-code multiplexing.

10.3 Key Code Design

We show the method to design the key code shown in Figure 10.2(a). Let $h_N(x, y)$ denote a designed key code. The value of N means the number of

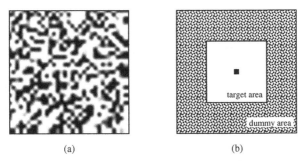

Fig. 10.2. (a) Designed key code and (b) its Fourier transform.

gray levels of the key code. For example, $h_2(x, y)$ means an optimally designed binary key code.

In order to design the key code, the energy function is introduced, given by

$$E = \sum_{\text{target area}} [|\text{FT}(h_N(x, y))| - C]^2, \tag{10.5}$$

where $FT(\cdot)$ denotes a Fourier-transform operation and C is a target value for the amplitude. The target area is shown in Figure 10.2(b). The center of the rectangle area is not included because the value of the center is much larger than the other area due to the characteristics of the Fourier transform. The key $h_N(x, y)$ that minimizes the energy function is the optimally designed key code. Note that the phase distribution of the Fourier transform of the key code is not included in the energy function. To find the key code that minimizes the energy function, a simulated annealing algorithm [17] is adopted.

10.4 Computer Simulations

10.4.1 Characteristics of the Key Code

We investigate the characteristics of the designed key code by computer simulations. Figure 10.3 shows the key code obtained after multiple iterations. The key consists of 128×128 pixels with nine gray levels. The number of gray levels is not limited to nine generally. However, because the photographic film used as a key code in our experiment does not have enough gray levels, the number is limited to nine. We set the target area to 64×64 pixels. The distributions of both the amplitude and the phase of the optimally designed key code are investigated. The Fourier transform of the key code is shown in Figure 10.4. We can see the almost uniform amplitude distribution (black and white denote 0 and 1.7, respectively) and uniformly random phase distribution (black and white denote $-\pi$ and π, respectively) within the target area. We set the mean

Fig. 10.3. Designed key code consisting of 128 × 128 pixels with nine gray levels.

(a)   (b)

Fig. 10.4. Fourier transform of the key code shown in Figure 10.3: (**a**) amplitude distribution (black and white denote 0 and 1.7, respectively) and (**b**) phase distribution (black and white denote −π and π, respectively).

value of the amplitude distribution to 1.0 so the maximum value is 1.7. The histograms of both the amplitude and the phase of the key code are shown in Figure 10.5. The amplitude distribution is not unity, but the shape of the histogram is much narrower than that of the conventional CGH key [4]. This means that the amplitude of the joint power spectrum is less modified than

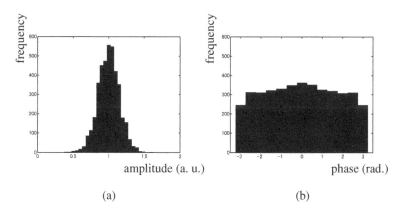

(a)   (b)

Fig. 10.5. Fourier transform of the key code shown in Figure 10.3: (**a**) amplitude distribution (black and white denote 0 and 1.7, respectively) and (**b**) phase distribution (black and white denote −π and π, respectively).

the conventional method. Although we do not include phase distribution in the energy function, the distribution is uniformly random. This means that the secure storage is carried out successfully [2].

10.4.2 Security of the Key Code

There are N^M candidates for a correct key code if M denotes the number of pixels of the key code with N gray levels. The key code shown in Figure 3 is one of 9^{16384} candidates. It is very difficult to find the correct key code by using a trial-and-error method of guessing the key. This is an approximate estimation because the Hamming distance between the key code and a counterfeit is not considered. However, if the characteristics of the key code (amplitude only, uniform Fourier amplitude distribution, and uniformly random Fourier phase distribution) are stolen, the risk of counterfeiting the key increases. We investigate whether the key code $h_N(x)$ to minimize the energy function E is determined as a unique solution. For several keys, that minimize the energy function, the Fourier phase distributions are calculated. Table 10.1 shows the standard deviations of argument of $H_{Ni}(u)H_{Nj}^*(u)$ for $i, j = 1, 2, \ldots, 5$, where $H_{Ni}(u)$ denotes the Fourier transform of the key code h_{Ni}. In this simulation,

Table 10.1. The standard deviations of argument of $H_{Ni}(u)H_{Nj}^*(u)$.

	$H_{N1}(u)$	$H_{N2}(u)$	$H_{N3}(u)$	$H_{N4}(u)$	$H_{N5}(u)$
$H_{N1}^*(u)$	0.00	1.37	1.14	0.96	1.14
$H_{N2}^*(u)$	1.37	0.00	1.35	1.40	1.30
$H_{N3}^*(u)$	1.41	1.35	0.00	1.43	1.35
$H_{N4}^*(u)$	0.93	1.40	1.43	0.00	1.44
$H_{N5}^*(u)$	1.41	1.30	1.35	1.44	0.00

we let the number of gray levels N be equal to nine. The standard deviation is large if two keys are not the same, and the standard deviation is equal to zero only if two keys are the same. From the simulations, we learn that the optimally designed key is not determined as a unique solution if one does not know the exact values of the key. The reason is that there are many degrees of freedom of the Fourier phase distribution of the key code because the distribution is not included in the energy function determined by Eq. (10.5). The existence of the multiple key codes hints that it is possible to record multiple images by key-code multiplexing.

10.4.3 Storage Performance

The data-storage performance using the optimally designed key code is illustrated. For rigorous analysis, we should include a Bragg selectivity and degeneracy for a volume hologram [18]. However, because we do not use a

Fig. 10.6. An original image to be recorded for computer simulation.

plane-wave reference but a random phase wave, it is hard to analyze the performance. Therefore, we show the computer simulations for a thick hologram to investigate the performance of our proposed secure storage system. The image consisting of 128×128 pixels to be recorded is shown in Figure 10.6. In the simulation, an input phase mask is bonded to the image to be recorded. The designed key code used in this simulation is the same as that shown in Figure 10.3. The correct readout image with the same key code is shown in Figure 10.7(a). The image is slightly different from the original image shown in Figure 10.6 due to the nonuniformity of the Fourier amplitude distribution as shown previously. The mean-square error between the original image and the readout image is 4.4×10^3. After binarization, the mean-square error is 5.0×10^3. This corresponds to a 7.7% bit-error rate. The image decrypted with an incorrect key code is shown in Figure 10.7(b). The image is quite different from the original image. The mean-square error between the original image and the readout image is 1.1×10^4. After binarization, the mean-square error is 1.7×10^4. This corresponds to 26% bit-error rate. The computer simulations confirm the performance of the proposed system.

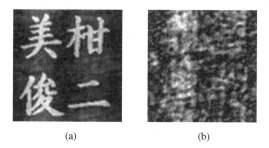

(a)   (b)

Fig. 10.7. Readout images with (**a**) a correct key code and (**b**) an incorrect key code by computer simulation.

10.5 Optical Experiments

10.5.1 Optical Experimental System

The secure optical data-storage system is demonstrated experimentally. The experimental setup is shown in Figure 10.8. A YAG laser (wavelength 532 nm) is used as the coherent light source. It is collimated by lenses L_1 (an objective lens) and L_2 with a focal length of 250 mm. A 10 mm $\times$ 12 mm $\times$ 2 mm LiNbO$_3$ crystal doped with 0.03 mol. % iron (LN) is used as a volume holographic recording medium. The image to be recorded and the key code recorded on a photographic film are placed side-by-side at input plane P_1. In the experiments, a phase mask is bonded to the image to be recorded. The extent of the Fourier spectrum of the mask corresponds to the target area of the Fourier spectrum of the key code. During the write-in process, shutters S_1 and S_2 are open. The collimated beam illuminates both the image to be recorded and the key code at plane P_1. They are Fourier-transformed by lens L_3 with a focal length of 50 mm, and the joint power spectrum is recorded into the crystal LN placed at P_2. In the readout process, shutter S_1 is closed and shutter S_2 is open. The joint power spectrum is illuminated by the Fourier transform of the key code placed at plane P_1. The readout image after Fourier transforming by lens L_4 with a focal length of 50 mm is captured by a CCD camera placed at P_3. The captured image is sampled with 512 $\times$ 480 pixels and is quantized to 8 bit gray levels by means of a frame-grabber board.

10.5.2 Encryption and Decryption

We encrypt the image shown in Figure 10.6 with a key code shown in Figure 10.3 optically. Both images consist of 128 $\times$ 128 pixels. The size of each pixel on the photographic film is 88 μm $\times$ 88μm, and the dimension of the images is 11.3 mm $\times$ 11.3 mm. The readout decrypted image with a correct key is shown in Figure 10.9. The readout decrypted image with an incorrect key is shown in Figure 10.10. From the experiment, we find the optical encryption using an optimized key is performed successfully.

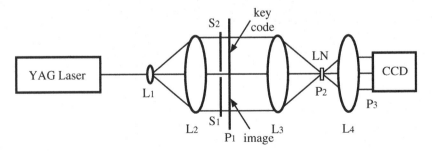

Fig. 10.8. Optical setup.

Fig. 10.9. A readout decrypted image with a correct key.

Fig. 10.10. A readout decrypted image with an incorrect key.

Figure 10.11 shows the binary image to be stored for quantitative evaluation. The key code shown in Figure 10.3 is used in the experiments. Both images consist of 128×128 pixels. The size of each pixel on the photographic film is the same as in the experiment above. The size of each cell of the image shown in Figure 10.11 is 3.2 mm × 3.2 mm. If we assume that one cell of the image denotes one bit, the image contains 256 bits of information. Figure 10.12(a) shows the binarized readout image with a correct key code.

In this case, the bit-error rate is 25%. Furthermore, we applied the spatial thresholding to the binarized readout image, and the resultant image is shown in Figure 10.12(b). In this case, the bit-error rate is 0%. The binarized readout image with an incorrect key code is shown in Figure 10.13. In this case, the bit-error rate is 50%.

Fig. 10.11. An original binary image for optical experiments.

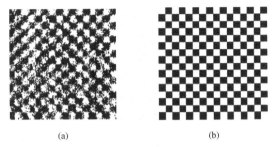

(a)   (b)

Fig. 10.12. Readout images with a correct key code: (**a**) after binarization and (**b**) after binarization and a spatial thresholding.

Fig. 10.13. Binarized readout image with an incorrect key code.

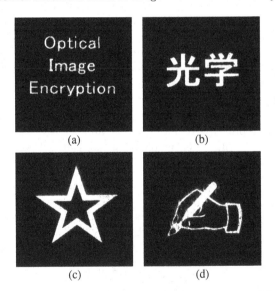

(a)   (b)

(c)   (d)

Fig. 10.14. Multiple images for angular multiplexing captured by CCD camera through the photorefractive crystal.

Fig. 10.15. Readout images from angular multiplexing holographic memory.

10.5.3 Secure Multiple Data Storage

Figure 10.14 shows the multiple images for angular multiplexing. The images are captured separately by a CCD camera through the photorefractive crystal. After multiple recording by angular multiplexing at 5° intervals, the readout images are as shown in Figure 10.15. Here one can see original images and no cross-talk images. The quality of the image is less than the original one. This is due to the nonuniformity of the Fourier amplitude distribution of the key code shown in Section 10.4. The readout image with an incorrect key code is a similar result, shown in Figure 10.13. Figures 10.16 and 10.17 show the experimental results for key-code multiplexing. Here also one can see original images and no cross-talk images. The quality of the image is less than the original one. This is due to nonuniformity for the same reason as for angular multiplexing. The readout image with an incorrect key code is a similar result, shown in Figure 10.13. The experimental results described above confirm the potential of our proposed optical data-storage system.

10.6 Conclusion

We have proposed a secure optical data-storage system with a random phase key code based on a configuration of a joint transform correlator. The method to design the key code has been proposed. Computer simulations and optical experiments confirm our proposed system.

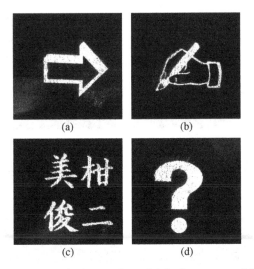

Fig. 10.16. Multiple images for key-code multiplexing captured by a CCD camera through the photorefractive crystal.

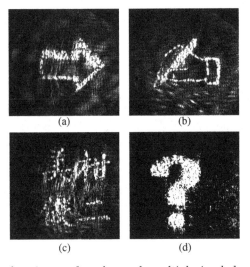

Fig. 10.17. Readout images from key-code multiplexing holographic memory.

Acknowledgments

The authors wish to thank Dr. Yoshiharu Morimoto and Mr. Shunji Mikan for their valuable comments and experimental support.

References

[1]   Li, H.-Y.S., Qiao, Y., and Psaltis, D., "An optical network for real-time face recognition." *Appl. Opt.* **32**, 5026–5035 (1993).

[2]   Réfrégier, P. and Javidi, B., "Optical image encryption based on input plane and Fourier plane random encoding." *Opt. Lett.* **20**, 767–769 (1995).

[3]   Javidi, B. and Nomura, T., "Securing information by use of digital holography." *Opt. Lett.* **25**, 28–30 (2000).

[4]   Nomura, T. and Javidi, B., "Optical encryption system with a binary key code." *Appl. Opt.* **39**, 4783–4787 (2000).

[5]   Tajahuerce, E., Mataoba, O., Verrall, S.C., and Javidi, B., "Opto-electronic information encryption with phase-shifting interferometry." *Appl. Opt.* **39**, 2313–2320 (2000).

[6]   Tajahuerce, E. and Javidi, B., "Encrypting three-dimensional information with digital holography." *Appl. Opt.* **39**, 6595–6601 (2000).

[7]   Unnikrishnan, G., Joseph, J., and Singh, K., "Optical encryption by double-random phase encoding in the fractional Fourier domain." *Opt. Lett.* **25**, 887–889 (2000).

[8]   Wang, B., Sun, C., Su, W., and Chiou, A.E.T., "Shift-tolerance property of an optical double-random phase-encoding encryption system." *Appl. Opt.* **39**, 4788–4793 (2000).

[9]   Heanue, J.F., Bashaw, M.C., and Hesselink, L., "Encrypted holographic data storage based on orthogonal-phase-code multiplexing." *Appl. Opt.* **34**, 6012–6015 (1995).

[10] Mok, F. H., "Angle-multiplexed storage of 5000 holograms in lithium niobate." *Opt. Lett.* **18**, 915–917 (1993).

[11] Matoba, O. and Javidi, B., "Encrypted optical storage with wavelength-key and random phase codes." *Appl. Opt.* **38**, 6785–6790 (1999).

[12] Tan, X., Matoba, O., Shimura, T., Kuroda, K., and Javidi, B., "Secure optical storage that uses fully phase encryption." *Appl. Opt.* **35**, 6689–6694 (2000).

[13] Goodman, J.W., *Introduction to Fourier Optics.* McGraw-Hill, New York (1996).

[14] Brown B.R., Lohmann, A.W., "Complex spatial filtering with binary masks." *Appl. Opt.* **5**, 967–969 (1966).

[15] Lohmann A.W., Paris, D.P., "Binary Fraunhofer holograms, generated by computer." *Appl. Opt.* **6**, 1739–1746 (1967).

[16] Hirokawa, K., Ukezono, N., Itoh, K., and Ichioka, Y., "Digital halftoning for computer generated holograms." *Proc. Photo-Opt. Instrum. Eng.* **2778**, 529–530 (1996).

[17] Aarts E. and Korst J., *Simulated Annealing and Boltzmann Machines.* (John Wiley & Sons, New York, 1989).

[18] Levene, M., Steckman, G.J., and Psaltis, D., "Method for controlling the shift invariance of optical correlators." *Appl. Opt.* **38**, 394–398 (1999).

11

Polarization Encoding for an Optical Security System

Takanori Nomura and Bahram Javidi

Summary. A polarization-encoding system for optical security verification is proposed. The polarization information is bonded to an identification card for security verification. As the polarization state cannot be imaged by an ordinary intensity-sensitive device such as a CCD camera, it can provide an additional degree of freedom in using optics to secure information. The expressions for polarization encoded input images are developed using Jones polarization calculus. A nonlinear joint transform correlator is used to provide an optical validation system. Computer simulations and optical experimental results are shown in support of the proposed method.

11.1 Introduction

Optical validation and security verification methods using optical-correlation systems have been proposed [1–5]. In some of these systems, the validation is based on correlation with a reference phase mask [2, 5]. In this chapter, we present an optical validation and security verification method that uses polarization encoding [6]. In this method, a gray-scale image such as a face or a fingerprint is bonded to a polarization-encoded mask. The polarization-encoded mask consists of randomly oriented linear polarizers. We call this composite image the polarization encoded image. The polarization-encoded image cannot be distinguished from the normal gray scale using an intensity-sensitive device such as a CCD camera. A nonlinear joint transform correlator (JTC) [7] is used to provide the verification system. Section 11.2 presents theoretically the polarization-encoded optical validation and security system using Jones matrix calculus [8]. Section 11.3 presents supporting computer simulation results. Results from an optical experiment are presented in Section 11.4.

11.2 Polarization-Encoding Optical Security Systems

In this section, we describe the polarization-encoding optical security system in detail. Let $f(x, y)$ denote a nonnegative and nonpolarized image to be identified. The image $f(x, y)$ is bonded to the polarization-encoded mask as shown in Figure 11.1. We call it a polarization-encoded image. In general, the polarization-encoded image looks the same as the original image because the polarization state cannot be seen by an intensity-sensitive detector. The polarization mask consists of small linear polarizers rotated at various angles from 0 to π.

In order to verify a polarization-encoded image, we optically compare the polarization-encoded image with the reference polarization mask. We use a nonlinear JTC optical system for verification. As shown in Figure 11.2, the polarization-encoded image and the reference polarization mask are placed side-by-side in the input plane of the correlator. The input-plane image, $h(x, y)$, can be written as

$$h(x, y) = f(x, y)p(x, y) + q(x, y), \tag{11.1}$$

where $p(x, y)$ and $q(x, y)$ denote the polarization-encoded mask and the reference polarization mask, respectively. Using the Jones matrices, the polarization-encoded mask and the reference polarization mask can be written as

$$\mathbf{P}(x, y) = \begin{pmatrix} \cos^2\alpha(x, y) & \sin\alpha(x, y)\cos\alpha(x, y) \\ \sin\alpha(x, y)\cos\alpha(x, y) & \sin^2\alpha(x, y) \end{pmatrix}, \tag{11.2}$$

$$\mathbf{Q}(x, y) = \begin{pmatrix} \cos^2\beta(x, y) & \sin\beta(x, y)\cos\beta(x, y) \\ \sin\beta(x, y)\cos\beta(x, y) & \sin^2\beta(x, y) \end{pmatrix}, \tag{11.3}$$

respectively. In Eqs. (11.2) and (11.3), $\alpha(x, y)$ and $\beta(x, y)$ denote the rotated angle of the linear polarizers for the polarization mask and the reference mask,

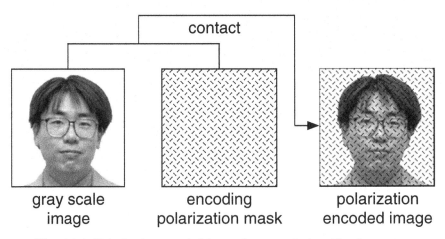

| gray scale | encoding | polarization |
| image | polarization mask | encoded image |

Fig. 11.1. Polarization-encoded image for an optical verification system.

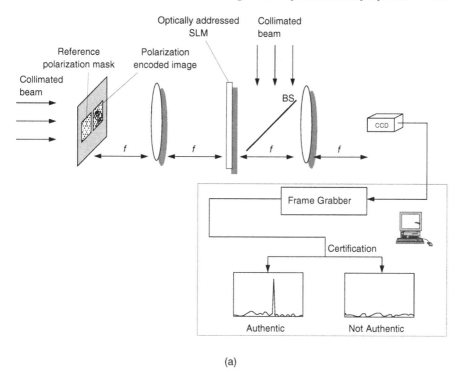

(a)

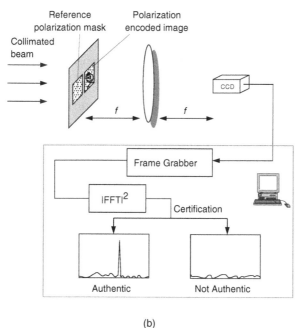

(b)

Fig. 11.2. Polarization-encoded optical verification system using a joint transform correlator: (**a**) all-optical system and (**b**) hybrid system.

respectively. To generalize, let us consider a light that is linearly polarized at an angle of $\pi/4$ relative to the x axis as the illuminating beam. In this case, the Jones calculus of the input image, $\mathbf{h}(x, y)$, can be written as

$$\mathbf{h}(x,y) = f(x,y)\mathbf{P}(x,y)\begin{pmatrix}1\\1\end{pmatrix} + \mathbf{Q}(x,y)\begin{pmatrix}1\\1\end{pmatrix}. \tag{11.4}$$

For simplicity, we have ignored the constant coefficient $1/\sqrt{2}$ of the Jones vector of the illuminating beam. A horizontal component, $h_H(x, y)$, and a vertical component, $h_V(x, y)$, of the optical field can be written as

$$h_H(x,y) = f(x,y)[\cos^2\alpha(x,y) + \sin\alpha(x,y)\cos\alpha(x,y)]$$
$$+[\cos^2\beta(x,y) + \sin\beta(x,y)\cos\beta(x,y)], \tag{11.5}$$

$$h_V(x,y) = f(x,y)[\sin\alpha(x,y)\cos\alpha(x,y) + \sin^2\alpha(x,y)]$$
$$+[\sin\beta(x,y)\cos\beta(x,y) + \sin^2\beta(x,y)]. \tag{11.6}$$

The input images are Fourier-transformed using a lens. Then the joint power spectrum of the polarization-encoded image and the reference polarization mask is captured by a CCD camera. The joint power spectrum is

$$|H(\xi,\eta)|^2 = |H_H(\xi,\eta)|^2 + |H_V(\xi,\eta)|^2, \tag{11.7}$$

where $H(\xi,\eta)$, $H_H(\xi,\eta)$, and $H_V(\xi,\eta)$ denote the Fourier transforms of $h(x, y)$, $h_H(x, y)$, and $h_V(x, y)$, respectively. Then the joint power spectrum is inverse Fourier-transformed by a computer. The joint power spectrum can be nonlinearly transformed to provide a high discrimination capability [9]. Finally we obtain the correlation between the polarization-encoded image and the reference polarization mask:

$$c(x,y) = f(x,y)[\cos^2\alpha(x,y) + \sin\alpha(x,y)\cos\alpha(x,y)] \tag{11.8}$$
$$\oplus[\cos^2\alpha(x,y) + \sin\alpha(x,y)\cos\alpha(x,y)]$$
$$+f(x,y)[\sin\alpha(x,y)\cos\alpha(x,y) + \sin^2\alpha(x,y)]$$
$$\oplus[\sin\alpha(x,y)\cos\alpha(x,y) + \sin^2\alpha(x,y)]$$
$$+[\cos^2\beta(x,y) + \sin\beta(x,y)\cos\beta(x,y)]$$
$$\oplus[\cos^2\beta(x,y) + \sin\beta(x,y)\cos\beta(x,y)]$$
$$+[\sin\beta(x,y)\cos\beta(x,y) + \sin^2\beta(x,y)]$$
$$\oplus[\sin\beta(x,y)\cos\beta(x,y) + \sin^2\beta(x,y)]$$
$$+f(x,y)[\cos^2\alpha(x,y) + \sin\alpha(x,y)\cos\alpha(x,y)]$$
$$\oplus[\cos^2\beta(x,y) + \sin\beta(x,y)\cos\beta(x,y)]$$
$$+f(x,y)[\sin\alpha(x,y)\cos\alpha(x,y) + \sin^2\alpha(x,y)]$$
$$\oplus[\sin\beta(x,y)\cos\beta(x,y) + \sin^2\beta(x,y)]$$
$$+f(x,y)[\cos^2\beta(x,y) + \sin\beta(x,y)\cos\beta(x,y)]$$

$$\oplus[\cos^2\alpha(x,y) + \sin\alpha(x,y)\cos\alpha(x,y)]$$
$$+f(x,y)[\sin\beta(x,y)\cos\beta(x,y) + \sin^2\beta(x,y)]$$
$$\oplus[\sin\beta(x,y)\cos\beta(x,y) + \sin^2\beta(x,y)],$$

where the $\oplus$ symbol denotes correlation. The first and second terms on the right-hand side of Eq. (11.8) correspond to autocorrelation of the polarization-encoded image. The third and fourth terms correspond to autocorrelation of the reference polarization mask. The fifth and the sixth terms are the cross correlation between the polarization-encoded image and the reference polarization mask. The rest of the terms denote the symmetry appearing in the cross correlation. The cross correlation and autocorrelation are spatially separated when the polarization-encoded image and the reference polarization mask are spatially separated. When the encoding polarization mask is the same as the reference polarization mask (that is, $\alpha(x,y) = \beta(x,y)$), a strong correlation is produced. When the two polarization masks are different, the correlation signal is lower than the autocorrelation signal peak. Thus, we can verify the image in terms of the correlation between the polarization-encoded mask and the reference polarization mask.

11.3 Computer Simulations

11.3.1 Correlation Peak and SNR Response Versus Error Rate of Polarization-Encoding Mask

The proposed verification system is based on a correlation peak between a polarization-encoded image and a polarization mask. Then we investigate two verification indices. They are a correlation peak and signal-to-noise ratio (SNR) response versus error rate of a polarization-encoding mask. The SNR is determined by

$$\text{SNR} = \frac{\text{correlation peak value}}{\Sigma(\text{correlation signals})}. \tag{11.9}$$

The dimension of the correlation signals is 64×64 pixels. We calculate the indices between two polarization-encoding masks, shown in Figure 11.3. One is a polarization mask A consisting of a combination of small linear polarizers. Another one is a polarization mask B consisting of a combination of small linear polarizers. The polarization state of some pixels of the mask B is not always the same as the polarization state of the corresponding pixel of the mask A. The number of pixels in both cases is 128×480 pixels. The gray levels in this image correspond to the rotated angles of the linear polarizer. White and black areas denote linear horizontal and vertical polarizations, respectively. The figures show an example of polarization masks consisting of 90 kinds of rotation angles of linear polarizers. The polarization-encoded image and the reference polarization mask are placed side-by-side at the JTC

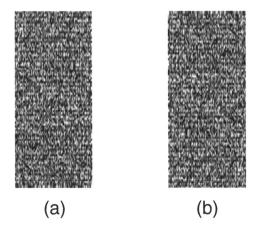

(a)   (b)

Fig. 11.3. Polarization masks used in investigation of correlation (**a**) peak response and (**b**) SNR response.

input plane. We change the rate (we may refer to "error rate") from 0% to 100%. We investigate 2, 4, and 90 kinds of rotated angles of the linear polarizer. We adopt here both conventional and binary JTCs. In the binary JTC, the joint power spectrum is binarized by its median value to provide a high discrimination capability [10].

Figures 11.4 and 11.5 show the correlation peak response versus the error rate and the SNR response versus the error rate, respectively. In the figures, the notation "C2" means that the polarization mask consists of two kinds of rotated angles and the correlation is obtained by a conventional JTC. The notation "B4" means that the polarization mask consists of four kinds of rotated angles and the correlation is obtained by a binary JTC. From the simulations, we find that the number of the rotation angles of the linear polarizer is not important. The main factor that determines the performance of the verification system is whether the joint power spectrum is binarized or not.

11.3.2 Verification

We show the performance of the proposed verification system by a computer simulation. The image used in the computer simulations is shown in Figure 11.6. The image consists of 128×128 pixels. A sample of a polarization mask is shown in Figure 11.7. The gray levels in this image correspond to the rotated angles of the linear polarizer. White and black areas denote linear horizontal and vertical polarizations, respectively. The polarization-encoded image and the reference polarization mask are placed side-by-side at the JTC input plane consisting of 512×512 pixels, as shown in Figure 11.8. The joint power spectrum and the correlation output obtained using a conventional JTC are shown in Figures 11.9 and 11.10, respectively. Figures 11.9(a) and 11.10(a) illustrate the case when the reference polarization mask is the same

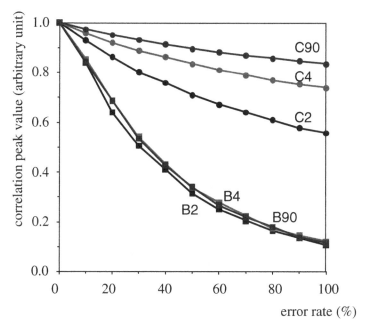

Fig. 11.4. Correlation peak response versus error rate of a polarization mask by computer simulations.

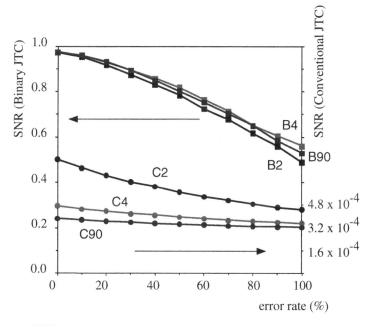

Fig. 11.5. SNR response versus error rate of a polarization mask by computer simulations.

Fig. 11.6. Gray-scale image consisting of 128 × 128 pixels for a computer simulation.

Fig. 11.7. Polarization mask consisting of a combination of small linear polarizers for a computer simulation.

Fig. 11.8. Input image for a joint transform correlator for computer simulations.

as the encoding polarization mask. Figures 11.9(b) and 11.10(b) illustrate the case when the reference mask is different from the encoding mask. Figure 11.10 shows only the area where the correlation signal is obtained and is normalized by the maximum value of Figure 11.10(a). It is clear that the correlator has verified the authentic polarization-encoded image and rejected the unauthorized image. Further simulations are done using a binary JTC. In these simulations, the joint power spectrum is binarized by its median value. Figures 11.11 and 11.12 show the joint power spectrum and the correlation results

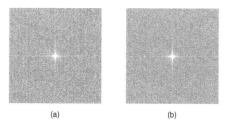

(a)   (b)

Fig. 11.9. Joint power spectra by computer simulation using a conventional joint transform correlator: (**a**) reference polarization mask identical to encoding polarization mask; (**b**) reference polarization mask different from encoding polarization mask.

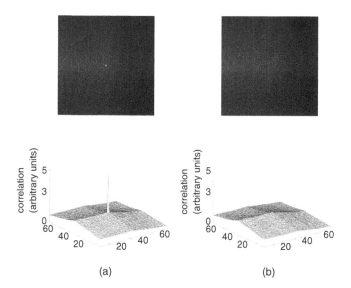

(a)   (b)

Fig. 11.10. Correlation results by computer simulation using a conventional joint transform correlator: (**a**) reference polarization mask identical to encoding polarization mask; (**b**) reference polarization mask different from encoding polarization mask.

using a binary JTC. Images in Figure 11.12 have been normalized by the maximum value of Figure 11.12(a).

11.4 Optical Experiments

11.4.1 Correlation Peak and SNR Response Versus Error Rate of Polarization-Encoding Mask

We show the correlation peak and SNR response versus error rate of the polarization-encoding mask using an all-optical system, as shown in Fig-

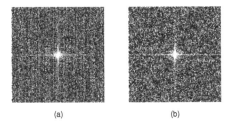

Fig. 11.11. Joint power spectra by computer simulation using a binary joint transform correlator: (**a**) reference polarization mask identical to encoding polarization mask; (**b**) reference polarization mask different from encoding polarization mask.

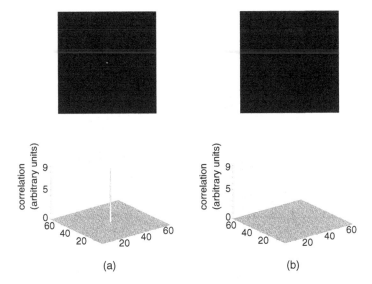

Fig. 11.12. Correlation results by computer simulation using a binary joint transform correlator: (**a**) reference polarization mask identical to encoding polarization mask; (**b**) reference polarization mask different from encoding polarization mask.

ure 11.2(a). A liquid-crystal panel consisting of 640 × 480 pixels is used for displaying two polarization masks. The He–Ne lasers are used as coherent light sources. The PAL–SLM (X-5641, Hamamatsu Photonics K.K.) is used for an optically addressed amplitude SLM. Lenses with focal lengths of 250 mm and 400 mm are used for first Fourier-transforming and second Fourier-transforming, respectively. Then the correlation result is sampled to 64 × 64 pixels and quantized to 8 bits by a frame grabber equipped in a personal computer.

Figures 11.13 and 11.14 show the correlation peak response versus error rate and SNR response versus error rate, respectively. The notation in the figures follows the notation in Figures 11.4 and 11.5. The experimental results demonstrate the possibility of the proposed system.

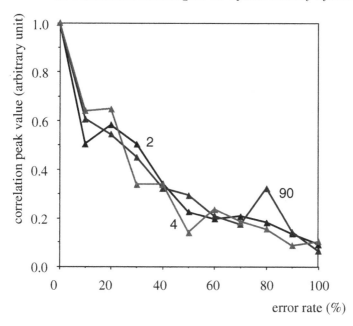

Fig. 11.13. Correlation peak response versus error rate of a polarization mask by optical experiments.

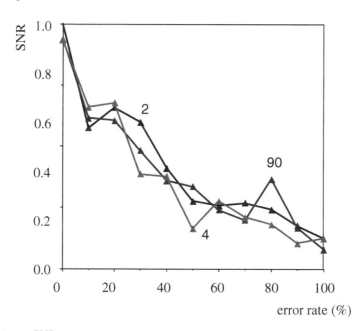

Fig. 11.14. SNR response versus error rate of a polarization mask by optical experiments.

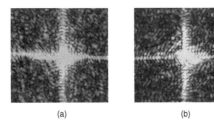

(a)   (b)

Fig. 11.15. Joint power spectra by optical experiments: (**a**) reference polarization mask identical to encoding polarization mask; (**b**) reference polarization mask different from encoding polarization mask.

11.4.2 Verification

We present optical experiments to demonstrate the proposed verification system. The experimental setup is the same as the system shown in Figure 11.2(b). A human face on photographic film is used as a gray-scale image. The dimensions of the input image are 6 × 6 mm. For simplicity, polarization masks consisting of 200 × 200 random binary (horizontal or vertical) linear polarizers arrays [11] are used. The arrays were made of two surface-relief-etched birefringent substrates joined face-to-face. Each square pixel is 30 mm. The He–Ne laser is used as a coherent light source. A lens with a focal length of 200 mm is used for optical Fourier-transforming. The joint power spectrum is captured by a CCD camera. Then it is sampled to 512 × 480 pixels and quantized to 8 bits by a frame grabber equipped in a personal computer. The digitized joint power spectrum is Fourier-transformed by using a fast Fourier-transform algorithm. Finally, by calculating the power of the Fourier transform, we obtain a correlation result. Figures 11.15 and 11.16 show the joint power spectrum and the correlation results. Figures 11.15(a) and 11.16(a) correspond to the case when the reference polarization mask is the same as the encoding polarization mask. Figures 11.15(b) and 11.16(b) correspond to the case when the reference mask is different from the encoding mask. The experimental results confirm the proposed verification system.

11.5 Conclusion

We have demonstrated an optical validation and security verification system using polarization-encoding of input images. The polarization-encoding can provide an additional degree of freedom to secure information using optical technologies. The polarization-encoding can be combined with a phase-encoding scheme to enhance the validation and security verification of the system. Computer simulations and optical experiments demonstrate the performance of the proposed system.

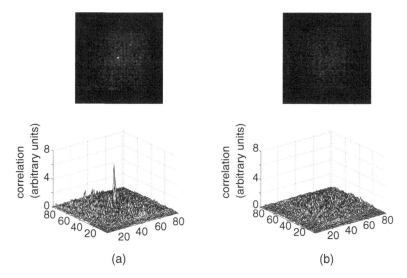

(a)   (b)

Fig. 11.16. Correlation results by optical experiments: (a) reference polarization mask identical to encoding polarization mask and (b) reference polarization mask different from encoding polarization mask.

Acknowledgments

We wish to thank Dr. Fang Xu and Prof. Yeshaiasu Fainman at UCSD for fabricating the polarization masks.

References

[1]   Li, H.-Y.S., Qiao, Y., Psaltis, D., "An optical network for real time face recognition." *Appl. Opt.* **32**, 5026–5035 (1993).

[2]   Javidi, B. and Horner, J.L., "Optical pattern recognition for validation and security verification." *Opt. Eng.* **33**, 1752–1756 (1994).

[3]   Réfrégier, P. and Javidi, B., "Optical image encryption using input plane and Fourier plane random encoding." *Opt. Lett.* **20**, 767–769 (1995).

[4]   Wilson, C.L., Watson, C.I., and Paek, E.G., "Combined optical and neural network fingerprint matching." *Proc. Soc. Photo-Opt. Instrum. Eng.* **3073**, 373–382 (1997).

[5]   Weber, D. and Trolinger, J., "Novel implementation of nonlinear joint transform correlators in optical security and validation." *Opt. Eng.* **38**, 62–68 (1999).

[6]   Javidi, B. and Nomura, T., "Polariation-encoding for optical security system." *Opt. Eng.* **39**, 2439–2443 (2000).

[7]   Weaver, C.S. and Goodman, J.W., "A technique for optical convolving two functions." *Appl. Opt.* **5**, 1248–1249 (1966).

[8]   Collett, H., "The Jones matrix calculus," in *Polarized Light*, Chap. 10, pp. 187–218 (Marcel Dekker, New York, 1993).

[9]   Javidi, B., "Nonlinear joint transform correlators," in B. Javidi and J. L. Horner, eds., *Real-Time Optical Information Processing*, Chap. 4, pp. 115–183 (Academic Press, New York, 1994).

[10]   Javidi, B. and Wang, J., "Binary nonlinear joint transform correlator with median an subset median thresholding." *Appl. Opt.* **30**, 967–976 (1991).

[11]   Ford, J.E., Xu, F., Urquhart, K., Fainman, Y., "Polarization-selective computer-generated-holograms." *Opt. Lett.* **18**, 456–458 (1993).

Stream Cipher Using Optical Affine Transformation

Jun Tanida and Toru Sasaki

Summary. In this chapter, a method for generating a sequence of pseudorandom patterns using two-dimensional affine transformation is presented. The method is called the pseudorandom pattern generation with affine transformation (PPGA). A parallel affine-transform feedback system is introduced as the platform of the PPGA. As an example of the application, a stream cipher is described. The procedure to implement the PPGA is explained with an evaluation of the results. Experimental result of optical implementation show the capabilities of the method.

12.1 Introduction

In recent years, information transferred over the Internet has increased tremendously. A variety of information from simple e-mail messages to secret codes on money and personal privacy is running everywhere. It can be said that a virtual society is growing over the internet. However, extension of such a virtual society is accompanied by increased crime perverting internet technologies so that demand for cipher technologies to guarantee the security of communication is increasing. In terms of implementation, we must consider the degree of importance and the required throughput on the cipher technology. For example, if safe transmission of medical images is the target, the highest level of security with a large throughput is required from the cipher technology. In such an application, current cipher technologies are not sufficient to satisfy such a demand.

As a solution for such requirements, cipher techniques based on optical methods are promising. In typical optical cipher techniques, the parallel nature of optical processing is fully utilized to achieve high processing throughput. Methods using phase images with random spatial distribution [1] and a stream cipher based on optical parallel processing [2] are good examples. In both cases, pseudorandom-number or pattern generation is an important process, which holds a key to increasing the processing throughput. For this problem, application of optical parallel processing is considered. A linear feedback shift register using a liquid-crystal spatial-light modulator and an optical implementation of a cellular automaton are proposed as effective methods for random pattern generation.

In this chapter, a method for generating a sequence of pseudorandom patterns using two-dimensional affine transformation is presented. The method is called (PPGA), is a technique for generating a sequence of images consisting of pseudorandom numbers (or pixels) in parallel. Since it generates one image at a time, the PPGA provides higher performance than a method generating a sequence of one-dimensional numbers. Moreover, the PPGA is suitable for optical implementation. Therefore, the PPGA is expected to be useful in a cipher system with a large capacity.

First, a parallel affine-transform feedback system, which is the platform of the PPGA, is introduced, and several examples of the possible processing by the system are presented. Then the procedure to implement the PPGA is explained and performance of the method is evaluated. Finally, some results from optical implementation are shown and the future issues regarding the PPGA are summarized.

12.2 Parallel Affine-Transform Feedback System

12.2.1 Architecture

PPGA is an architecture of optical computing based on iterative processing by optical transformations. The system generates a series of images using two-dimensional affine transformations. A two-dimensional affine transformation consists of a linear transformation and a translation,

$$\mathbf{x}' = \begin{bmatrix} a & b \\ c & d \end{bmatrix} \mathbf{x} + \begin{bmatrix} e \\ f \end{bmatrix}, \tag{12.1}$$

where $\mathbf{x}$ and $\mathbf{x}'$ indicate the positions on the input and output planes, respectively.

Figure 12.1 shows the processing procedure of the parallel affine-transform feedback system. In this system, two-dimensional affine transformations are used as the fundamental operation. The operations are performed iteratively

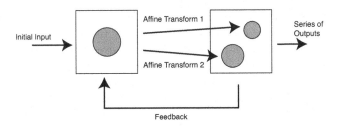

Fig. 12.1. Parallel affine-transform feedback system.

with feedback of the transformed images. A series of output images or a resulting image after a sufficient number of iterations are utilized.

A notable point of the system is that multiple affine transformations over all pixels on the input image are executed in parallel, while a typical digital processor achieves the same task pixel-by-pixel. In addition, an affine transformation requires real-valued operations on the coordinates, which consume time and computational resources. On this point, the optical implementation gives an effective solution for the processing.

12.2.2 Fractal Generation

The original form of the parallel affine-transform feedback system was presented as an optical processing system called an optical fractal synthesizer [5]. It differs from the parallel affine-transform feedback system in that the optical fractal synthesizer utilizes only contractive affine transformations, which is a restriction of the basic principle. Since the processing manner of the optical fractal synthesizer is identical to the parallel affine-transform feedback system and useful to understanding its operations, we explain the optical fractal synthesizer briefly.

The optical fractal synthesizer generates fractal shapes [6] based on an iterated function system [7]. An iterative function system is a function system consisting of a set of contractive transformations. For the case of N affine transformations, the iterative function system is expressed as

$$W(A) = \bigcup_{l=1}^{N} w_l(A), \quad A \subset \mathbf{R}^2, \tag{12.2}$$

where A indicates a set of points in a two-dimensional real number space $\mathbf{R}^2$. Note that A is a pattern on a two-dimensional image plane, and $w_l(A)$ is a contractive affine transformation onto a set of points A representing as

$$w_l(A) = \left\{ \mathbf{x}'; \mathbf{x}' = \begin{bmatrix} a_l & b_l \\ c_l & d_l \end{bmatrix} \mathbf{x} + \begin{bmatrix} e_l \\ f_l \end{bmatrix}, \ \mathbf{x} \in A \right\}, \tag{12.3}$$

where $\mathbf{x}$ and $\mathbf{x}'$ are two-dimensional vectors. The elements of the matrix satisfy $|a_l d_l - b_l c_l| < 1$.

When the iterated function system is applied to an arbitrary set of points, the set follows the equation

$$B_{i+1} = W(B_i), \quad B_i \subset \mathbf{R}^2, \tag{12.4}$$

where i is the iteration number. After a sufficient number of iterations, a set of points B_∞ satisfying the equation is

$$B_\infty = W(B_\infty), \quad B_\infty \subset \mathbf{R}^2. \tag{12.5}$$

The set B_∞ is unchanged by the transformation of the iterative function system. This set of points is called the attractor and has a complicated shape, or *fractal*, according to the parameter set $\{a_l, \cdots, f_l; l = 1, \cdots, N\}$ in Eq.(12.3).

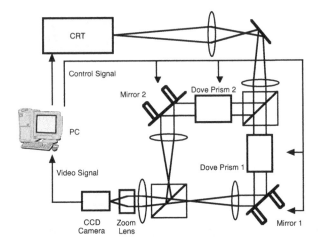

Fig. 12.2. Experimental setup of an optical fractal synthesizer.

12.2.3 Optical Fractal Synthesizer

The optical fractal synthesizer is an optical processor on which the recurrence formula in Eq. (12.4) is implemented. Figure 12.2 shows an optical setup of the experimental system. In the system, an initial image corresponding to B_0 is displayed on the CRT. The image is duplicated by the first beam splitter, and an affine transformation is applied to each of the duplicated images. The affine transformation is achieved with the dove prisms for image rotation and the mirrors for image translation. The second beam splitter combines the two transformed images. Finally, the zoom lens contracts the combined image, which is detected by the CCD camera. The detected image corresponds to B_1. Repeating the same process, we can perform the operation in Eq. (12.4). Note that all the processes of the optical system are achieved in parallel so that high processing throughput is obtained.

Figure 12.3 shows a series of images transformed step-by-step. The initial image is a rectangle pattern of B_0. By increasing the iteration, a finer pattern structure of pattern is obtained, which is a kind of fractal pattern. The images B_7 and B_8 are almost identical. The reason is that the resolution points of the optical system are finite, so the generated pattern is not a *pure* fractal. In spite of this limitation, the optical fractal synthesizer is an interesting processor based on optical iterative processing.

12.2.4 Control of the Optical Fractal Synthesizer

An affine transform performed by the experimental optical fractal synthesizer is described as

$$w_l(A) = \{\mathbf{x}'; \mathbf{x} = S(s_l)R(\theta_l)M(j_l)\mathbf{x} + \mathbf{t}_l, \mathbf{x} \in A\}, \quad l = 1, 2. \tag{12.6}$$

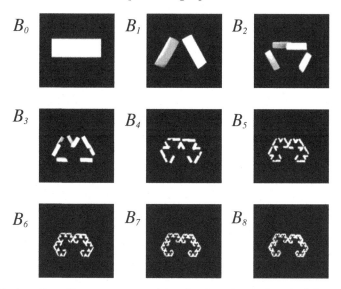

Fig. 12.3. A series of images generated by the experimental optical fractal synthesizer.

$S(s)$, $R(\theta)$, and $M(j)$ are two-dimensional matrices for contraction, rotation, and reflection, respectively,

$$S(s) = \begin{bmatrix} s & 0 \\ 0 & s \end{bmatrix}, \tag{12.7}$$

$$R(\theta) = \begin{bmatrix} \cos\theta & -\sin\theta \\ \sin\theta & \cos\theta \end{bmatrix}, \tag{12.8}$$

$$M(j) = \begin{bmatrix} j & 0 \\ 0 & 1 \end{bmatrix}, \quad j \in \{1, -1\}, \tag{12.9}$$

where s is the contraction ratio, θ is the rotation angle, and j is a reflection parameter where $j = 1$ for an erect image and $j = 0$ for the reflected one with respect to the y axis. $\mathbf{t}_l$ is a vector specifying translation. Hereafter, the affine parameters of the experimental optical fractal synthesizer are denoted by $\{s_l, \theta_l, j_l, \mathbf{t}_l; l = 1, 2\}$.

12.2.5 Extension to a Parallel Affine-Transform Feedback System

The optical fractal synthesizer only uses contract transformations in the iterative function system. However, removing this restriction, we can extend its functionality. The parallel affine-transform feedback system is the extended architecture. On the system, not only fractal patterns but also randomly varying patterns can be obtained by specifying the parameters in Eq. (12.3). In the following sections, the parallel affine-transform feedback system is utilized as the platform for the processing.

12.3 Pseudorandom Pattern Generation with Affine Transformation

12.3.1 Stream Cipher

The stream cipher is a method of encryption using secret keys for variable lengths of messages [2, 4]. The secret-key encryption is a class of encryption methods in which a common key is kept by the sender and the receiver. As long as the key is not leaked, the secret of the message is preserved. To handle variable lengths of messages, the corresponding length of a random-number sequence is used as the key. Owing to the high throughput of algorithms for random-number generation, the stream cipher is promising for treating a large volume of information.

Figure 12.4 shows the processing procedure of the stream cipher. Usually, a message is treated as a collection of 1-bit data. In our method, a message is assumed to be two-dimensional image, considering the capability of random image generation of the PPGA. In the encryption phase, the sending image and the key image are added, and a modular operation with modulus n applied on it, where n is the number of gray levels of the image. For the key image, an image composed of randomly varying pixels is used. Since the encrypted image looks like an image consisting of randomly varying pixels, others cannot recognize the contents of the message. In the decryption phase, the key image is subtracted from the encrypted image, and a modular operation with modulus n is applied. Note that synchronization of the key images between the sender and the receiver is required to retrieve the correct information. In addition, an effective method of random pattern generation for the key images is crucial to achieve high-throughput transmission.

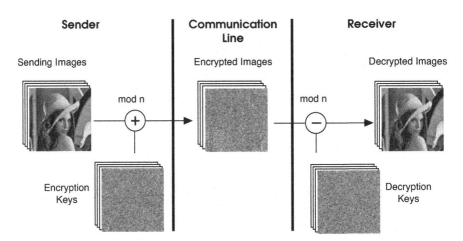

Fig. 12.4. Stream cipher for a sequence of images.

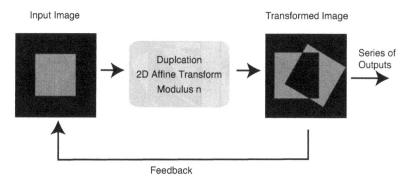

Fig. 12.5. Procedure of the PPGA.

12.3.2 Pseudorandom Pattern Generation

The PPGA was developed to meet the demand on a stream cipher for a large volume of data. The processing procedure is similar to that of the optical fractal synthesizer. Figure 12.5 depicts the procedure. The input image is duplicated, and each of the images is transformed with different affine parameters. Then the transformed images are added, and a modular operation with modulus n is applied to the added image. By repeating the same procedure with the feedback image, we can obtain a series of output images.

On the PPGA, affine transforms composed of rotation, magnification, and reflection are employed. A set of affine transforms are represented as

$$w_l : \mathbf{x}' = S(s_l)R(\theta_l)M(j_l)\mathbf{x} + \mathbf{t}_l, \quad l = 1, \cdots, N, \tag{12.10}$$

where l is the identifier of the affine transform and N is the total number of affine transforms. θ_l is the rotation angle, s_l is the magnification ratio, and j_l is the reflection parameter introduced in Eq.(12.9). $S(s_l)$, $R(\theta_l)$, and $M(j_l)$ are operators for magnification, rotation, and reflection given by Eqs.(12.7)–(12.9). $\mathbf{t}_l$ is a two-dimensional vector specifying translation.

To apply the PPGA to random pattern generation for the stream cipher, the parameters for the affine transforms and the iteration number can be used to specify patterns of the key image. Since the PPGA is based on a deterministic process, we can reproduce a specific pattern using the same parameters. Therefore, these values are called *key parameters*. For a secure cipher method, estimation of the keys must be difficult. From this point, the space of the key parameters of the PPGA should be large. As explained in the next subsection, pattern distribution generated by the PPGA is sensitive to a change of the key parameters. As a result, the number of possible keys becomes large and resistance against exhaustive attacks can be obtained. In addition, if key parameters are changed for individual iterations, key estimation will be made very difficult.

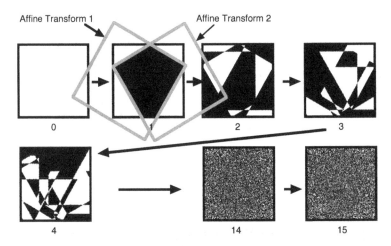

Fig. 12.6. A series of random patterns generated by the PPGA.

Figure 12.6 shows a result of computer simulation of the PPGA. In this case, the generated image consists of 256×256 pixels with 256 gray levels. The modulus of the modular operation was set as 256. The iteration number is indicated in the figure. After 14 iterations, an image with randomly distributed patterns is successfully generated.

Figures 12.7(a)-(c) are the patterns resulting from three different key parameters. Table 12.1 summarizes the key parameters used in the generation. They seem to be composed of well-randomized patterns. To evaluate the difference between the patterns generated by different parameters, subtraction images for Figures 12.7(a) and 12.7(b) and Figures 12.7(a) and 12.7(c) are calculated as shown in Figures 12.7(d) and 12.7(e), respectively. These results indicate that patterns with different key parameters are sufficiently different from each other. Figure 12.7(f) shows patterns generated when the magnification ratio of the affine transform is less than 1. In this case, randomized patterns are localized within a region surrounded by a fractal shape.

12.3.3 Evaluation of Security Strength

Principle Verification

Performance of the stream cipher implemented by the PPGA is evaluated by computer simulation. A standard image Lena, of 256×256 pixels with 256 gray levels is used as the original image. The modulus of the modular operation is set as 256. Simulation results are shown in Figure 12.8. It is confirmed that the original image is well-hidden in the encrypted image and correctly retrieved after decryption.

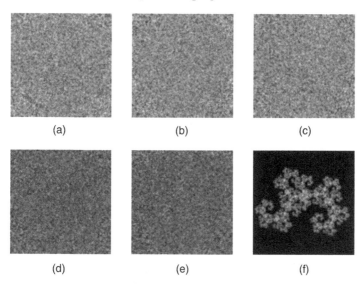

Fig. 12.7. A variation of random patterns generated by the PPGA: (**a**)–(**c**) random patterns with key parameters in Table 12.1; (**d**) subtraction image between (**a**) and (**b**); (**e**) subtraction image between (**a**) and (**c**); and (**f**) random pattern with contract affine transforms.

Table 12.1. Key parameters for generating patterns of Figures 12.7(a)–(c) and 12.7(f).

Case	l	s_l	θ_l	$\mathbf{t}_l$	Iteration
(a)	1	1.2	60	$(50, 0)$	19
	2	1.2	120	$(-50, 0)$	19
(b)	1	1.2	60	$(50, 0)$	18
	2	1.2	120	$(-50, 0)$	18
(c)	1	1.201	60	$(50, 0)$	19
	2	1.201	120	$(-50, 0)$	19
(f)	1	0.7	60	$(50, 0)$	19
	2	0.7	120	$(-50, 0)$	19

Possible Key Number

To evaluate the security strength of the method, several tests are executed. First, the effect of key-parameter variance is studied. Evaluation is achieved by decryption using key parameters that are slightly different from those used in encryption. Figure 12.9 shows images decrypted from the encrypted image of Figure 12.8(c) with key images generated by different key parameters. Table 12.2 summarizes the key-parameter variance from the correct one. Even if the difference of between the key parameters is quite small, the original image is

Fig. 12.8. Simulation result of a stream cipher by PPGA: (**a**) original image, (**b**) key image, (**c**) encrypted image, and (**d**) decrypted image.

not retrieved for any cases. Note that not only the affine parameters but also the iterative numbers can be utilized as the key parameters.

As shown by the experiment, a small variance of the key parameters provides sufficiently different key patterns for encryption. This suggests that a large key space can be obtained by a combination of the key parameters in the PPGA. For example, if key parameters for N affine transformations are chosen according to a rule of Table 12.3, the number of possible key number becomes $(7.2 \times 10^8)^N$. To improve the security strength against exhaustive attacks, increasing of the number of affine transforms N is quite effective.

Table 12.2. Key-parameter variance for decryption.

Case	Parameter	Variance
(a)	s_1	+0.01
(b)	θ_1	+0.1
(c)	$\mathbf{t}_1$	$(+1, 0)$
(d)	iteration	+1

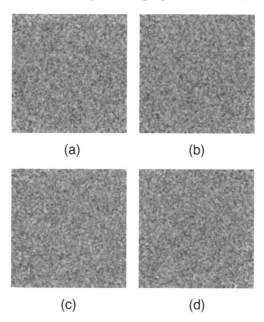

Fig. 12.9. Decrypted images using slightly different key parameters.

Table 12.3. An example of a key-selection rule.

Key parameter	Selection rule	Total number
s_l	every 0.01 over 1.1–1.3	20
θ_l	every 0.1 degree over 0–360°	3600
$\mathbf{t}_l$	any point in 100×100 pixels	10,000

Evaluation of Randomness

When the PPGA is applied to the stream cipher, randomness of the distribution of the generated image is important for information security. However, the mechanism of the PPGA has not been studied, and evaluation of the randomness is not sufficient so that, by comparing with the linear feedback shift register (LFSG) [8], which is a typical system of random pattern generation, the characteristics of the PPGA are investigated. Then the randomness is evaluated with correlation functions.

The LFSR is a random pattern generator consisting of K shift registers connected in tandem. Figure 12.10 shows the schematic diagram of the LFSR. The LFSR returns the resulting value of a modular operation with modulus

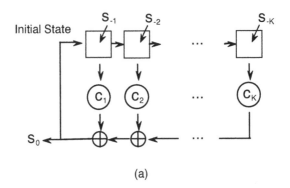

(a)

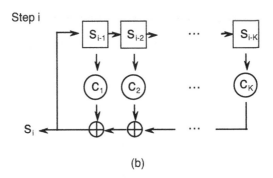

(b)

Fig. 12.10. Linear feedback shift register.

2 on the weighted sum of all the registers as

$$s_i = \sum_{\substack{k=1 \\ (\text{mod } 2)}}^{K} c_k s_{i-k}, \tag{12.11}$$

where k is the identifier of registers, c_k is the weight on each register, $\sum_{(\text{mod } 2)}$ indicates a sum of modulus 2, i is the iteration number, and $\{s_{-1}, s_{-2}, \cdots, s_{-K}\}$ are supplied as the initial conditions. As shown in Figure 12.10, s_i is generated from $\{s_{i-1}, s_{i-2}, \cdots, s_{i-K}\}$ in one step. In the following step, the value of each register is transferred to the next one and s_i is fed back as the input of the shift register. Repeating the same operation, a sequence of random number $s_i(i = 1, 2, \cdots)$ can be obtained.

On the other hand, patterns generated by the PPGA are represented as

$$f_{i+1}(\mathbf{x}) = \sum_{\substack{l=0 \\ (\text{mod } n)}}^{N} f_i\left(A_l^{-1}(\mathbf{x} - \mathbf{a}_l)\right), \tag{12.12}$$

Table 12.4. Comparison between the LFSR and the PPGA.

Item	LFSR	PPGA
data value	register value	pixel value
data container	shift register	pixel
processing	data transfer by shift register	data transfer by affine transform
	modular operations	modular operations
data paths	single and obvious	multiple and complicated

where $\mathbf{x}$ is a vector indicating the pixel position and $f_i(\mathbf{x})$ shows the value of pixel $\mathbf{x}$ at step i. A_l and $\mathbf{a}_l$ are a linear transform matrix and a translation vector of affine transformation, respectively, and l is the identifier of N affine transforms.

Correspondence between the operations by the LFSR and the PPGA are summarized in Table 12.4. As is seen from the table, the LFSR and the PPGA have many corresponding points. A notable difference is that the PPGA has much more flexible data-transfer paths, whereas the LFSR has fixed ones. The PPGA obeys the same computational process as the LFSR, but the data-transfer paths of the PPGA are hidden, which makes it more difficult to infer the keys than in the LFSR.

Evaluation with Correlation

Randomness of the patterns generated by the PPGA is evaluated by means of correlation functions. If the pattern values are distributed randomly, autocorrelation of the pattern shows a single narrow peak only at the origin, and if two patterns have different randomness, cross correlation between the patterns shows a flat distribution.

To evaluate the degree of randomness, Peak to correlation energy (PCE) [9] is adopted. PCE is a measure of peak narrowness, which is defined as

$$PCE = \frac{|f(x_{peak}, y_{peak})|^2}{E}, \tag{12.13}$$

$$E = \int_{-\infty}^{\infty} \int_{-\infty}^{\infty} |f(x,y)|^2 dx dy, \tag{12.14}$$

where $f(x,y)$ is the correlation function of the target pattern, and (x_{peak}, y_{peak}) is the position of the correlation peak. The denominator of PCE is the energy of the correlation function. Since the energy depends on the numbers of pixels and gray levels, patterns to be compared by PCE must be identical in those values.

Figures 12.11(a) and (b) show autocorrelation functions of the pattern of Figure 12.8(b) and a reference pattern generated by the LFSR, respectively. Narrow peaks are obtained for both cases. Figure 12.11(c) indicates the cross

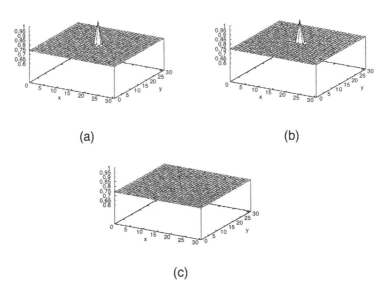

(a)   (b)

(c)

Fig. 12.11. Correlation on generated patterns: (**a**) and (**b**), autocorrelation of the patterns generated by the PPGA and the LFSR; (**c**) cross correlation between the patterns with different iteration numbers by the PPGA.

correlation between the patterns generated by the PPGA with the same affine parameters except the iteration numbers (one is 18 and the other is 19). In this case, no peak appears, which shows that no similar part exists in the two patterns. PCE values of these correlation functions are summarized in Table 12.5. The result indicates that the pattern generated by the PPGA has the same degree of randomness as that generated by the LFSR. In addition, a low value of PCE for the patterns generated by different iteration numbers suggests the validity of using the iteration number as a key parameter.

Table 12.5. PCE for correlation peaks in Figure 12.11.

Case	PCE
(a)	2.72×10^{-5}
(b)	2.74×10^{-5}
(c)	1.54×10^{-5}

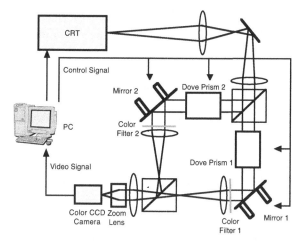

Fig. 12.12. Experimental setup of an optical parallel affine-transform feedback system.

12.4 Optical Implementation of PPGA

12.4.1 Optical System for Accurate Processing

Correct operation of the PPGA is verified by an experimental optical affine feedback system. As is well-known, even a slight difference of the initial states develops into quite different states after a sufficient number of affine transformations. This indicates that incomplete optical implementation with low accuracy results in a failure of patterns reproduction. To overcome the problem, we must realize optical transformations with high accuracy for the PPGA.

The optical system is shown in Figure 12.12, which is a modified version of the optical fractal synthesizer in Figure 12.2. In the optical system, we develop two notable techniques to improve the precision of the operation. One is spatial division of the processing plane, and the other is wavelength multiplexing.

The spatial division is a technique in which data points on a processing image are divided into multiple groups and each group is processed time-sequentially to prevent cross talk from the neighboring pixels. Using this technique, we can obtain an accurate affine transformation even if the point-spread function of the optical system is not sufficiently sharp.

The wavelength multiplexing is introduced to compensate for intensity differences between two optical paths. In the original setup of the optical fractal synthesizer, the affine-transform images associated with different optical paths are just overlapped and the intensity summation is detected by the CCD camera. Although this is a simple implementation of image summation, nonuniformity of intensity over the transformed image causes unrecoverable errors. To avoid this problem, two different wavelengths are assigned to the individual paths, and the image summation is performed by the computer after detection with the color CCD camera. Due to the flexibility of digital processing, nonuniformity of the optical signals can be compensated.

12.4.2 Experimental Result

To simplify the modular operation in the algorithm, binary images are used in the system. Namely, black-and-white optical distribution is utilized for the images to be encrypted and the key image generated by the PPGA. In this case, the modular operation can be altered by a logical exclusive OR operation. Although only binary images are treated by this configuration, bit plane decomposition enables us to achieve gray and color image encryption.

An experimental result of a stream cipher by the optical parallel affine-transform feedback system is shown in Figure 12.13. The original image consists of 200×200 pixels with 8 bit gray levels. After being decomposed into eight bit-planes, each binary image is encrypted with different key images. For generation of the key images, a set of affine parameters shown in Table 12.6 is used with different iteration numbers from 13 to 20. As seen from Figure 12.13, the original information cannot be inferred from the encrypted images. For the decryption keys, the same affine parameters and iteration numbers are employed. The final result shows that the stream cipher system works correctly and that the PPGA can be applied to the stream cipher.

12.5 Feature Issues

There are several issues with a stream cipher using the PPGA for practical usage. From a theoretical point of view, the size of the key space is an important measure indicating security strength. Since the key parameters of the PPGA are analog values, the quantized resolution determines the size of the key space. However, as the resolution step becomes finer, more precise adjustment and more severe control of the optical implementation are required. To derive the optimum conditions, further studies are important.

Adaptability to the current Internet technology is another issue. Although light waves carry a huge amount of information over the Internet, most of the terminal devices process the information after light-to-electron conversion. If the method presented is applied directly to raw optical signals, it is expected

Table 12.6. Affine parameters for pattern generation.

l	s_l	θ_l	$\mathbf{t}_l$
1	1.2	60	$(75, 0)$
2	1.2	120	$(-75, 0)$

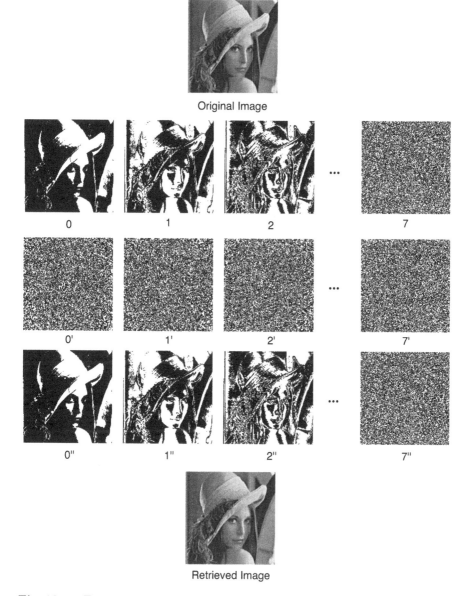

Original Image

0   1   2   ···   7

0'   1'   2'   ···   7'

0"   1"   2"   ···   7"

Retrieved Image

Fig. 12.13. Experimental result of a stream cipher by optical implementation of the PPGA. The original image is decomposed into eight bit-planes, and each bit-plane image is encrypted with a key image generated by the PPGA. The correct image is retrieved as a weighted sum of the decrypted images.

to be an effective technique for optical communication. Time-to-space and space-to-time conversion techniques [10] are promising for the purpose.

Packaging of the optical system is a common issue for optical information systems. Recently, several interesting techniques for optical system packaging have been presented in the area of optical interconnection [11, 12]. Using these techniques, the parallel affine-transform feedback system can be embodied with compact hardware. Introduction of such special hardware makes it difficult to replicate the original system and increases security strength.

12.6 Summary

As an example of information security techniques based on optical processing, a method for generating a sequence of pseudorandom patterns using two-dimensional affine transformation called PPGA has been described. A parallel affine-transform feedback system is introduced as the platform of the PPGA. Then the procedure to implement the PPGA is explained with an evaluation of the results. Experimental results of optical implementation show the capabilities of the method presented. It is expected that the PPGA will open a new field of optical security and contribute progress in information science.

References

[1]   P. Réfrégier and B. Javidi, "Optical image encryption based on input plane and Fourier plane random encoding," *Opt. Lett.* **20**, 767–769 (1995).

[2]   M. Madjarova, M. Kakuta, M. Yamaguchi, and N. Ohyama, "Optical implementation of the stream cipher based on the irreversible cellular automata algorithm," *Opt. Lett.* **22**, 1624–1626 (1997).

[3]   S. Zhang and M. Karim, "High-security optical integrated stream ciphers," *Opt. Eng.* **38**, 20–24 (1999).

[4]   T. Sasaki, H. Togo, J. Tanida, and Y. Ichikoka, "Stream cipher based on pseudorandom number generation with optical affine transformation," *Appl. Opt.* **39**, 2340–2346 (2000).

[5]   J. Tanida, A. Uemoto, and Y. Ichioka, "Optical fractal synthesizer: concept and experimental verification," *Appl. Opt.* **32**, 653–658 (1993).

[6]   B.B. Mandelbrot, *The Fractal Geometry of Nature* (Freeman, San Francisco, 1983).

[7]   M.F. Barnsley, *Fractals Everywhere* (Academic, Boston, 1988).

[8]   J.L. Massey, "Shift-register synthesis and BCD decoding," *IEEE Trans. Information Theory* **IT-15**, 122–127 (1969).

[9]   B.V.K.V. Kumar and L. Hassebrook, "Performance measures for correlation filters," *Appl. Opt.* **29**, 2997–3006 (1990).

[10]   T. Konishi and Y. Ichioka, "Ultrafast image transmission by optical time-to-two-dimensional-space-to-time-to-two-dimensional-space conversion," *J. Opt. Soc. Am. A* **16**, 1076–1088 (1999).

[11] J. Jahns and B. Acklin, "Integrated planar optical imaging system with high interconnection density," *Opt. Lett.* **18**, 1594–1596 (1993).

[12] G. Li, D. Huang, E. Yuceturk, P.J. Marchand, S.C. Esener, V.H. Ozguz, and Y. Liu, "Three-dimensional optoelectronic stacked processor by use of free-space optical interconnection and three-dimensional VLSI chip stacks," *Appl. Opt.* **41**, 348–360 (2001).

13

Applications of Digital Holography for Information Security

Takanori Nomura, Enrique Tajahuerce, Osamu Matoba, and Bahram Javidi

Summary. Secure optical storage based on a configuration of a joint transform correlator (JTC) using a photorefractive material is presented. A key code designed by an optimized algorithm so that its Fourier transform has a uniform amplitude distribution and a uniformly random phase distribution is introduced. Original two-dimensional data and the key code are placed side-by-side at the input plane. Both of them are stored in a photorefractive material as a joint power spectrum. The retrieval of the original data can be achieved with the same key code. We can record multiple two-dimensional data in the same crystal by angular multiplexing and/or key-code multiplexing.

13.1 Introduction

Optical information-processing techniques have proved to be a real alternative to purely electronic processing in security, encryption, and pattern-recognition applications [1–13]. This is partially due to recent advances in optoelectronic devices and components, such as detectors, modulators, optical memories, and displays. Now, in general, it is easy to transfer information from electronic to optical domains and vice versa at high speeds. In this way, it is possible to combine the advantages of both approaches to develop more efficient security applications. This fact is especially relevant when securing information codified in the form of two-dimensional (2-D) or three-dimensional (3-D) images because, in these cases, optical systems are unavoidable.

One of the approaches to securing information by optical means consists in the use of random phase-encoding techniques [1–3,14–35]. In these methods, images or holograms are transformed, by using random phase masks, into noise distributions. Therefore, the original information remains encrypted and can be recovered only by means of a random phase mask acting as the key. However, in general, the resulting encrypted data contain both phase and amplitude and, thus, must be recorded and stored holographically [36]. This necessity makes it difficult to transmit the encrypted information over conventional communication channels. Also, if not digitized, or converted in some way, the original information must be reconstructed optically.

Fully complex information may be stored and communicated digitally by using digital holography [37–44]. With this technique, holograms are captured by an intensity-recording device, such as a CCD camera, and reconstructed in a computer. The information contained in the digital hologram may also be reconstructed optically after transmission by using computer-generated holograms. This method has been applied in holographic interferometry, wavefront sensing, and other optical metrology applications. Among the different techniques used to achieve digital holograms, a way to efficiently utilize the CCD capabilities is to use digital phase-shifting interferometry [45–50]. This phase-measurement technique can be more precise than using off-axis holography and can also be implemented in real time.

In this work, we review different methods based on digital holography for securing information stored as 2-D or 3-D images [51–53]. Digital holography is used to record the complex amplitude distribution of the diffraction patterns generated by images illuminated with coherent light. Encryption is performed using random phase codes. These techniques avoid analog holographic recording and the corresponding chemical or physical development. Also, they allow digital transmission of the information through conventional communication channels. The encryption is performed immediately and directly on the complex information, but decryption requires no more computation than a usual image-reconstruction procedure. Consequently, there is a potential speed advantage over fully digital encryption techniques. After decryption, electronic reconstruction with a one-step FFT procedure or optical reconstruction methods can be applied, depending on the specific applications. Computer decryption is less secure because the phase key is stored electronically, but no manual focusing adjustment is required. Therefore, these techniques combine the high space-bandwidth product and speed of optical processing with the advantages of electronic transmission, storage, and post-processing.

In the first approach described, a 2-D image is encrypted optically by double-random phase encoding with uniformly distributed random phase keys in both the input and Fourier planes of an optical processor architecture [51]. The method is similar to the first double-random phase-encoding technique proposed in the literature, but the final complex encrypted information is recorded digitally. A CCD camera is used for direct recording of the encrypted hologram with an off-axis digital holographic technique based on a Mach–Zehnder interferometer architecture. The amplitude distribution necessary for decryption, the key, is measured using the same optical system. This method allows the encrypted data to be stored, transmitted, and decrypted digitally. Furthermore, the system can be used for secure video storage, and transmission optical experiments are shown to illustrate the proposed method.

As a second approach, we describe in detail an optical encryption technique that uses phase-shifting interferometry to record the fully-complex encrypted information of a 2-D image, again with two random phase masks [52]. Both

Fourier and Fresnel domain optical encryption are achieved by using one random phase mask attached to the input in the object beam and another phase mask at a variable position in the reference beam of a Mach–Zehnder phase-shifting interferometer system. Information about the complex key to be used in the decryption process is also obtained by phase-shifting interferometry. As in the previous method, the encrypted information can be transmitted through digital communication lines, while the key can be transmitted either electronically or by making controlled copies of the reference phase mask used in the encryption procedure. Decryption can thus be performed either electronically or optoelectronically.

In the previous encryption techniques, the information to be secured or verified was encoded as a 2-D image using amplitude and phase. However, the ability of digital holograms to reconstruct in the computer the amplitude distribution generated by 3-D objects has recently been reported [48,49]. This idea has been applied to optically secure 3-D information [53]. In this approach, a phase-shifting interferometer records the phase and amplitude information generated by a 3-D object at a plane located in the Fresnel diffraction region by using an intensity-recording device. This information is encrypted optically with the Fresnel diffraction pattern generated by a random phase mask and stored electronically. The key is also electronically recorded by phase-shifting interferometry. The decrypted image can be obtained digitally or optically by using the complex field associated with the key.

In Section 13.2, we describe the technique to encrypt a 2-D object with random phase codes in the input and Fourier planes by using digital holography with an off-axis interferometer architecture. Section 13.3 describes how digital phase-shifting interferometry can be adapted to optoelectronically encrypt the Fraunhofer or a Fresnel diffraction pattern of a 2-D input. In Section 13.4, we show how to extend this technique to the encryption of 3-D scenes. Also, it is shown that it is possible to generate different perspectives of the input object. Finally, Section 13.5 summarizes the conclusions.

13.2 Securing Information by Use of Off-Axis Digital Holography

As mentioned in the previous section, the main motivation for developing security methods by using digital holography is to avoid conventional holographic recording, as holographic plates or photorefractive crystals, in order to permit digital transmission and postprocessing of the complex encrypted information. Random phase codes are used to modify the Fraunhofer or Fresnel diffraction patterns of the input object, but the encrypted complex amplitude distribution is recorded digitally with an electronic camera by using interferometric techniques. In this way, decryption can be performed digitally or optically.

Two random phase codes, one attached to the object and the other located in the Fourier plane, are used to encrypt the input image in the first reported

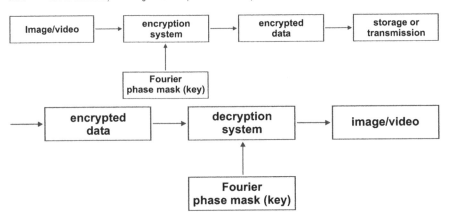

Fig. 13.1. Secure image/video–data-storage/transmission system that uses a combination of double-random phase encryption and a digital holographic technique: (a) transmitter–encoder, (b) receiver–decoder.

approach, by Javidi and Nomura [51]. Figure 13.1 sketches out this secure image/video storage/transmission method. A random phase mask is attached to the input image and a second is located in the Fraunhofer plane of an optical processor. The encrypted information is then recorded with off-axis digital holography instead of using conventional holographic materials. The interference of the encrypted diffraction pattern with a planar reference beam constitutes the Fourier hologram to be recorded by a CCD camera and stored in the computer. Digital information about the Fourier phase mask is obtained in the same way by removing the input object and the first phase mask. Decryption is then performed digitally in the computer by multiplying the extracted encrypted hologram and the Fourier phase-mask hologram followed by an inverse Fourier transformation. The system is well-adapted to work in real time but has space-bandwidth limitations due to the use of off-axis holography.

The proposed optical system for implementing this technique is shown in Figure 13.2. It consists of a Mach–Zehnder interferometer architecture with a He–Ne laser used as a coherent light source. The lower arm of the interferometer is the optical path where encryption is performed with the optical processor and the random phase masks. The upper arm constitutes the reference wave. At the CCD camera, the hologram is created by the interference between the encrypted data and the slightly inclined reference plane wave.

Let $t(x,y)$, $a(x,y)$, and $H(x,y)$ denote the complex amplitude distribution of the image to be encrypted, the input random phase mask, and the Fourier random phase mask, respectively. Variables x and y are spatial coordinates, and u and v denote the coordinates in the Fourier domain. As shown in Figure 13.2, the input random phase mask, $a(x,y) = \exp[i\phi_O(x,y)]$, is bonded with the image $t(x,y)$ to be encrypted. The resultant product of the two images is

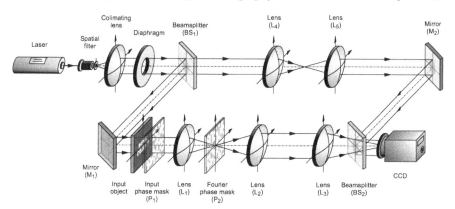

Fig. 13.2. Optical experiment setup for double-random phase-mask encryption with off-axis digital holography.

Fourier-transformed by lens L_1 and is multiplied by the Fourier phase mask $H(x, y) = \exp[i\phi_F(x, y)]$ at plane P_2. The $4f$ optical system constituted by lenses L_2 and L_3 images this complex distribution onto the CCD camera. The Fourier hologram is recorded by using the interference of this encrypted diffraction pattern with a planar reference beam, with amplitude distribution $R(x, y)$ at the output plane, slightly inclined with respect to the optical axis. The reference wave passes through the $4f$ optical system of lenses L_4 and L_5 to keep the spatial coherence. The hologram is then registered by a CCD camera and digitized into the computer.

Aside from constant factors, the complex amplitude distribution at plane P_2, $U_F(x, y)$, after the second phase mask, is given by the Fourier transformation

$$U_F(x, y) = \exp[i\phi_F(x, y)$$
$$\times \iint\limits_{-\infty}^{\infty} t(x', y') \exp[i\phi_O(x', y')] \exp\left[-i\frac{2\pi}{\lambda f}(xx' + yy')\right] dx'\, dy' \quad (13.1)$$

where f is the focal length of lens L_1, and λ is the wavelength of the laser beam. In Eq. (13.1) we assume that the object is located at a distance f from lens L_1. In convolution form, Eq. (13.1) can be written as

$$U_F(x, y) = \left[\tilde{t}\left(\frac{x}{\lambda f}, \frac{y}{\lambda f}\right) \otimes \tilde{a}\left(\frac{x}{\lambda f}, \frac{y}{\lambda f}\right)\right] \times H(x, y), \quad (13.2)$$

where $\tilde{t}(x, y)$ and $\tilde{a}(x, y)$ are the Fourier transforms of $t(x, y)$ and $a(x, y)$, respectively, and $\otimes$ denotes a convolution operation. The image of $U_F(x, y)$ through the lens doublet $L_1 - L_2$ interferes with the reference $R(x, y)$ providing the hologram. The irradiance distribution at the output plane $I_{\mathrm{E}}(x, y)$ is

$$I_{\mathrm{E}}(x, y) = \left| U_{\mathrm{F}}\left(\frac{x}{M}, \frac{y}{M}\right)\right|^2 + U_{\mathrm{F}}\left(\frac{x}{M}, \frac{y}{M}\right) R^*(u, v) + U_{\mathrm{F}}^*\left(\frac{x}{M}, \frac{y}{M}\right) R(u, v),$$
(13.3)

where M is the lateral magnification between the plane P_2 and the output plane, and the symbol $*$ denotes complex conjugation. Inasmuch as we can know the first and second terms on the right-hand side of Eq. (13.3) a priori by obtaining the power spectrum of the encrypted data and reference beam, we can get the following holographic data, $I_{\mathrm{E}}'(u, v)$:

$$I_{\mathrm{E}}'(x, y) = U_{\mathrm{F}}\left(\frac{x}{M}, \frac{y}{M}\right) R^*(x, y) + U_{\mathrm{F}}^*\left(\frac{x}{M}, \frac{y}{M}\right) R(x, y).$$
(13.4)

This irradiance distribution constitutes the encrypted information. To obtain the key, we can also obtain the holographic data of the Fourier phase mask alone, $I_{\mathrm{K}}'(u, v)$, given by

$$I_{\mathrm{K}}'(x, y) = H\left(\frac{x}{M}, \frac{y}{M}\right) R^*(x, y) + H^*\left(\frac{x}{M}, \frac{y}{M}\right) R(x, y).$$
(13.5)

Once these patterns are recorded, decryption is performed in the following way. Because the reference beam is a slightly inclined planar wave, we can extract the first term on the right-hand side of Eq. (13.4) and the second term on the right-hand side of Eq. (13.5) by Fourier-transforming the holographic data to obtain the encrypted data and the Fourier phase mask, respectively. By multiplying the extracted encrypted data and the Fourier phase mask, we obtain

$$U_d(x, y) = U_{\mathrm{F}}\left(\frac{x}{M}, \frac{y}{M}\right) R^*(x, y) \times H^*\left(\frac{x}{M}, \frac{y}{M}\right) R(x, y)$$

$$= \tilde{t}\left(\frac{x}{\lambda f M}, \frac{y}{\lambda f M}\right) \otimes \left(\frac{x}{\lambda f M}, \frac{y}{\lambda f M}\right),$$
(13.6)

where we have used that $|H(x, y)|^2$ is equal to a constant because the phase mask has only phase value. Finally, by using an inverse Fourier transformation, we can obtain the decrypted data $D(x, y)$ as

$$D(x, y) = \int\!\!\!\int_{-\infty}^{\infty} U_d(x', y') \exp\left[i\frac{2\pi}{\lambda f}(xx' + yy')\right] dx'dy' = t(Mx, My)a(Mx, My).$$
(13.7)

In Eq. (13.7), we have assumed that the inverse Fourier transformation is performed with the same scale factor as that in Eq. (13.1), as is the case if this operation is performed optically, again using the lens L_1. The intensity of Eq. (13.7) produces a scaled version of the original image because $t(x, y)$ is a positive real-valued function and $a(x, y)$ is phase only.

Fig. 13.3. Gray-level pictures of the irradiance corresponding to the digitally reconstructed input images to be encrypted.

For computer reconstruction, the inverse discrete Fourier transformation is evaluated numerically with the equation,

$$|t(m',n')|^2 = \left| \sum_{m'=0}^{N-1} \sum_{n'=0}^{N-1} U_{\mathrm{D}}(m,n) \exp\left[i\frac{2\pi}{N}(mm' + nn') \right] \right|^2, \qquad (13.8)$$

where m and n are the discrete spatial coordinates in the CCD plane and m', n' are those corresponding to the object plane. In Eq. (13.8), it is assumed that the number of pixels in both orthogonal directions of the CCD is the same, denoted by N. Equation (13.8) can be calculated through a fast Fourier-transform algorithm (FFT) [54].

In the experimental verification of this method, the input mask was a random phase mask with a correlation length of less than 10 mm, while the Fourier phase mask was a lens. A lens was used in this preliminary experiment because of the lack of a sufficient space–bandwidth product of both the optical system and the CCD camera. Lens L_1 has a numerical aperture of 0.10 mm, lenses L_2 and L_3 each have a numerical aperture of 0.14 mm, and lenses L_4 and L_5 each have a numerical aperture of 0.17 mm. The hologram was captured by a CCD camera with 512×480 pixels and quantized to 8 bits of gray levels by means of a frame-grabber board. The CCD array had dimensions 6.4×4.8 mm. Once the encrypted hologram was recorded, the input image, the input phase mask, and lens L_1 were removed to record the decrypting key. This is just the hologram of the Fourier phase mask.

Figure 13.3 shows the input images to be decrypted in the experiment constituted by two words. These images were obtained by digital off-axis holography with an input phase mask but without the Fourier phase mask. The reason for the noise in the image is the scattering produced by the thickness of the input random phase mask and the limitation on the numerical aperture of the lens L_1.

Digital holograms of the encrypted data and the Fourier phase mask are shown in Figure 13.4. The digitally reconstructed encrypted images are shown in Figure 13.5(a) and 13.5(b).

These images were obtained by inverse Fourier transforming of the digital hologram of the encrypted data in Figure 13.3(a) and 13.3(b), respectively, but without using the key. Thus, the original images cannot be recognized. However, digitally reconstructed images that have been decrypted with the key (i.e., the hologram of the Fourier phase mask), are shown in Figure 13.6.

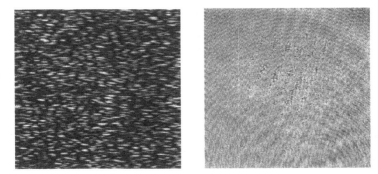

Fig. 13.4. Digital holograms of (**a**) the encrypted data and (**b**) the Fourier phase mask.

Fig. 13.5. Images that have been digitally reconstructed by: (**a**) and (**b**), applying an inverse Fourier transformation to the digital hologram of the encrypted data, thus obtaining an unsuccessful reconstruction; (**c**) and (**d**), decrypting by using the digital hologram of both the encrypted data and the Fourier phase mask.

Now the original images are clearly reconstructed. The experimental results demonstrate the feasibility of the proposed method.

13.3 Securing Information by Use of Phase-Shift Digital Holography

On-axis digital holography has also been applied to encryption. Both Fourier and Fresnel domain optical encryption have been achieved by using two random phase masks and a Mach–Zehnder phase-shift interferometer architecture [52]. In this case, one phase mask is attached to the object, which is Fourier-transformed or free-space propagated. The second phase mask is located at an arbitrary position in the reference beam that is interfered on-axis with the previous diffraction pattern. Phase shift with retarder plates allows one to obtain the phase and amplitude of the encrypted data, which can be transmitted digitally. To decrypt, information about the reference mask is obtained also by phase-shift interferometry, and decryption is performed in the computer. This technique is more efficient than that using off-axis interferometry and still has a great level of security, as it uses two random phase codes.

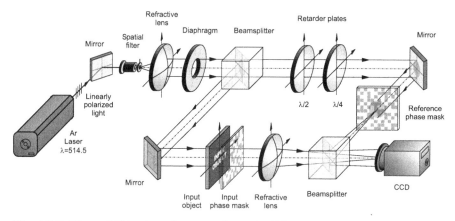

Fig. 13.6. Phase-shifting interferometer for optical Fourier encryption of 2-D images with two random phase masks.

Other digital holographic methods have been applied to encrypt 2-D input images [55,56]. For example, in the method of Lai and Neifeld [55], only one random phase mask is used. The phase code is located in the object beam of an interferometer architecture at an arbitrary distance from the input, so the phase mask modifies a Fresnel diffraction pattern of the input object and is propagated again toward the output plane. This approach is similar to the method of Matoba and Javidi but without the phase mask at the input plane [23]. The complex information associated with the encrypted image is recorded also with a CCD camera using phase-shift interferometry. In this case, phase shifting is obtained using a liquid-crystal variable retarder.

13.3.1 Fourier Encryption with On-Axis Digital Holography

The outline of an optical encryption system for encrypting Fraunhofer diffraction patterns of a 2-D input object is depicted in Figure 13.6 [52]. It is based on a Mach–Zehnder interferometer. An argon laser beam, expanded and collimated, is divided by a beam splitter into two plane wave fronts traveling in different directions. These are the object and the reference beams. After reflecting in a mirror, the object beam travels through the input transparency, whose amplitude transmittance, $t(x, y)$, contains the data to be encrypted. A refractive lens produces a representation of the Fourier transform of the input into the CCD detector through the second beam splitter. A first random phase mask, $a(x, y) = \exp[i\phi_O(x, y)]$ is attached to the input object. Therefore, aside from constant factors, the complex amplitude distribution of the Fraunhofer diffraction pattern at the output plane, $U_H(x, y)$, is given by

$$U_H(x, y) = \iint\limits_{-\infty}^{\infty} t(x', y') \exp[i\phi_O(x', y')] \exp\left[-i\frac{2\pi}{\lambda f}(xx' + yy')\right] dx'dy', \quad (13.9)$$

where f is the focal length of the refractive objective, and λ is the wavelength of the laser beam. In the preceding equation, we assume that the input is located at a distance $d = f$ from the refractive lens. The complex amplitude distribution is the Fourier transform of the product of the input transmittance modified by the input random phase mask. Thus, in principle, by measuring the amplitude and phase of the complex distribution $U_H(x, y)$, the amplitude of the input function, $t(x, y)$, may be recovered by computing the inverse Fourier-transform intensity. Note that the phase mask attached to the input not only improves security but also allows one to adapt the intensity distribution of the Fraunhofer diffraction pattern in Eq. (13.9) to the dynamic range of the detector.

The parallel reference beam passes through two phase retarders of one-quarter and one-half wave plate, is reflected by a mirror, and is modified by a second random phase mask. The system is aligned in such a way that, without the phase mask, the reference beam generates a plane wave traveling perpendicular to the CCD sensor after reflecting in the second beam splitter. The light provided by the argon laser is linearly polarized. In this way, by suitably orienting the phase retarders, the phase of the reference beam can be shifted to different values. Let us assume that the phase of the parallel beam after the second retarder plate is φ_o when the fast axis of both plates is aligned with the direction of polarization of the incident light. In this way, by aligning successively the different slow and fast axes of the phase retarders with the direction of polarization of the incident light, different phase values $\varphi_o + \alpha_p$ with $\alpha_1 = 0$, $\alpha_2 = -\pi/2$, $\alpha_3 = -\pi$, and $\alpha_4 = -3\pi/2$ can be produced.

The reference phase mask has a random phase distribution $\phi_R(x, y)$ and is placed at a distance $L = L_1 + L_2$ from the CCD detector, as shown in Figure 13.6. The complex amplitude distribution of the reference beam at the output plane can be calculated using the Fresnel–Kirchhoff integral with the expression

$$U_R(x, y; \alpha_p) = \exp(i\alpha_p)\exp\left[i\frac{\pi}{\lambda L}(x^2 + y^2)\right]$$

$$\times \int\int_{-\infty}^{\infty} \exp\left[i\phi_R(x', y')\right]\exp\left[i\frac{\pi}{\lambda L}(x'^2 + y'^2)\right]$$

$$\times \exp\left[-i\frac{2\pi}{\lambda L}(xx' + yy')\right]dx'dy', \tag{13.10}$$

where α_p denotes the relative phase changes introduced by the retarder plates on the reference beam. Note that other constant phase factors have been omitted in the preceding equation. The intensity pattern recorded by the linear intensity recording device, such a CCD camera, is then given by

$$I_P(x, y, \alpha_p) = |U_H(x, y) + U_U(x, y; \alpha_p)|^2, \tag{13.11}$$

with U_H and U_R given by Eqs. (13.9) and (13.10), respectively. Because Eq. (13.10) provides a random-noise-like phase and amplitude distribution, the image provided by the CCD will also look like a random intensity distribution.

The complex light field at the output plane is evaluated with digital phase-shifting interferometry by recording four intensity patterns with the reference beam phase shifted by the previous values of α [45,48]. Denoting the complex amplitude distribution at the output plane generated by the object and the reference beams with

$$U_H(x,y) = A_H(x,y) \exp[i\phi_H(x,y)]$$

and

$$U_R(x,y;\alpha_p) = A_R(x,y) \exp[i(\phi_R(x,y) + \alpha_p)],$$

respectively, we can rewrite Eq. (13.11) as

$$I_p(x,y) = [A_H(x,y)]^2 + [A_R(x,y)]^2$$
$$+ 2A_H(x,y)A_R(x,y) \cos[\phi_H(x,y) - \phi_R(x,y) - \alpha_p]. \quad (13.12)$$

It is straightforward to show that this phase-shifting interferometry technique provides the encrypted phase

$$\phi_E(x,y) = \phi_H(x,y) - \phi_R(x,y) = \arctan\left[\frac{I_4(x,y) - I_2(x,y)}{I_1(x,y) - I_3(x,y)}\right]. \quad (13.13)$$

The amplitude of the encrypted image is just the product of the amplitudes

$$A_E(x,y) = A_H(x,y)A_R(x,y)$$
$$= \frac{1}{4}\left\{[I_1(x,y) - I_3(x,y)]^2 + [I_4(x,y) - I_2(x,y)]^2\right\}^{1/2}. \quad (13.14)$$

Note that it is not possible to recover the complex amplitude distribution $U_H(x,y)$ generated by the object beam without knowledge of the random-like functions $\phi_R(x,y)$ and $A_R(x,y)$. Thus, it is not possible to obtain the input function, $t(x,y)$, by an inverse Fourier transformation. The input information is encrypted such that it can only be decrypted with the reference complex amplitude distribution $U_R(x,y;0)$ or using the same reference phase mask $\phi_R(x,y)$ located in the proper 3-D position, which is acting as the key.

The effect of the encryption is similar to that achieved by the original double phase-encoding method with digital holography reviewed in Section 13.2. The phase and amplitude given by Eqs. (13.13) and (13.14) can be simply understood as the phase and amplitude of the product of the Fourier complex amplitude, $U_O(x,y)$, with a second random complex amplitude distribution that is the complex conjugate of $U_R(x,y;0)$. By simply imaging the reference phase mask over the output detector instead of using free-space propagation, the encrypted image would be that achieved by the double phase method with the system in Figure 13.1. However, with this new technique, we improve

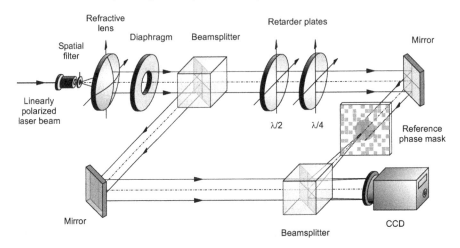

Fig. 13.7. Optical system to measure the key.

security by extending the encryption to the Fresnel domain. Note that, in this method, not only the phase but also the amplitude of the diffraction pattern of the input is modified. Besides, the three-dimensional position of the reference phase mask acts also as a key.

In order to decrypt the information obtaining the original complex amplitude distribution $U_O(x, y)$, we must also use the phase-shifting interferometry technique to achieve the key complex distribution $U_R(x, y; 0)$ (i.e., the Fresnel diffraction pattern generated by the reference phase mask at the output plane). By simply removing the input transparency, the input phase mask, and the Fourier-transforming lens in the optical system, as is shown in Figure 13.7, the phase and the amplitude of the Fresnel diffraction pattern generated by the reference random phase mask can be measured. In this case, the intensity at the output plane of the system in Figure 13.7 is given by

$$I_p'(x, y) = |A_C \exp(i\phi_C) + U_R(x, y, \alpha_p)|^2, \tag{13.15}$$

where A_C and ϕ_C are the constant amplitude and phase of the object beam at the output plane. The phase, $\phi_K(x, y)$, and amplitude, $A_K(x, y)$, provided by the phase-shifting interferometric technique are now

$$\phi_K(x, y) = \phi_C - \varphi_R(x, y) = \arctan\left[\frac{I_4'(x, y) - I_2'(x, y)}{I_1'(x, y) - I_3'(x, y)}\right] \tag{13.16}$$

and

$$A_K(x, y) = A_C A_R(x, y) = \frac{1}{4}\left\{[I_1'(x, y) - I_3'(x, y)]^2 + [I_4'(x, y) - I_2'(x, y)]^2\right\}^{1/2}. \tag{13.17}$$

Parameters ϕ_C and A_C in the previous equations are only constant factors and thus can be simply substituted by 0 and 1, respectively. Thus, Eqs. (13.16) and

(13.17) allow us to obtain directly the phase $\varphi_R(x, y)$ and amplitude $A_R(x, y)$. These functions act as the key for decryption of the information contained in Eqs. (13.13) and (13.14).

The decrypted complex distribution, $U_D(x, y)$, can be obtained digitally by simply combining Eqs. (13.13) and (13.14) with Eqs. (13.16) and (13.17). This allows us to obtain the phase

$$\phi_D(x, y) = \phi_E(x, y) - \phi_K(x, y), \tag{13.18}$$

and the amplitude

$$A_D(x, y) = \begin{cases} \dfrac{A_E(x, y)}{A_K(x, y)} & \text{if } A_K(x, y) \neq 0 \\ 0 & \text{otherwise.} \end{cases} \tag{13.19}$$

These phase and amplitude functions reproduce the phase and amplitude of the diffraction pattern $U_H(x, y)$.

By proper propagation of $U_D(x, y)$, using the Fresnel–Kirchhoff integral, it is possible to obtain the intensity distribution in any other plane of the incident light beam within the paraxial approximation. In particular, the input object intensity can be recovered by an inverse Fourier transformation. Reconstruction can be implemented optically or digitally.

As in the previous section, for direct computer reconstruction, the inverse discrete Fourier transformation

$$|t(m', n')|^2 = \left| \sum_{m'=0}^{N-1} \sum_{n'=0}^{N-1} U_D(m, n) \exp\left[i\frac{2\pi}{N}(mm' + nn') \right] \right|^2 \tag{13.20}$$

is evaluated numerically, where, again, m, n, m', and n' are discrete spatial coordinates. If we consider only the horizontal transversal direction, we have $x = m\Delta x$ and $x' = m'\Delta x'$, Δx and $\Delta x'$ being the spatial resolutions in the CCD plane and the input plane, respectively.

The optical system in Figure 13.6 was experimentally constructed using an argon laser emitting a vertical polarized light beam with $\lambda = 514.5$ nm. The retarder plates were $\lambda/2$ and $\lambda/4$ wave plates optimized for the preceding wavelength. The four-step phase shifting was performed manually by rotating the phase retarders. However, other techniques can be applied to achieve phase shifts in real time, such as with piezoelectric mirrors or liquid-crystal retarders [49]. The Fourier transform lens L in the object beam had a focal length $f = 200$ mm and a numerical aperture NA = 0.1. The input image, with the input phase mask bonded, was located at a distance f from L. The input information was encoded as a binary image in a black-and-white transparency. The reference phase mask was located in the reference beam of the interferometer at a distance $L = 300$ mm from the CCD. The different interferograms were registered by a CCD camera, sampled with 480×480 pixels, and quantized to 8 bits of gray levels using a frame grabber.

Fig. 13.8. Image of the input object for Fourier encryption obtained by digital holography with the phase-shift interferometer in Figure 13.6 without the reference phase mask.

Figure 13.8 shows a gray-level picture of the computer-reconstructed image of the input to be encrypted with the input phase mask bonded. It was obtained recording its Fraunhofer diffraction pattern by phase-shifting interferometry but without the reference phase mask. Once the phase and amplitude at the CCD plane were obtained, an inverse FFT algorithm was applied to recover the input image. The encrypted phase and amplitude distributions after locating the reference phase mask as indicated in Figure 13.6 are shown in Figures 13.9(a) and 13.9(b), respectively, as gray-level pictures. By inverse Fourier transformation via computer of the complex amplitude distribution associated to these images, without using a proper decryption key, we obtain the picture in Figure 13.9(c). This result shows that only a random-like intensity pattern is obtained when information about the key is not available.

Fig. 13.9. Result of the 2-D Fourier encryption experiment: (**a**) and (**b**) are gray-level pictures of the encrypted phase and amplitude of the input object; (**c**) shows an unsuccessful attempt at direct reconstruction from the encrypted hologram without the proper key.

In order to decrypt the previous information, the phase and amplitude functions that characterize the key were experimentally measured, also by phase-shift interferometry, using the optical system in Figure 13.7 and applying Eqs. (13.16) and (13.17). These distributions are shown in Figure 13.9(a) and 13.9(b) as gray-level pictures. By correcting the images in Figure 13.8(a) and 13.8(b) with the proper decryption key, as is stated mathematically in Eqs. (13.18) and (13.19), it is possible to reconstruct the original input image by inverse Fourier transformation in the computer. A gray-level picture of the decrypted image, almost identical to the original one, is shown in Figure 13.9(c).

13.3.2 Fresnel Encryption with On-Axis Digital Holography

The idea developed in the previous section can be modified in different ways by changing the optical setup in Figure 13.6. For example, the reference random phase mask can either be imaged or Fourier-transformed at the output plane instead of free-space propagated as commented previously. Also, the optical transformation applied to the object beam can be similarly modified. This section explains how to modify the previous system into a lensless architecture by recording a Fresnel diffraction pattern of the input image instead of the Fraunhofer diffraction pattern. This modification allows a more compact and versatile optical system and improves security. Now, in order to decrypt the information, it is also necessary to know the position of the input object along the optical axis and the wavelength of the light beam.

The optical security system based on the encryption of a Fresnel diffraction pattern of the input object is depicted in Figure 13.10 [52]. It is based also in a Mach–Zehnder interferometer. The argon laser beam, expanded and collimated, is divided by a beam splitter into two plane wave fronts. After reflecting in a mirror, the object beam is diffracted by the input transparency, with transmittance $t(x, y)$, which has the first random phase mask attached, with phase $\phi_O(x, y)$. The light diffracted by the input object and the first phase mask travels through the second beam splitter and is recorded by the CCD detector. Aside from constant factors, now the complex amplitude distribution of the Fresnel diffraction pattern obtained at the output plane, $U_H(x, y)$, is given by

$$U_H(x, y) = \exp\left[i\frac{\pi}{\lambda d}(x^2 + y^2)\right] \int\limits_{-\infty}^{\infty}\!\!\int t(x', y') \exp[i\phi_O(x', y')]$$

$$\times \exp\left[i\frac{\pi}{\lambda d}(x'^2 + y'^2)\right] \exp\left[-\frac{2\pi}{\lambda d}(xx' + yy')\right] dx'dy', \quad (13.21)$$

where d is the distance from the input object to the output plane. By measuring the amplitude and phase of the complex distribution $U_H(x, y)$, the amplitude of the input function, $t(x, y)$, may be recovered. This time, the

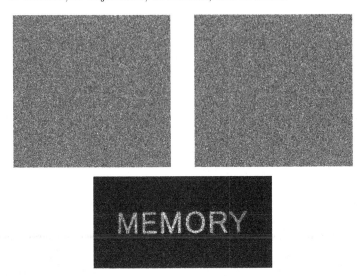

Fig. 13.10. Result of the 2-D decryption experiment after Fourier encryption: (**a**) and (**b**) are gray level pictures of the key phase and amplitude; (**c**) shows the decrypted image when the encrypted hologram in Figure 13.9(**a**) and 13.9(**b**) is corrected with the proper key.

reconstruction is performed by computing an inverse Fresnel transform operation either optically or numerically. In this technique, the first random phase mask modifies and smoothes the intensity of the Fresnel diffraction pattern of the input image recorded by the CCD.

The technique used to encrypt and record this information is similar to that used in the previous section. Phase-shifting interferometry is employed again to record the phase and amplitude of the Fresnel diffraction pattern at the output plane. A parallel reference beam is interfered on-axis with the Fresnel pattern. By simply rotating the retarder plates located in the reference beam, it is possible to obtain four interference patterns with the phase of the reference shifted in intervals of $\pi/2$. The combination of these interference patterns would provide the phase and amplitude of the diffraction pattern in Eq. (13.21). As before, in order to achieve secure encryption, a second random phase mask is located in the reference beam. The Fresnel diffraction pattern provided by this reference phase mask at the output plane is given again by Eq. (13.10). The encrypted intensity patterns are then given by Eqs. (13.11) and (13.12), where now the complex distribution $U_H(x, y)$ is that given by Eq. (13.21). In this way, Eqs. (13.13) and (13.14) again provide the encrypted phase and amplitude of the input signal. Both images will look like random distributions due to the action of the input and reference phase masks. Without information about the reference phase code, it is very difficult to recover the original input data.

The key functions are obtained also by phase-shifting interferometry, removing the input function and the input phase mask from the optical system, as is shown in Figure 13.7. After recording the corresponding interference patterns, denoted by Eq. (13.15), application of Eqs. (13.16) and (13.17) provides the phase and amplitude of the complex encryption key $U_K(x, y)$. Only by having this complex key is it possible to decrypt the information stored digitally. This is done by using Eqs. (13.18) and (13.19), which provide the decrypted complex distribution $U_D(x, y)$. Aside from constant factors, this distribution coincides with the Fresnel diffraction pattern of the input image with the phase code attached, $U_H(x, y)$. Note that, instead of digitally sending the key function $U_K(x, y)$, it is also possible to send a copy of the reference phase mask itself and information about its three-dimensional position to the authorized remote site. In this case, a second optical system similar to that in Figure 13.7 will be used to recover the digital key after transmission.

Again, the reconstruction of the original information, $t(x, y)$, can be implemented optically or by computer. In this case, we need information not only about the location of the input image in the optical path of the interferometer but also about the wavelength of the incident light and the pixel size of the CCD. Optical reconstruction requires us to codify the Fresnel diffraction pattern of the input as a computer-generated hologram. By simple free-space propagation, the intensity of the original image will be recovered from the hologram. Computer reconstruction can be performed by applying an inverse discrete Fresnel transform algorithm to simulate free-space propagation. Thus, the input information can be retrieved from the equation

$$|t(m, n)|^2 = \left| \sum_{m=0}^{N-1} \sum_{n=0}^{N-1} U_D(m', n') \exp\left[-i\pi \frac{\Delta x^2}{\lambda d} (m'^2 + n'^2) \right] \right.$$
$$\left. \times \exp\left[i\frac{2\pi}{N}(mm' + nn') \right] \right|^2, \tag{13.22}$$

where we have assumed again that the size of the pixels in both transversal directions is the same. Evaluation of the previous expression can be performed in the computer in a short time by using an FFT algorithm.

In order to verify the behavior of this Fresnel encrypting technique, the optical system in Figure 13.11 was constructed using optical elements similar to those employed in the previous section. We again used the line with wavelength $\lambda = 514.5$ nm of an argon laser to illuminate the system. The input image, with the input phase mask bonded, was located at a distance $d = 350$ mm from the output plane. The input information to be encrypted, a set of alphabet characters, was encoded again in a black-and-white transparency. The encrypting reference phase mask was located in the reference beam at a distance $L = 300$ mm from the CCD camera. The phase mask is a commercially available plastic diffuser of randomly varying thickness with a correlation length of 6 μm in both the x and y directions. Interferograms with different phase shifts, achieved by rotating the retarder plates manually,

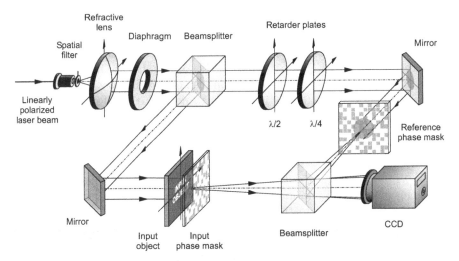

Fig. 13.11. Phase-shifting interferometer for optical Fresnel encryption of 2-D images with two random phase masks.

Fig. 13.12. Image of the input object for Fresnel encryption obtained by digital holography with the lensless phase-shift interferometer in Figure 13.11 without the reference phase mask.

were registered by the monochrome CCD camera, with 480×480 pixels, and digitized to 256 gray levels.

In Figure 13.12, we show a gray-level picture of the image to be encrypted generated by phase-shift interferometry. To obtain this image, the Fresnel diffraction pattern at the output plane was recorded with the system in Figure 13.11 but without using the reference phase mask. An inverse Fresnel transform algorithm was then applied to recover the original image in the computer. Note that the phase mask attached to the input does not prevent one from obtaining the original intensity. Scattering in the reconstructed image is produced by the limited size and dynamic range of the detector.

The previous image was encrypted by locating the second phase mask in the reference beam of the interferometer. Figures 13.13(a) and 13.13(b) show gray-level pictures of the encrypted phase and amplitude distributions, respectively. In order to simulate an attempt at reconstruction without the proper key, we apply an inverse Fresnel transformation of the complex amplitude distribution associated with these images. The result is shown in Figure

Fig. 13.13. Result of the 2-D encryption experiment with Fresnel encryption: (**a**) and (**b**) are gray level pictures of the encrypted phase and amplitude of the input object; (**c**) is an attempt at direct reconstruction from the encrypted hologram.

13.13(c). As expected, only a random-like intensity pattern is obtained and the original image cannot be recognized at all.

To recover the original information, the key phase and amplitude functions were first obtained by using the optical system in Figure 13.7 (i.e., by removing the input object and input phase mask in Figure 13.11). The key complex distribution is shown in Figures 13.14(a) and 13.14(b) as gray-level pictures representing the phase and amplitude, respectively. By decrypting the images in Figures 13.13(a) and 13.13(b) with the proper key, the Fresnel diffraction pattern of the input, with the phase mask attached, is obtained. After an inverse Fresnel transformation in the computer, it is possible to reconstruct the intensity of the original input image. The result is shown in Figure 13.14(c). Note that the decrypted image is almost identical to the original one.

13.4 Three-Dimensional Object Encryption with Digital Holography

The ideas described in the previous section have also been applied to secure information codified in a 3-D scene [53]. The method is based on the ability of digital holograms to reconstruct in the computer images of 3-D objects focused at different planes [49,50]. Optical encryption is performed by recording the interference of a Fresnel diffraction pattern of the 3-D input with a Fresnel

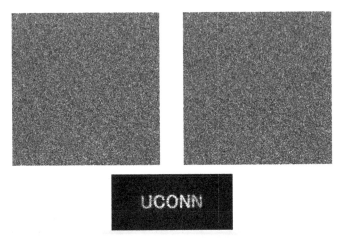

Fig. 13.14. Result of the 2D decryption experiment: (**a**) and (**b**) are gray level pictures of the phase and amplitude of the key; (**c**) shows the decrypted image when the encrypted hologram is corrected with the proper key.

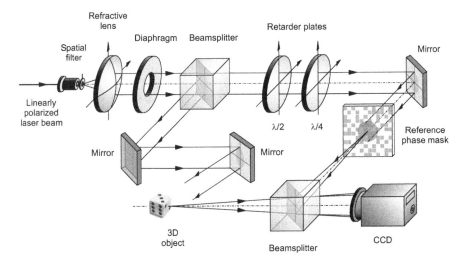

Fig. 13.15. Phase-shifting holographic system to encrypt 3-D information.

pattern generated by a random phase mask. Digital decryption in the computer is only possible with information about the reference phase mask, which can be obtained also with interferometric techniques. The reconstruction of the 3-D object can be performed optically by encoding the decrypted digital hologram on a spatial-light modulator, as has been proved recently [57,58].

The outline of the optical system to encrypt 3-D information is shown in Figure 13.15. It is based on a Mach–Zehnder interferometer architecture. The first beam splitter (BS) divides a collimated light beam from an argon laser into the reference and the object beams. The object beam illuminates the 3-D input after reflecting in a couple of mirrors (M). Assuming that the incident light is diffracted only once, we can describe the opaque object as a 3-D continuum distribution of point sources with relative amplitude $U_O(x, y, z)$, where x and y are transversal coordinates and z is the paraxial distance to the output plane. In this way, the complex amplitude distribution at the output plane, $U_H(x, y)$, can be evaluated with the 3-D integral of superposition

$$U_H(x,y) = A_H(x,y) \exp\left[i\phi_H(x,y)\right] = \frac{1}{i\lambda} \int\!\!\!\int\!\!\!\int_{-\infty}^{\infty} U_O(x',y';z)\frac{1}{z} \exp\left(i\frac{2\pi}{\lambda}z\right)$$

$$\times \exp\left\{i\frac{\pi}{\lambda z}\left[(x-x')^2 + (y-y')^2\right]\right\} dx'dy'dz, \qquad (13.23)$$

where λ is the wavelength of the incident light. In Eq. (13.19), $A_H(x, y)$ and $\phi_H(x, y)$ are the amplitude and phase, respectively, of the complex amplitude distribution at the output plane generated by the object beam. Note that in this approximation, neglecting secondary diffraction, the object can be considered also as a continuum of 2-D images at different distances z to the output plane. In the previous equation, some constant factors have been neglected.

The parallel reference beam travels through the retarder plates and is diffracted by a random phase mask. The phase retarders are again one-quarter, $\lambda/4$, and one-half, $\lambda/2$, wave plates. They modulate the phase of the reference beam with four phase-shift values α_p by aligning the fast and slow axes of the retarders. The phase mask, with a random phase distribution $\Phi_R(x, y)$, is located at an arbitrary distance $L = L_1 L_2$ from the detector and generates a complex field for each α_p, given again by Eq. (13.10).

The output intensity results from the coherent superposition of the diffraction patterns in Eqs. (13.23) and (13.10) and can be written as in Eqs. (13.11) or (13.12), where now functions $U_H(x, y)$, $A_H(x, y)$, and $\phi_H(x, y)$ are those defined in Eq. (13.23). Recording four interference patterns $I_p(x, y)$ with the reference phase shifted by the four values α_p, it is straightforward to show from Eq. (13.12) that the measured phase is given by Eqs. (13.13) and (13.14).

The amplitude $A_E(x, y)$ and phase $\phi_E(x, y)$ constitute the encrypted hologram of the 3-D object. Without knowledge of the functions $A_R(x, y)$ and $\phi_R(x, y)$, which act as keys for the decryption, it is not possible to recover images of the 3-D object by inverse Fresnel propagation. To obtain the key functions, again the optical system in Figure 13.7 is used. The reference and object parallel beams impinge on the axis into the output plane. In this way, we record four other intensity patterns $I_p'(x, y)$ that, by applying Eqs. (13.16) and (13.17), provide the phase $\phi_K(x, y) = \phi_C \Phi_R(x, y)$ and the amplitude $A_K(x, y) = A_C A_R(x, y)$ generated by the phase mask. As in the previous sections, parameters ϕ_C and A_C are the constant phase and amplitude of the object beam and can be substituted by 0 and 1, respectively. Decryption is performed by combining the encrypted hologram, determined by $A_E(x, y)$

and $\phi_E(x, y)$, with the key functions $A_K(x, y)$ and $\phi_K(x, y)$. As an alternative, instead of using expressions similar to those in Eqs. (13.18) and (13.19), decryption can be done directly from the intensity measurements using

$$\phi_D(x, y) = \arctan\left[\frac{(I_4 - I_2)(I_1' - I_3') - (I_1 - I_3)(I_4' - I_2')}{(I_4 - I_2)(I_4' - I_2') - (I_1 - I_3)(I_1' - I_3')}\right] \tag{13.24}$$

and

$$A_D(x, y) = \left[\frac{(I_1 - I_3)^2 + (I_4 - I_2)^2}{(I_1' - I_3')^2 + (I_4' - I_2')^2}\right]^{1/2}. \tag{13.25}$$

For simplicity, in Eqs. (13.24) and (13.25), the spatial dependence of $I_p(x, y)$ and $I_p'(x, y)$ has been omitted. The previous functions, $A_D(x, y)$ and $\phi_D(x, y)$, constitute the amplitude and phase of the decrypted Fresnel digital hologram, $U_D(x, y) = A_D(x, y) \exp[i\phi_D(x, y)]$, of the 3-D object. By free-space propagation of $U_D(x, y)$, the amplitude distribution of the input object can be reconstructed.

After decryption, a direct digital reconstruction of the input object can be performed by computing a Fresnel integral numerically. Let $U_D(m, n)$ be the discrete amplitude distribution of the decrypted digital hologram, where m and n are discrete coordinates along the orthogonal directions x and y, respectively. In this way, $x = m\Delta x$ and $y = n\Delta y$, Δx and Δy being the resolution of the CCD detectors. The discrete complex amplitude distribution $U_O(m', n')$ at a plane orthogonal to the digital hologram, located at a distance d, is given, aside from constant factors, by the discrete Fresnel transformation

$$U_O(m', n') = \exp\left[\frac{-i\pi}{\lambda d}(\Delta x'^2 m'^2 + \Delta y'^2 n'^2)\right] \sum_{m'=0}^{N_x - 1}\sum_{n'=0}^{N_y - 1} U_D(m, n)$$
$$\times \exp\left[\frac{-\pi}{\lambda d}(\Delta x^2 m^2 + \Delta y^2 n^2)\right] \exp\left[-i2\pi\left(\frac{m'm}{N_x} + \frac{n'n}{N_y}\right)\right]. \tag{13.26}$$

In Eq. (13.26), m' and n' are discrete coordinates at the reconstruction plane, $\Delta x'$ and $\Delta y'$ denote the resolution at this plane, and N_x and N_y are the number of pixels of the detector along the x and y axes, respectively.

It is important to note that different regions of the hologram record light arising from different perspectives of the 3-D object. Therefore, as in conventional analog holography, we can reconstruct different views of the 3-D object by using different windows in the decrypted hologram, as shown in Figure 13.16, and illuminating them with tilted plane waves. To this end, let us consider that the input object to be encrypted was located at a distance d from the CCD. To reconstruct a particular view of the decrypted 3-D object, first we must consider only a rectangular window, $U_D'(m, n; a_x, a_y)$, of the digital hologram $U_D(m, n)$ centered at the proper location. Parameters a_x and a_y

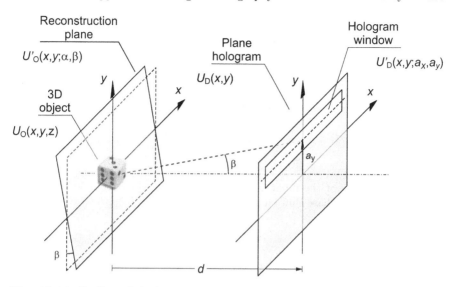

Fig. 13.16. Outline of the location of the window in the digital hologram used to generate different vertical perspectives of the 3-D object numerically.

denote the coordinates of the center of the window. In this way, the information contained in this region of the hologram corresponds to a direction of observation that subtends angles α and β with the optical axis given by

$$\alpha = \frac{a_x \Delta x}{d} \quad \text{and} \quad \beta = \frac{a_y \Delta y}{d}. \tag{13.27}$$

Second, we must consider that the hologram is illuminated by a parallel light beam directed toward the 3-D object (i.e., a light beam tilted by angles ? and ? with respect to the optical axis given by Eq. (13.27)). In this way, the perspective of the object will remain centered in the reconstruction of the hologram. The angular range achieved will be limited only by the size of the detector.

From the considerations above, the rectangular region of the hologram to be used for reconstruction must be defined as

$$U_D'(m, n; a_x, a_y) = U_D(m, n)\mathrm{rect}\left(\frac{m - a_x}{b_x}, \frac{n - a_y}{b_y}\right) \exp\left[i2\pi(a_x m + a_y n)\right],$$
$$\tag{13.28}$$

where $\mathrm{rect}(x, y)$ is the rectangle function and b_x and b_y denote its transversal size. The linear phase factor in Eq. (13.28) simulates the effect of a tilted plane wave incident on the hologram. Now, the discrete complex amplitude distribution $U_O'(m, n; \alpha, \beta)$ at a plane located at a distance d and tilted by angles α and β with respect the hologram plane can be computed using the

Fig. 13.17. Image of the input 3-D object to be encrypted obtained by digital holography with the phase-shift interferometer in Figure 13.15 without the reference phase mask.

equation

$$U'_O(m', n'; \alpha, \beta)$$
$$= \exp\left[\frac{-i\pi}{\lambda d}\left(\Delta x'^2 m'^2 + \Delta y'^2 n'^2\right)\right] \sum_{m'=0}^{N_x-1} \sum_{n'=0}^{N_y-1} U'_D\left(m, n; \frac{\alpha d}{\Delta x}, \frac{\beta d}{\Delta y}\right)$$
$$\times \exp\left[\frac{-i\pi}{\lambda d}\left(\Delta x^2 m^2 + \Delta y^2 n^2\right)\right] \exp\left[-i2\pi\left(\frac{m'm}{N_x}, \frac{n'n}{N_y}\right)\right]. \quad (13.29)$$

Note that the introduction of the linear phase factor in Eq. (13.28) is equivalent to a tilted plane wave illuminating the hologram in Eq. (13.29).

As in previous sections, the operation in Eq. (13.29) can be efficiently computed using a fast Fourier-transform algorithm. Thus, different perspectives can be generated at high speed. Points on the surface of the object at distances z from the hologram different from d will appear defocused in the reconstructed image. Nevertheless, the planes of reconstruction can also be changed easily in the computer starting from the same digital hologram. Note also that the field of focus can be increased by diminishing the size of the hologram window at the expense of a reduction of resolution.

In the experimental verification, we performed the encryption of a 3-D object with the phase-shifting interferometer depicted in Figure 13.1. A megapixel camera with pixel size equal to 9×9 μm was used to record the interference patterns between the object and reference beams with a dynamic range equivalent to 256 gray levels. The object has the shape of a cube with 10 mm lateral size. In Figure 13.17, we show the axial view of the object generated from a nonencrypted digital hologram using Eq. (13.26). The object was located at a distance $d = 570$ mm from the CCD, and no phase mask was used at the reference beam.

In Figure 13.18, we show the amplitude and phase of the encrypted hologram when the random phase mask was located as shown in Figure 13.15. The phase mask is again a plastic diffuser of randomly varying thickness with

Fig. 13.18. Gray-level representation of: (**a**) the encrypted amplitude and (**b**) the encrypted phase of the 3-D object in Figure 13.17 by using a random reference phase mask.

Fig. 13.19. (**a**) Result of decryption of the encrypted information contained in Figure 13.18 by using the proper key; (**b**) incorrect decryption using a wrong phase key.

a correlation length of 6 μm in both the x and y directions. The amplitude and phase key generated by the random phase mask at the output plane, obtained by using the optical system in Figure 13.7, are similar to those already shown in Figures 13.14(a) and 13.14(b). The decrypted image shown in Figure 13.19(a) was generated with Eq. (13.26) using the entire decrypted hologram. For comparison, in Figure 13.19(b) we show a decryption performed without the correct key. Noise in the decrypted image arises from errors in the evaluation of the Fresnel diffraction pattern generated by the phase mask in the CCD. These errors are due to the limited size of the detector pixels and the gray-level quantization.

Finally, in Figure 13.20 we show different perspectives of the decrypted 3-D object reconstructed from the decrypted digital Fresnel hologram. We used a rectangular window to limit the hologram to a vertical size of 256 pixels and horizontal size equal to the size of the CCD. The size reduction is necessary to reconstruct different perspectives from different regions of the hologram. But this also reduces the resolution and, consequently, the quality of the images with respect to that obtained using the whole digital hologram. Thus, we only reduce the size of the window in the direction for which we want to achieve different views of the 3-D object. In this case, the window was displaced only in the vertical direction, generating an angular difference of 0.7° between the different perspectives labeled (a), (b), and (c) in Figure 13.20. Any other view angle, limited only by the size of the CCD, can be reconstructed in the same way.

13.5 Conclusions

We have described several optoelectronic techniques based on digital phase-shifting holography for security applications. Information codified as both 2-D and 3-D images is optically processed and captured electronically. These techniques combine the high space-bandwidth product and speed of optical processing with the advantages of electronic transmission, storage, and post-processing. Optical experiments have been shown to illustrate the proposed methods.

In the first approach, encryption of 2-D images into a random complex pattern is performed using two random phase masks in an off-axis interferometer architecture. Electronic decryption can be achieved using a one-step FFT reconstruction procedure but only having the proper key. Alternatively, decryption can be performed optically by encoding the digital hologram as a computer-generated one and displaying it on a spatial light modulator.

Second, we have described the process of encryption of 2-D images into a random complex pattern also by using two random phase codes but in a phase-shifting interferometer architecture. Both Fraunhofer and Fresnel diffraction patterns of the input are efficiently encrypted. The complex key to be used in the decryption process is also obtained by phase-shift interferometry. In this case, the 3-D position of the reference random phase mask also acts as a key. This method allows us to use more efficiently the reduced resolution of CCD cameras.

Fig. 13.20. Different perspectives of the 3-D object obtained after decryption. Angles of view are: **(a)** $\beta = 0.7°$, **(b)** $\beta = 0°$, and **(c)** $\beta = -0.7°$.

A similar technique has been described to encrypt information encoded as a 3-D image. Once decrypted, views of the 3-D object can be reconstructed at different axial distances and with different perspectives in the computer. Alternatively, the reconstruction of the 3-D object can be performed optically by encoding the decrypted digital hologram on a spatial-light modulator.

Acknowledgments

E. Tajahuerce acknowledges partial financial support by the agreement between the Universitat Jaume I and the Fundació Caixa Castelló-Bancaixa (Ref. P1-1B2002-29), Spain.

References

[1]   B. Javidi and J.L. Horner, "Optical pattern recognition for validation and security verification," *Opt. Eng.* **33**, 1752–1756 (1994).

[2]   P. Réfrégier and B. Javidi, "Optical image encryption based on input plane and Fourier plane random encoding," *Opt. Lett.* **20**, 767–769 (1995).

[3]   F. Goudail, F. Bollaro, B. Javidi, and P. Réfrégier, "Influence of a perturbation in a double phase-encoding system," *J. Opt. Soc. Am A* **15**, 2629–2638 (1998).

[4]   H.-Y. Li, Y. Qiao, and D. Psaltis, "Optical network for real-time face recognition," *Appl. Opt.* **32**, 5026–5035 (1993).

[5]   C.L. Wilson, C.I. Watson, and E.G. Paek, "Combined optical and neural network fingerprint matching," *Proc. SPIE* **3073**, 373–382 (1997).

[6]   A. Pu, R. Denkewalter, and D. Psaltis, "Real-time vehicle navigation using a holographic memory," *Opt. Eng.* **36**, 2737–2746 (1997).

[7]   P. Lalanne, H. Richard, J.C. Rodier, P. Chavel, J. Taboury, K. Madani, P. Garda, and F. Devos, "2D generation of random numbers by multimode fiber speckle for silicon arrays of processing elements," *Opt. Commun.* **76**, 387–394 (1990).

[8]   N. Yoshikawa, M. Itoh, and T. Yatagai, "Binary computer-generated holograms for security applications from a synthetic double-exposure method by electron-beam lithography," *Opt. Lett.* **23**, 1483–1485 (1998).

[9]   J.F. Heanue, M.C. Bashaw, and L. Hesselink, "Encrypted holographic data storage based on orthogonal-phase-code multiplexing," *Appl. Opt.* **34**, 6012–6015 (1995).

[10] R.L. van Renesse, *Optical Document Security* (Artech House, Boston, 1998).

[11] J.L. Horner and B. Javidi, eds., *Optical Engineering, Special Issue on Optical Security*, Vol. 38 (SPIE, Bellingham, WA, 1999).

[12] B. Javidi and T. Nomura, "Polarization encoding for optical security systems," *Opt. Eng.* **39**, 2439–2443 (2000).

[13] B. Hennelly and J.T. Sheridan, "Optical image encryption by random shifting in fractional Fourier domains," *Opt. Lett.* **28**, 269–271 (2003).

[14] T.F. Krile, M.O. Hagler, W.D. Redus, and J.F. Walkup, "Multiplex holography with chirp-modulated binary phase-coded reference-beam masks," *Appl. Opt.* **18**, 52–56 (1979).

[15] J.E. Ford, Y. Fainman, and S.H. Lee, "Array interconnection by phase-coded optical correlation," *Opt. Lett.* **15**, 1088–1090 (1990).

[16] C. Denz, G. Pauliat, G. Roosen, and T. Tschudi, "Volume hologram multiplexing using a deterministic phase encoding method," *Opt. Commun.* **85**, 171–176 (1991).

[17] H. Lee and S.K. Jin, "Experimental study of volume holographic interconnects using random patterns," *Appl. Phys. Lett.* **62**, 2191–2193 (1993).

[18] R.K. Wang, I.A. Watson, and C.R. Chatwin, "Random phase encoding for optical security," *Opt. Eng.* **35**, 2464–2469 (1996).

[19] Y.H. Kang, K.H. Kim, and B. Lee, "Volume hologram scheme using optical fiber for spatial multiplexing," *Opt. Lett.* **22**, 739–741 (1997).

[20] B. Javidi, A. Sergent, G. Zhang, and L. Guibert, "Fault tolerance properties of a double phase encoding encryption technique," *Opt. Eng.* **36**, 992–998 (1997).

[21] F. Goudail, F. Bollaro, B. Javidi, and P. Réfrégier, "Influence of a perturbation in a double phase-encoding system," *J. Opt. Soc. Am A* **15**, 2629–2638 (1998).

[22] B. Javidi and E. Ahouzi, "Optical security system with Fourier plane encoding," *Appl. Opt.* **37**, 6247–6255 (1998).

[23] O. Matoba and B. Javidi, "Encrypted optical memory system using three-dimensional keys in the Fresnel domain," *Opt. Lett.* **24**, 762–764 (1999).

[24] O. Matoba and B. Javidi, "Encrypted optical storage with angular multiplexing," *Appl. Opt.* **38**, 7288–7293 (1999).

[25] P.C. Mogensen and J. Glückstad, "Phase-only optical encryption," *Opt. Lett.* **25**, 566–568 (2000).

[26] G. Unnikrishnan, J. Joseph, and K. Singh, "Optical encryption by double-random phase encoding in the fractional Fourier domain," *Opt. Lett.* **25**, 887–889 (2000).

[27] Z. Zalevsky, D. Mendlovic, U. Levy, and G. Shabtay, "A new optical random coding technique for security systems," *Opt. Commun.* **180**, 15–20 (2000).

[28] T. Nomura and B. Javidi, "Optical encryption using a joint transform correlator architecture," *Opt. Eng.* **39**, 2031–2035 (2000).

[29] B. Zhu, S. Liu, and Q. Ran, "Optical image encryption based on multifractional Fourier transforms," *Opt. Lett.* **25**, 1159–1161 (2000).

[30] E. Tajahuerce, J. Lancis, B. Javidi, and P. Andres, "Optical security and encryption with totally incoherent light," *Opt. Lett.* **26**, 678–680 (2001).

[31] S. Liu, Q. Mi, and B. Zhu, "Optical image encryption with multistage and multichannel fractional Fourier-domain filtering," *Opt. Lett.* **26**, 1242–1244 (2001).

[32] X. Tan, O. Matoba, T. Shimura, and K. Kuroda, "Improvement in holographic storage capacity by use of double-random phase encryption," *Appl. Opt.* **40**, 4721–4727 (2001).

[33] H.T. Chang, W.C. Lu, and C.J. Kuo, "Multiple-phase retrieval for optical security systems by use of random-phase encoding," *Appl. Opt.* **41**, 4825–4834 (2002).

[34] N.K. Nishchal, J. Joseph, and K. Singh, "Optical phase encryption by phase contrast using electrically addressed spatial light modulator," *Opt. Commun.* **217**, 117–122 (2003).

[35] T. Nomura, S. Mikan, Y. Morimoto, and B. Javidi, "Secure optical data storage with random phase key codes by use of a configuration of a joint transform correlator," *Appl. Opt.* **42**, 1508–1514 (2003).

[36] H.J. Caulfield, ed., *Handbook of Optical Holography* (Academic, London, 1979).

[37] L. Onural and P.D. Scott, "Digital decoding of in-line holograms," *Opt. Eng.* **26**, 1124–1132 (1987).

[38] U. Schnars, "Direct phase determination in hologram interferometry with use of digitally recorded holograms," *J. Opt. Soc. Am A* **11**, 2011–2015 (1994).

[39] U. Schnars and W.P.O. Jüptner, "Direct recording of holograms by a CCD target and numerical reconstruction," *Appl. Opt.* **33**, 179–181 (1994).

[40] G. Pedrini, Y.L. Zou, and H.J. Tiziani, "Digital double-pulsed holographic interferometry for vibration analysis," *J. Mod. Opt.* **40**, 367–374 (1995).

[41] Y. Takaki, H. Kawai, and H. Ohzu, "Hybrid holographic microscopy free of conjugate and zero-order images," *Appl. Opt.* **38**, 4990–4996 (1999).

[42] J.C. Marron and K.S. Schroeder, "Three-dimensional lensless imaging using laser frequency diversity," *Appl. Opt.* **31**, 255–262 (1992).

[43] U. Schnars, T.M. Kreis, and W.P.O. Jüptner, "Digital recording and numerical reconstruction of holograms: reduction of the spatial frequency spectrum," *Opt. Eng.* **35**, 977–982 (1996).

[44] E. Cuche, F. Bevilacqua, and C. Depeursinge, "Digital holography for quantitative phase-contrast imaging," *Opt. Lett.* **24**, 291–293 (1999).

[45] J.H. Bruning, D.R. Herriott, J.E. Gallagher, D.P. Rosenfeld, A.D. White, and D.J. Brangaccio, "Digital wavefront measuring interferometer for testing optical surfaces and lenses," *Appl. Opt.* **13**, 2693–2703 (1974).

[46] K. Creath, "Phase-measurement interferometry techniques," in E. Wolf, ed., *Progress in Optics*, Vol. 26, pp. 349–393 (North-Holland, Amsterdam, 1988).

[47] J. Schwider, "Advanced evaluation techniques in interferometry," in E. Wolf, ed., *Progress in Optics*, Vol. 28, pp. 271–359 (North-Holland, Amsterdam, 1990).

[48] I. Yamaguchi and T. Zhang, "Phase-shifting digital holography," *Opt. Lett.* **22**, 1268–1270 (1997).

[49] T. Zhang and I. Yamaguchi, "Three-dimensional microscopy with phase-shifting digital holography," *Opt. Lett.* **23**, 1221–1223 (1998).

[50] I. Yamaguchi, J. Kato, S. Ohta, and J. Mizuno, "Image formation in phase-shifting digital holography and applications to microscopy," *Appl. Opt.* **40**, 6177–6186 (2001).

[51] B. Javidi and T. Nomura, "Securing information by means of digital holography," *Opt. Lett.* **25**, 29–30 (2000).

[52] E. Tajahuerce, O. Matoba, S.C. Verrall, and B. Javidi, "Optoelectronic information encryption using phase-shifting interferometry," *Appl. Opt.* **39**, 2313–2320 (2000).

[53] E. Tajahuerce and B. Javidi, "Encrypting three-dimensional information with digital holography," *Appl. Opt.* **39**, 6595–6601 (2000).

[54] J.W. Cooley and J.W. Tukey, "An algorithm for the machine calculation of complex Fourier series," *Math. Comput.* **19**, 297–301 (1965).

[55] S. Lai and M.A. Neifeld, "Digital wavefront reconstruction and its application to image encryption," *Opt. Commun.* **178**, 283–289 (2000).

[56] R. Arizaga, R. Henao, and R. Torroba, "Fully digital encryption technique," *Opt. Commun.* **221**, 43–47 (2003).

[57] O. Matoba, T.J. Naughton, Y. Frauel, N. Bertaux, and B. Javidi, "Real-time three-dimensional object reconstruction by use of a phase-encoded digital hologram," *Appl. Opt.* **41**, 6187–6192 (2002).

[58] O. Matoba and B. Javidi, "Optical retrieval of encrypted digital holograms for secure real-time display," *Opt. Lett.* **27**, 321–323 (2002).

14

Gait-Based Human Identification Using Appearance Matching

A. Kale, N. Cuntoor, B. Yegnanarayana, A.N. Rajagopalan, and R. Chellappa

Summary. In this chapter, we present an appearance-based approach for recognizing human gait. Given the gait video of an individual, the images are binarized and the width of the outer contour of the silhouette of that individual is obtained for each image frame. Several gait features are derived from this basic width vector. Temporally ordered sequences of the feature vectors are then used to represent the gait of a person. While matching the feature templates for recognition, dynamic time-warping (DTW), which is a nonlinear time-normalization technique, is used to deal with naturally occurring changes in the walking speeds of individuals. The performance of the proposed method is tested on indoor as well as outdoor gait databases, and the efficacy of different gait features and their noise resilience is studied. The experiments also demonstrate the effect of change in the viewing angle and frame rate of data capture on the accuracy of gait recognition.

14.1 Introduction

Gait refers to the style of walking of an individual. Often in surveillance applications it is difficult to get face or iris information at high enough resolution for recognition. Studies in psychophysics [1] indicate that humans have the capability of recognizing people from even impoverished displays of gait, indicating the presence of identity information in gait. From early medical studies [2, 3], it appears that there are 24 different components to human gait and that, if all the measurements are considered, gait is unique. It is interesting, therefore, to study the utility of gait as a biometric.

Approaches to gait recognition can be broadly classified as being model-based or model-free. Examples of the first kind include the work of Cunado et al. [4], who extract gait signature by fitting the movement of the thighs to an articulated pendulum-like motion model. The idea is somewhat similar to an early work by Murray [2], who modeled the hip rotation angle as a simple pendulum, the motion of which was approximately described using simple harmonics. In [5], two sets of activity-specific static and stride parameters are extracted for different individuals. The expected confusion for each set is computed to guide the choice of parameters under different imaging conditions (that is, indoor vs. outdoor, side view vs. angular view, etc.). The set of stride parameters (four), which is smaller than the set of static parameters (two), is found to exhibit greater resilience to viewing direction. In a recent work, Lee and Grimson [6] fit ellipses to different parts of the binarized silhouette of a person. The parameters of these ellipses (such as the location of the centroid, eccentricity, etc.) are used as features to represent the gait of that person.

Examples of the model-free approach to gait include the work of Little and Boyd [7], who extract frequency and phase features from moments of the motion image derived from optical flow and use template matching to recognize different people by their gait. Huang et al. [8] also use optical flow to derive the motion-image sequence corresponding to a gait cycle. Principal components analysis (PCA) is then carried out to derive what are called eigengaits for recognition. Cutler et al. [9] use image self-similarity as a gait feature. PCA is used to reduce dimensionality, and the K nearest-neighbor rule is applied for classification.

A gait cycle corresponds to one complete cycle from rest (standing) position to right foot forward to rest to left foot forward to rest position. The movements within a cycle consist of the motion of the different parts of the body such as head, hands, legs, and so forth. The characteristics of an individual are reflected not only in the dynamics and periodicity of a gait cycle but also in the size and shape of that individual. Given the video of an unknown individual, we wish to use gait as a cue to find who among the N individuals in the database the person is. For a normal walk, gait sequences are repetitive and exhibit nearly periodic behavior. As gait databases continue to grow in size, it is conceivable that identifying a person only by gait may be difficult. However, gait can still serve as a useful indexing tool that allows us to narrow the search down to a considerably smaller set of potential candidates.

In a pattern-classification problem, choice of the feature as well as the classifier is important. We choose the width of the outer contour of the silhouette of a person as the basic image feature since it contains structural as well as dynamical information on the person's gait. As will be shown, the outer contour contains sufficient information for recognizing gait. From the raw width vector, different low-dimensional features are derived. These include the smoothed and down-sampled width vector, the eigensmoothed width vector, and the velocity feature vector. Temporally ordered sequences of these feature vectors are used for compactly representing the person's gait. Typically, 5–10 contiguous half-cycles of gait data per subject may be available, and the number of frames per cycle ranges from 8 to 20. The amount of training data is inadequate for adopting statistical model-based approaches such as the Markov model, as it may not be possible to reliably estimate the parameters of the model. Hence, template matching is adopted for comparing the probe and the reference sequences of the temporally ordered feature vectors. Typically, gait cycles when taken at different times tend to be unequal in length due to changes in walking speeds of the individuals. Hence, a classifier based on direct template matching is not appropriate. To deal with this issue, dynamic time warping (DTW) is employed for matching gait sequences. The DTW concept has been used quite successfully by the speech community for text-dependent speaker recognition/verification [10]. DTW uses an optimum time expansion/compression function for producing nonlinear time normalization so that it is able to deal with misalignments and unequal sequence lengths of the probe and the reference gait sequences. Importantly, DTW can be used even with limited training data.

The performance of our approach is tested on four standard gait databases; namely, the University of Maryland (UMD) database, the MIT database, the Carnegie Mellon University (CMU) database, and the University of Maryland angular-walk database (UMD3). These databases contain video sequences of individuals walking in a normal manner along certain predefined paths. The UMD and CMU databases have both frontal and side views, the MIT database has only side-view sequences, and the UMD3 database is not restricted to side or frontal views. The proposed DTW-based approach is equally applicable to side as well as frontal views, and the results are analyzed to determine the factors that contribute to accuracy of gait recognition. The idea is to study the efficacy of the features derived from the basic width vector and their resilience to noise in the estimate of the width vector. Our experiments also reveal the effect of differences in walking speeds and the relevance of different body parts for gait recognition. Low frame rate during data capture and methods to mitigate its effect are discussed. The effect of changes in viewing angle on gait recognition is examined.

The organization of the chapter is as follows. In Section 14.2, the basic width vector, its extraction, and its relevance for the gait problem are explained. Section 14.3 discusses different low-dimensional features derivable from the basic width vector for gait representation. In Section 14.4, we describe how gait sequences can be matched using dynamic time warping. Section 14.5 briefly reviews the gait databases used in our studies. Experimental results are discussed in Section 14.6. Conclusions and suggestions for further studies are given in Section 14.7.

14.2 The Width Vector

An important issue in gait-based human identification is the extraction of salient features that will effectively capture gait characteristics. In order to be insensitive of the color and texture of clothing and illumination, it is reasonable to consider the binarized silhouettes of a subject. Given the image sequence of a subject, background subtraction [11] is applied to the image sequence. The resultant motion image is binarized to get the silhouettes (see Figure 14.1). A bounding box is placed around the part of the motion image that contains the outer contour of the moving person. The width along each row of the silhouette, computed as the difference in the locations of the

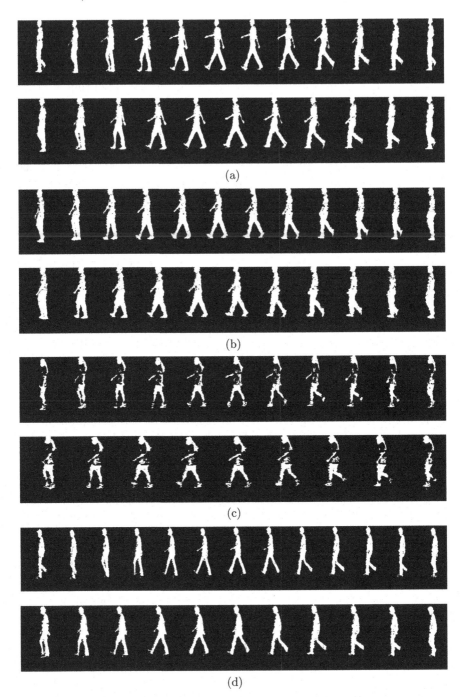

Fig. 14.1. Silhouettes for a full gait-cycle corresponding to (**a**) Person 1 (**b**) Person 2 (**c**) Person 3 (**d**) Person 4.

rightmost and the leftmost boundary pixel in that row, is stored. The physical structure of the subject as well as the swing of the limbs and other details of the body are retained in the width vector thus derived, but the pose information is lost. For the four individuals shown in Figure 14.1, an overlay of the width vectors derived from the silhouettes is given in Figure 14.2. The width-vector plots clearly bring out the fact that there is relatively more swing in the middle region (corresponding to hands) and in the end region (corresponding to feet) as compared to other parts of the body. However, these plots do not depict the temporal information. To bring out the temporal effects explicitly, the width vectors are plotted as a sequence of gray-level patterns in Figure 14.3 for the same four individuals. In this figure, about five full gait cycles are shown. The vertical axis corresponds to the index of the width vector, while the horizontal axis corresponds to the frame number. The total number of frames for each plot was fixed at 110 for uniformity. We call this the *temporal* plot of the width vector. It is clear that the width vector is approximately periodic and gives the extent of movement of different parts of the body. It varies with time in a quasiperiodic fashion, with the time taken for the completion of a half-cycle treated as the period.

Every component of the width vector contributes to the gait signature of a subject. The brighter a pixel, the larger is the value of the width vector in that position. The top part in each of the plots corresponds to the head region, the middle part corresponds to the torso, and the bottom part corresponds to the foot region of the individual. Note that the intensity variations in the torso and leg region (corresponding to the hand and leg swings) are larger than in the head region. The extent of the dark region near the top and the bottom of the temporal-width plots reflects the height differences among the individuals.

A comparison of the temporal-width plots across individuals reveals interesting gait information. The structural or static differences among people is clearly revealed by the extent of the nonzero portions in these plots. One can also decipher dynamic information. For instance, in Figure 14.3, the hand motion in the case of Person 1 is more pronounced as compared with Person 2. Note that the intensities are brighter for Person 1 in the middle region compared with Person 2. There also exist visible differences in the brightness gradients or velocities for different people (i.e., the velocity profile for each individual is different). There are differences in the repetition frequency that may be useful to distinguish people at a gross level. For instance, Person 3 has 12 half-gait cycles for the same number of frames compared with 10 half-gait cycles for Person 1 and 10 half-gait cycles for Person 2. For most female walkers, the frequency was found to be higher compared with their male counterparts. Thus, considering a gait (half)-cycle as a unit of observation, we can capture both spatial and temporal characteristics of an individual's gait signature over several cycles.

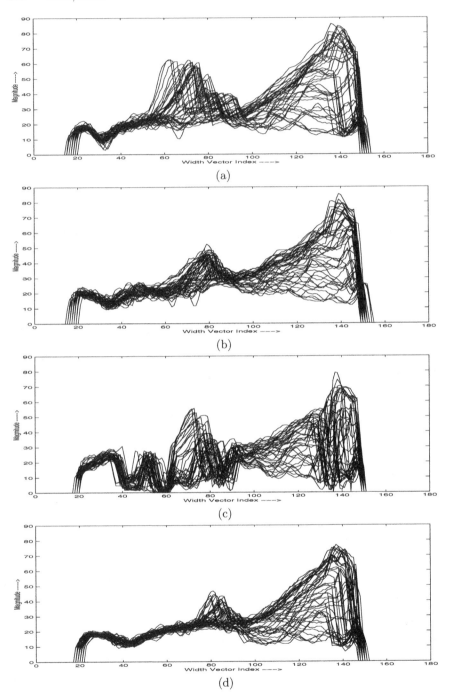

Fig. 14.2. Width overlay plots for (**a**) Person 1 (**b**) Person 2 (**c**) Person 3 (**d**) Person 4.

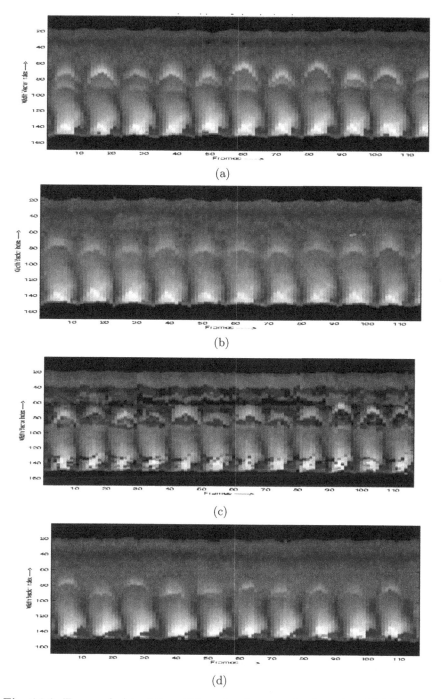

Fig. 14.3. Temporal plot of the width vector for several gait cycles of (**a**) Person 1 (**b**) Person 2 (**c**) Person 3 (**d**) Person 4.

14.3 Gait Representation

As discussed, the width vector captures important characteristics about an individual's gait. However, the dimension of the raw width vector can be quite high (as large as 640). In this section, we enumerate different low-dimensional but effective features that can be derived from the basic width vector. The idea is to arrive at a compact representation that exploits redundancy in the gait data for dimensionality reduction. Features that are obtained directly from the basic width vector are called direct features, and these include the smoothed and down-sampled versions of the width vector. On the other hand, features derived using an eigenanalysis of the width vector across frames are called eigenfeatures. Corresponding to every frame of the gait cycle of an individual, one can extract any of the feature vectors discussed here. The gait of the individual is then represented by temporally ordering the feature vectors in accordance with the image frames in the gait cycle of that individual. We now describe each of the features in detail.

14.3.1 Direct Features

Examples of direct features are shown in Figure 14.6.

1. *The raw width vector.* This is the basic width vector discussed in Section 14.2 and is derived directly from the video sequence.
2. *3-point smoothed width vector.* This is derived from the raw width vector by smoothing it with a 3-point mean filter.
3. *5-point, 11-point, and 21-point smoothing and down-sampling by a factor of 4, 8, and 16, respectively.* The motivation behind smoothing and down-sampling stems from the fact that the original width vector has redundancies. Hence, it should be possible to discriminate reasonably well using lower-dimensional features. Also, the original raw width vector is derived by simple and quick preprocessing. Hence, it is usually noisy, and smoothing helps to mitigate the effect of noise.
4. *Difference feature vector.* This is useful to study the effect of dynamics for gait-based human identification. There are many different ways to extract dynamic information from the width vector. One would be simply to take the difference of successive raw width vectors across frames, thus preserving only the changes that occur between frames, during a gait cycle followed by smoothing and down-sampling. There is a trade-off between smoothing and extracting dynamic information. Although some degree of smoothing is required to counter noise, too much smoothing can alter the dynamic information. Obviously, most of the structural information, such as girth of the person, is lost when we go to the velocity domain. It is to be expected that neither dynamic nor structural information, in isolation, will be sufficient to capture gait. As will be shown later, both are necessary and cannot be decoupled.

14.3.2 Eigenfeatures

From the temporal-width plots, we note that, although the width vector changes with time within a gait cycle, there is a high degree of correlation among the width vectors across frames. Most changes occur in the hand and in the leg regions. The rest of the body parts do not undergo any significant changes during a gait cycle. Hence, it is reasonable to expect that gait information in the width vector can be derived with much fewer coefficients. This is what we attempt to do in the eigenanalysis of gait.

All the width vectors corresponding to several training gait cycles of an individual are used to construct the covariance or scatter matrix for that individual. Principal components analysis is then performed to derive the eigenvectors for this data.

Given the width vectors $\{W(1), \cdots, W(N)\}$, for N frames $W(.) \in R^M$, we compute the eigenvectors $\{V(1), \cdots, V(M)\}$ corresponding to the eigenvalues of the scatter matrix arranged in descending order and reconstruct the corresponding width vectors using $(m < M)$ most significant eigenvectors as

$$W_r(i) = \left(\sum_{j=1}^{m} w_j V(j) \right) + \overline{W},$$

where $w_j = \langle W(i), V(j) \rangle$ and $\overline{W} = (W(1) + \cdots + W(N)/N)$.

Note that every individual has his/her own eigengait space. Reconstruction of the width vector using only the first and second eigenvectors is shown in Figure 14.5(f). The consequences of approximating the width vector using only the first two eigenvectors are that the effect of noise is suppressed and the width vector is smooth. The first two eigenvectors capture the physical structure of the person as well as the typical arm and leg swings and serve as a freeze-frame representation of the gait sequence.

14.4 Matching Gait Sequences using DTW

Since only limited training data are usually available for gait analysis, a template-matching approach is adopted in this chapter for recognition. The input to the gait-recognition algorithm is a sequence of image frames. Each image frame can be compactly represented by any of the feature vectors discussed in Section 14.3 to derive a template that consists of a temporally ordered sequence of feature vectors for representing gait. The number of frames in the reference gait sequence and the probe gait sequence depends on the number of gait cycles available. The larger the number of gait cycles used for matching, the better we can expect the performance to be. Typically, the number of frames in the reference and probe data will differ. Moreover, the reference and probe gait data are seldom synchronized. Therefore, direct matching of a

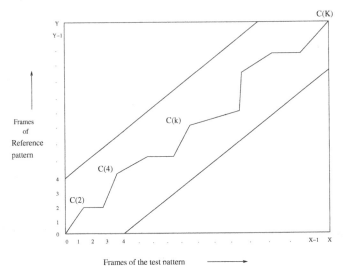

Fig. 14.4. A typical dynamic time-warping path.

probe gait sequence with a reference gait sequence should be avoided. In this work, a pattern-matching method based on a dynamic programming paradigm is used for dealing with this situation. Dynamic time warping (DTW) [10] can be used to compensate for the variability in the speed of walking which in turn reflects the number of frames for each gait cycle. A distance metric (usually the Euclidean distance) is computed between the two feature sets representing the gait data. A decision function is arrived at by integrating the metric over time. The DTW method was originally developed for isolated word recognition [12] and later adapted for text-dependent speaker verification [10]. The gait problem is analogous to text-based speaker identification/verification wherein different individuals utter the same text and differ only in the characteristics of their utterance [10].

Consider the x–y plane shown in Figure 14.4, where the x and the y axes represent the frame numbers of the probe and the reference patterns, respectively. Assume that the first frame of the reference and probe sequence are both indexed as 1. Let the last frames of the reference and probe sequences be indexed as X and Y, respectively. The match between the two sets can be represented by a sequence of K points $C(1), C(2), \ldots, C(k), \ldots, C(K)$, where $C(k) = (x(k), y(k))$, $x(k)$ is a frame index of the probe sequence, and $y(k)$ is a frame index of the reference sequence. Here, $C(k)$ represents mapping of the time axis of the probe sequence onto that of the reference sequence. The sequence

$$F = C(1), C(2), \ldots, C(k), \ldots, C(K)$$

is called the warping path.

The process of time normalization uses certain constraints depending on the problem at hand. Some of the constraints are the following [13].

1. *Endpoint constraints.* The fixed end points of the patterns lead to a set of constraints on the warping function. They are of the form

$$x(1) = 1, \ y(1) = 1, \ x(k) = X, \ \text{and} \ y(k) = Y \qquad (14.1)$$

(i.e., the first and the last frames of the probe sequence should be matched with the first and the last frames of the reference, respectively). Under these constraints, a typical warping would look like the one shown in Figure 14.4.

2. *Monotonicity constraint.* This constraint requires feature vectors belonging to a probe sequence (or the reference sequence) to be matched in monotonically increasing order. This maintains the temporal order of the sequence during time normalization. This is accomplished by ensuring that

$$x(k-1) \leq x(k) \ \text{and} \ y(k-1) \leq y(k). \qquad (14.2)$$

3. *Local continuity constraints.* This constraint can be used to ensure use of each probe frame during normalization and also to ensure that no more than one frame is skipped in the reference. This is achieved by setting the constraints

$$x(k) - x(k-1) = 1 \ \text{and} \ y(k) - y(k-1) \leq 2. \qquad (14.3)$$

4. *Global path constraints.* This constraint restrains the extent of compression or expansion. The warping path can be constrained by defining a region around which the warping path is allowed to traverse. The region between the two parallel lines marked in the x–y plane in Figure 14.6 defines the allowable grid or the region through which the warping path can traverse. The warping path can be limited to a band around the diagonal so that the search time for the optimal path is significantly reduced. This is reasonable, as it is unlikely that the probe sequence of the genuine case would deviate significantly from its reference.

After deciding the local and global constraints, the DTW algorithm is applied as follows.

1. *Local distance computation.* This involves computing the distance between the feature vectors representing the probe and reference frames. The distance between the probe and reference feature vectors $\mathbf{t_x}$ and $\mathbf{r_y}$ can be computed using

$$d(C(k)) = |\mathbf{t_x} - \mathbf{r_y}|$$

so that the total distance along the path F is given by

$$E(F) = \sum_{k=1}^{K} d(C(k)) \qquad (14.4)$$

The distance $E(F)$ is the similarity score, which can be used at the decision-making stage.

2. *Cumulative distance computation.* The cumulative distance $D(x(k), y(k))$ is the minimum distance to reach $(x(k), y(k))$ at the kth stage starting from $(1, 1)$. This distance is the sum of all the local distances of points through which the warping path passes to reach $(x(k), y(k))$ from $(1, 1)$. Initialize the cumulative distance $D(1, 1)$ to $L(1, 1)$, where $L(1, 1) = d(C(1))$. For all the other points lying within the global region of search, compute the cumulative distance using the local path constraints (i.e., the point $(x(k), y(k))$ can only be reached from the points $((x(k) - 1), y(k))$ or $((x(k) - 1), (y(k) - 1))$ or $((x(k) - 1), (y(k) - 2))$. If the distance at stage k is obtained as

$$D(x(k), y(k)) = \min \begin{cases} D((x(k) - 1), y(k)) + L(x(k), y(k)) \\ D((x(k) - 1), (y(k) - 1)) + L(x(k), y(k)) (14.5) \\ D((x(k) - 1), (y(k) - 2)) + L(x(k), y(k)) \end{cases}$$

then $D(X, Y)$ gives the distance between the probe and the reference sequences.

3. *Backtracking.* Using the cumulative distance matrix and the local path constraints, the warping path F can be obtained by backtracking.

14.4.1 Implementation details

Use of dynamic time warping relies on similar start and end points for the probe and gallery. Hence, prior to applying the DTW algorithm, it is necessary to parse the video sequence into cycles. A simple way to achieve this is using the width feature explained in Section 14.2. From Figure 14.5, it is easy to see that $N(t) = (\sum_{j=1}^{M} I(j))(t)$, where $I(j)$ denotes the width at row j of the image; that is, the sum of widths will show a periodic variation (see Figure 14.5). The troughs of the resulting wave form correspond to the rest positions during the walk cycle, while the crests correspond to the part of the cycle where the hands and legs are maximallly displaced. A half-cycle consists of frames between two successive troughs of a sum-of-intensities plot. In general, the exact trough may be hard to determine in the presence of noise, and picking a frame in its neighborhood is usually adequate. Given a video sequence, the width feature thus provides a natural parsing in terms of half-cycles. In our experiments, four half-cycles from the probe sequence are matched with four half-cycles from the gallery sequence.

Euclidean distance was chosen as the local distance. A global constraint of $\max(0.1 N_{gallery}, 0.1 N_{probe})$ was chosen. Using the similarity matrix resulting from the matching algorithm, the cumulative match characteristic is as explained in [14]. Essentially, the cumulative match characteristic is a plot of the number of times the right person occurs in the top n matches, where $n < G$ and G denotes the number of people in the gallery as a function of n.

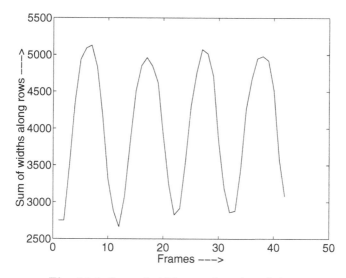

Fig. 14.5. Sum of widths as a function of time.

14.5 Data

The UMD Database. Gait data were captured with the outdoor video research (OVR) facility at the University of Maryland (`http://degas.umiacs.umd.edu/hid`). Cameras located at right angles and at a height of 4.5 above the ground were used for data capture. The UMD data set consists of 44 individuals. There are 38 male and 6 female subjects with different ethnicity, physical build, and other characteristics. The individuals were asked to walk normally along a T-shaped path. Because the cameras are orthogonally placed and the path is T-shaped, a side view and a frontal view could be captured simultaneously as the person walked. The videos for each individual were recorded with intervals ranging from about half a day to a maximum of one month. The size of the bounding box containing the person was 170×138. Labeling was done to mark the sex, test/training sequences, direction of walk, number of gait cycles, and left/right foot forward.

The CMU Database. The CMU database (`http://hid.ri.cmu.edu`) consists of different views of indoor gait sequences of 25 individuals walking on a treadmill. Different situations such as a person walking at a slow pace, fast pace, and while carrying a ball in his/her hands are considered. Even though it is an indoor database, it serves as a good test bed for evaluating performance under variations in walking speed and also to assess the relative importance of body parts. For example, in the sequence where the person is carrying the ball, there is very little upper-body movement. The size of each image frame in this database is 640×486 pixels.

The MIT Database. The MIT database (`http://www.ai.mit.edu/people/llee/HID/intro.htm/`) consists of side views of gait sequences of 25 subjects

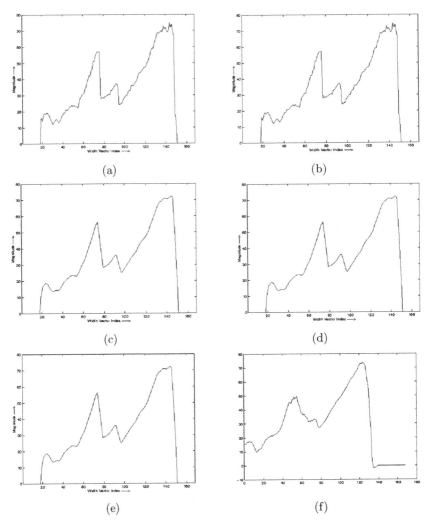

Fig. 14.6. Width plots for an arbitrary frame. (**a**) Unsmoothed raw feature. (**b**) 3-point smoothed feature. (**c**) 5-point smoothed feature. (**d**) 11-point smoothed feature. (**e**) 21-point smoothed feature. (**f**) Reconstructed width vector using the first two eigenvectors.

captured indoors and collected on four different days. There are four to eight segments per subject per day with two to five contiguous half-cycles per segment. The number of subjects on a particular day varies from 4 to 16. As a result, the degree of overlap across days is less. The frame rate of the camera is only 15 frames per second (as compared with 25 for the UMD database), and the image size is 128×104 pixels. A lower frame rate leads to sparse temporal sampling of the gait information and must be tackled appropriately. This database provides a good test bed for evaluating training and test databases captured on different days.

The UMD3 Database. The UMD3 database consists of outdoor gait sequences of 12 people walking along straight-line paths at azimuth angles of $0°$, $15°$, $30°$, $45°$, and $60°$ from the side view. The data were captured at the rate of 25 frames per second by a tripod-mounted camera. Two sequences of each person were captured, with a time interval of at least 1 hr between them. One of the sequences was used as the reference and the other was used as the probe. This database provides a test bed for assessing gait-recognition performance as a function of the viewing angle. The image size for zero azimuth is 210×105 pixels.

14.6 Experimental Results

In this section, we present the results of our approach to gait recognition on all the databases described in Section 14.5. Each of the databases has its own interesting features, and our exercise reveals several interesting facts about the gait-based human-identification problem. The UMD database has gait sequences for side as well as frontal views. Hence, the performance of all the features discussed in Section 14.3 was first studied for this database to get a feel for what might be a good feature vector for gait representation and recognition. The method described in Section 14.4 was used for recognition.

14.6.1 Results for the UMD Database

For this database, there were approximately 40 frames, corresponding to four half-cycles in the gallery and probe. The recognition results for the direct features are given in Table 14.1. From this table, we note that all the features have done well despite the presence of noise in the estimate of the basic width vector. The experiment also shows that dimensionality reduction is possible to a great extent by exploiting the redundancy in the gait data. In the experiments to come, we choose the 5-point smoothed 42-dimensional width vector, which is between the two extremities of no smoothing and very heavy smoothing, as the direct feature for performance analysis.

It must be mentioned here that similar to hand dominance (right/left-handedness), foot dominance (right/left-leggedness) also exists in individuals.

Table 14.1. UMD database: Cumulative match scores (CMS) for the first five ranks for different directly derived feature vectors.

Feature\Rank	1	2	3	4	5
Unsmoothed 168 dim feature	79.0	81.4	83.2	86.0	86.0
3-point smoothed 168 dim feature	79.0	81.4	83.7	86.0	86.0
5-point smoothed 42 dim feature	76.7	83.7	83.7	88.3	88.7
11-point-smoothed 21 dim feature	76.7	83.7	83.7	83.7	90.0
21-point-smoothed 11 dim feature	79.0	86.0	86.5	88.3	90.0

In practice, it is difficult to align the reference and the probe sequences accurately with respect to heel strike. The result of ignoring heel-strike information (left foot forward or right foot forward first) in gait analysis is reflected in Table 14.2. Suppose there are five half-cycles in both reference and probe sequences for a particular subject. The first four half-cycles of the two sequences are matched to generate a matrix of similarity scores. Then, the reference sequence is matched with a probe sequence shifted by a half-cycle to generate another matrix of similarity scores. Of the two shifted probe sequences, only one can provide a better match if the subject exhibits foot dominance. The two similarity scores are combined using the minimum-error criterion. The improvement in recognition when heel strike is taken into account can be noted from Table 14.2.

For the eigenfeatures, the performance results obtained by combining different numbers of eigenvectors are given in Table 14.3. Again, four half-cycles of each subject were used for matching. Note that by using just the first two eigenvectors an accuracy of 80% is achievable. Other eigenvectors are noisy and, in fact, tend to lower the accuracy. Hence, we have used only the first two eigenvectors for computing the eigenfeatures in our experiments. It is also of interest to study the effect of the number of half-cycles on matching. Intuitively, one would expect to obtain a better match with a larger number of half-cycles (i.e., with a longer gait sequence). When the number of half-cycles is reduced systematically from four to one, the corresponding results are given in Table 14.4. While reducing the number of half-cycles, the number of frames was kept approximately constant by linearly interpolating the components of

Table 14.2. UMD database: CMS for the first five ranks for the case of two full cycles having a relative phase shift of one half-cycle.

Feature\Rank	1	2	3	4	5
Set of four half-cycles 1(a)	68.1	77.2	84.0	84.0	84.0
Set of four half-cycles shifted by one 1(b)	70.4	79.5	81.8	86.3	86.3
Minimum of 1(a) and 1(b)	79.0	81.4	83.2	86.0	86.0

Table 14.3. CMS for the UMD database using different eigenfeatures.

Feature\Rank	1	2	3	4	5
Eigenvector 1	73	75	80	80	84
Eigenvectors 1, 2	80	87	90	90	91
Eigenvectors 1, 2, 3	68	80	84	84	84
Eigenvectors 1, 2, 3, 4	73	77	84	84	84
Eigenvectors 1, 2, 3, 4, 5	70	73	79	82	84
Eigenvector 2	58	63	67	72	74
Eigenvectors 2, 3	61	64	67	82	76
Eigenvectors 2, 3, 4	61	65	68	68	72
Eigenvectors 2, 3, 4, 5	68	68	72	74	74
Eigenvector 3	51	54	60	62	62
Eigenvectors 3, 4	55	60	68	68	72
Eigenvector 4	22	28	35	42	42
Eigenvector 5	20	24	36	40	44

Table 14.4. Effect of length of gait sequence for the UMD database: CMS for the first five ranks.

Feature\Rank	1	2	3	4	5
Four half-cycles	80.0	87.0	90.0	90.0	91.0
Two half-cycles	71.0	80.0	82.0	84.0	87.0
One half-cycle	65.0	67.0	77.0	83.0	87.0

the width vector. As expected, the performance degrades as the length of the sequence is reduced. A comparison of the results of our method using two eigenvectors with the results of Lee and Grimson [6] and Bobick and Johnson [5] is given in Table 14.5.

To assess the relative discriminability of the structural component of the width vector versus its dynamic component, we computed the difference of the width vectors corresponding to successive frames in the gait sequence. Eigendecomposition of the velocity vector was carried out, and the vector reconstructed using the first two eigenvectors was computed. Table 14.6 shows the results obtained by considering only the velocity profile as the feature

Table 14.5. Comparison of different algorithms on the UMD database.

	CMS Scores	
Experiment	Rank 1	Rank 5
Our approach(using two eigenvectors)	80	91
GaTech [5]	17	42
MIT [6]	41	58

Table 14.6. UMD database: Cumulative match scores for the velocity profile.

Feature\Rank	1	2	3	4	5
Smoothed and differenced 168 dim feature	41.9	51.6	61.2	70.9	74.1
Eigendecompostion of velocity profile	56	75	76	80	83
Eigendecompostion of width vector and successive difference of projected weights	32.0	40.0	51.0	61.0	65
Eigendecompostion of width vector and using 2nd eigenvector	58.0	63.0	67.0	72.0	74.0

of interest. Alternatively, the width vector could first have been subjected to eigendecomposition and the successive differences of the projected values used for matching. Note that the accuracy drops significantly if only the velocity information is used. Clearly, for gait-based human identification, both structural and dynamic information are important.

Frontal Gait

Here, the aim is to recognize humans using the frontal gait information as a cue. As a subject walks toward the camera, an apparent increase in the size of the subject and in the swing of the arms and legs can be observed. The outer contour of the subject carries information that is useful in discriminating among subjects. Figure 14.7(a) shows the variation of the width vector (that reflects in the outer contour) as a function of time. The size of the width vector grows, indicating that the subject is approaching the camera. We retain the last four half cycles of each sequence as shown in Figure 14.7(b), and the raw width vector itself is used for matching. The results are tabulated in Table 14.7. To account for the apparent change in the size of the subject, we normalize the width vectors by computing an appropriate scale factor. The positions of the head and the feet (top and bottom pixels) are identified in each frame and smoothed using a median filter. To the resulting sinusoidal patterns shown in Figure 14.7(c), two straight lines, one to the top and one to the bottom, are fit. The distance between the two lines in each frame is used to compute the normalizing factor. The normalized plot is shown in Figure 14.7(d). The recognition rates obtained before and after normalization are summarized in Table 14.7. Note that the accuracy for the frontal gait sequences is lower as compared with the side-view case. This is because the swing is less pronounced now, and the dynamics are not as effectively captured as in the case of the sideview. For the UMD database, because we have two orthogonal cameras taking pictures simultaneously, we get both frontal and side-view gait sequences. The improved results obtained by combining the evidence from both the views at the decision level are given in Table 14.7.

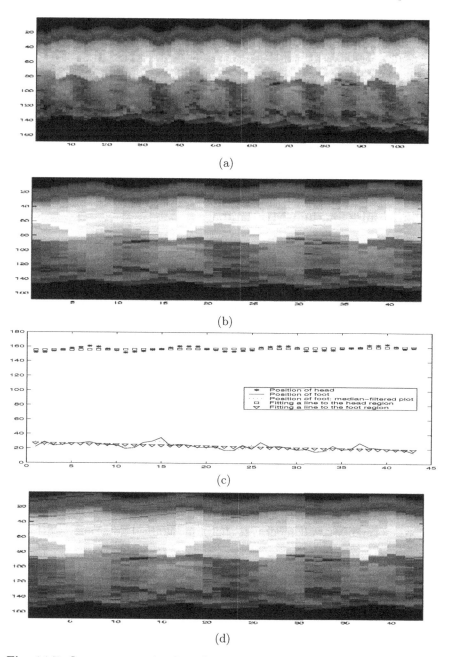

Fig. 14.7. Outer contour plot for a frontal gait sequence in the UMD datasbase. (**a**) Subject walking towards the camera (**b**) 4 half-cycles chosen for recognition (**c**) The position of head and foot in each frame is located to compute the normalizing factor. Two straight lines are then fit to the trajectories of the head and foot. (**d**) The 4 half-cycles normalized.

Table 14.7. CMS results for the frontal gait sequences of the UMD database.

Feature\Rank	1	2	3	4	5
Raw width vector (without normalization)	66.0	67.0	74.0	81.0	86.0
Normalization in the vertical direction	69.0	74.0	86.0	90.0	90.0
Normalization in both vertical and horizontal directions	71.0	84.0	87.0	87.0	87.0
Frontal + Sideview (decision based on fusion)	85.0	87.0	92.0	94.0	95.0

14.6.2 Results for the CMU database

The following experiments were performed on this database: (i) Train on slow walk and test on slow walk, (ii) train on fast walk and test on fast walk, (iii) train on slow walk and test on fast walk, and (iv) train on walk carrying a ball and test on walk carrying a ball. For this database, the number of frames corresponding to four half-cycles varied from 55 for slow walk to about 75 for fast walk. The results are given in Table 14.8.

We note that the eigensmoothed feature performs better than the directly smoothed feature. This can be attributed to the fact that eigensmoothing exploits spatiotemporal redundancy, unlike direct smoothing, which uses only spatial smoothing. When the reference is the slow-walk sequence and the probe is the fast-walk sequence, the performance is found to be inferior to that for the case when the reference and probe are both slow-walk sequences. DTW is known to perform poorly [15] when the ratio of reference sequence length to probe sequence length is less than 0.5 or more than 2. In the CMU data set, this ratio for the worst case was 1.36. From Table 14.7, we note that the DTW-based method is reasonably robust to changes in walking speed. In fact, in our experiments, the value of the ratio for one of the mismatched cases was 1.15. A few frames in the gait cycles of this incorrectly recognized person under slow- and fast-walk modes are shown in Figures 14.8(a) and 14.8(b). As

Table 14.8. CMU database: Recognition results for different experiments.

Experiment	Feature	Rank				
		1	2	3	4	5
Slow vs. Slow	Directly smoothed feature	70.8	83.3	87.5	95.8	95.8
	Eigensmoothed feature	95.8	95.8	95.8	95.8	100.0
Fast vs. Fast	Directly smoothed feature	83.3	83.3	83.3	83.3	87.5
	Eigensmoothed feature	95.8	95.8	95.8	95.8	100.0
Fast vs. Slow	Directly smoothed feature	54.1	75.0	87.5	87.5	87.5
	Eigensmoothed feature	75.0	83.3	83.3	83.3	87.5

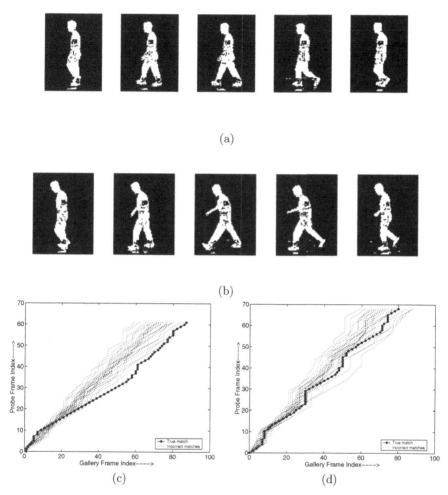

Fig. 14.8. Sample images of the same subject corresponding to (**a**) slow-walk and (**b**) fast-walk (Notice the change in posture and body dynamics) (**c**) Warping path for person with largest reference sequence-length to probe sequence-length ratio (**d**) Warping path for the person in (**a**) and (**b**).

is apparent from the Figure 14.8, the posture as well as hand swings for the person are quite different in cases of fast walk and slow walk. Figures 14.8(c) and 14.8(d) show the warping paths for the person with the highest ratio and the incorrectly recognized person, respectively. Note that the warping path for the correctly recognized person is much more regular as compared with that of the incorrectly recognized individual. Hence, it is the change in the posture

Table 14.9. CMS for the ball experiment in the CMU database using the eigens-moothed feature.

Experiment\Rank	1	2	3	4	5
Ball vs. Ball	95.45	100.00	100.00	100.00	100.00

and body dynamics of the person rather than the mismatch in the length of the reference and probe sequences that was responsible for the mismatch.

Finally, we consider the case when the person is walking with a ball in his hand. In this situation, most of the gait dynamics is confined to the leg region. For this experiment, we observe from Table 14.9 that the recognition performance is very good (person identified correctly within the top two matches). This experiment suggests that for the purpose of recognition certain parts of the body may be more preferable than the others. Incidentally, this has been noted in kinesiology research as well [16].

14.6.3 Results for the MIT Database

The evaluation scheme for the MIT database is as follows. For training, data collected on days 2, 3, and 4 were used. Data collected on day 1 were used for testing. Overall, there are eight subjects. We extract two half-cycles (which is the maximum number of half-cycles common to all the subjects in the database) and build the directly smoothed width vector. Since the frame rate is 15 frames/second, on the average, there are only 7–8 frames per half-cycle, so there were about 28 frames in four half-cycles. We recognize that the length of the gait sequence is not adequate. As discussed earlier, the larger the number of half-cycles, the better would be the possibility of obtaining a correct match. Nevertheless, four out of the eight people were correctly recognized. Since the frame rate is low for this database, it results in coarse sampling in the temporal domain and presents a problem while matching. When we used linear interpolation between frames to partially compensate for the coarse sampling, the recognition rate increased to five matches out of eight. When the eigenfeatures were tried on this database, the results were not very different.

14.6.4 Results for the UMD3 database

The UMD3 database contains gait data corresponding to different azimuth angles. The gait sequences were captured at different times at angles $\theta = 0°$, $15°$, $30°$, $45°$, and $60°$ with respect to the side view. To account for the foreshortening effect as a person walks at nonzero azimuths, the height of each individual was computed from the zero-azimuth width plots, and this was used to normalize the height corresponding to nonzero azimuth angles. Figure 14.9 shows the width plots for one of the individuals in the database for each

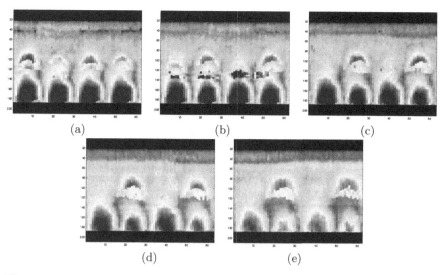

Fig. 14.9. Width profile of an individual for different azimuth angles (**a**) $0°$ (**b**) $15°$ (**c**) $30°$ (**d**) $45°$ and (**e**) $60°$.

of the angles above. There were approximately 60 frames corresponding to four half-cycles for the UMD3 database. It can be seen that for large azimuth angles, the hand swings appear only in alternate half-cycles. This is because the body occludes the presence of hands in the alternate half-cycles. Another observation is that the range of variation in the width amplitude is small for large azimuth angles. Therefore, recognition performance can be expected to be more sensitive to the presence of noise in the gait sequence for higher values of θ.

Two full cycles having a relative shift of one half-cycle were used as the gallery sequence (similar to the first UMD database) and two full cycles captured at a later time were chosen as the probe sequence. Both the 5-point directly smoothed feature and the eigensmoothed feature were tested for recognition, and the results are given in Table 14.10. In Table 14.10, L and R refer to two full gait cycles with a relative shift of one half-cycle, while F refers to the case when we accept the smaller of the two entries corresponding to L and R. It is clear that $\theta = 0$, which is the side view, provides the best viewing direction for gait recognition. Since the variation in the amplitude of the width vector was small for large θs, eigensmoothing results in better performance. For smaller values of θs when the width vector components are relatively large, eigensmoothing does not seem to yield any significant improvement over direct smoothing.

Table 14.10. UMD3 database: CMS for the first three ranks for different azimuth angles using the direct and the eigen-feature vectors.

Angle from side view	Sequence chosen	Rank (direct)			Rank (eigen)		
		1	2	3	1	2	3
0°	L	75.00	91.67	100.00	91.67	91.67	100.00
	R	75.00	83.33	91.67	83.33	83.33	91.67
	F	91.67	91.67	100.00	91.67	91.67	100.00
15°	L	66.67	83.33	100.00	91.67	91.67	100.00
	R	58.33	75.00	83.33	91.67	91.67	100.00
	F	91.67	91.67	100.00	91.67	91.67	100.00
30°	L	66.67	83.33	91.67	83.33	83.33	91.67
	R	25.00	50.00	50.00	58.33	91.67	100.00
	F	83.33	83.33	91.67	83.33	83.33	100.00
45°	L	33.33	41.67	50.00	41.67	66.67	91.67
	R	66.67	75.00	83.33	83.33	91.67	100.0
	F	50.00	83.33	100.00	58.33	83.33	100.00
60°	L	33.33	50.00	58.33	58.33	75.00	91.67
	R	41.67	58.33	58.33	75.00	83.33	91.67
	F	58.33	83.33	91.67	75.00	83.33	91.67

14.7 Conclusions

In this chapter, we have proposed a new approach to gait recognition using the dynamic programming paradigm. The method was tested on five different databases using several features. The width of the outer contour of the binarized silhouette of a person was used as the gait feature. Different feature vectors were derived from the basic width vector using either direct smoothing and down-sampling or by way of eigenanalysis for gait representation and matching. During the matching process, a dynamic time-warping algorithm was used to effectively handle unequal lengths of the reference and the probe gait sequences. Various situations such as changes in walking speed, sensitivity to viewing angle, different floor characteristics, low frame rate, and others, were considered. The performance of our method was found to be quite satisfactory on all the databases. The eigensmoothed feature performed the best, followed by the directly smoothed feature vector. The velocity profile of the width vector is not sufficient to capture gait characteristics when used in isolation. Our experiments confirm that the side view is the optimal one for capturing gait characteristics. As the azimuth angle increases, the recognition accuracy falls. The method is capable of recognizing even frontal gait but with lesser accuracy as compared with that of the side view.

It would be interesting to analyze whether the warping-path characteristics of the DTW algorithm can be used to determine the direction of motion. Also, a study that combines evidence from different component-level features of the human body for the purpose of recognition deserves investigation.

References

[1]   J. Cutting and L. Kozlowski, "Recognizing friends by their walk: Gait perception without familiarity cues," *Bull. Psychonom. Soc.*, **9**, 353–356 (1977).

[2]   M.P. Murray, A.B. Drought, and R.C. Kory, "Walking patterns of normal men," *J. Bone J. Surg.*, **46A**, (2):335–360 (1964).

[3]   M.P. Murray, "Gait as a total pattern of movement," *Am. J. of Phys. Med.*, **46**, pp. 290–332 (1967).

[4]   D. Cunado, J.M. Nash, M.S. Nixon, and J.N. Carter, "Gait extraction and description by evidence-gathering," *Proceedings of the International Conference on Audio and Video Based Biometric Person Authentication*, pp. 43–48 (1995).

[5]   A.F. Bobick and A. Johnson, "Gait extraction and description by evidence-gathering," *Proc. of the IEEE Conference on Computer Vision and Pattern Recognition*, 423–430 (IEEE, New York, 2001).

[6]   L. Lee and W.E.L. Grimson, "Gait analysis for recognition and classification," *Proceedings of the IEEE Conference on Face and Gesture Recognition*, pp. 155–161. (IEEE, New York, 2002).

[7]   J. Little and J. Boyd, "Recognizing people by their gait: The shape of motion," *Videre* **1**, (2), 1–32 (1998).

[8]   P.S. Huang, C.J. Harris, and M.S. Nixon, "Recognizing humans by gait via parametric canonical space," *Arti. Intell. Eng.* **13**, (4), 359–366 (1999).

[9]   R. Cutler, C. Benabdelkader, and L.S. Davis, "Motion based recognition of people in eigengait space," *Proceedings of the IEEE Conference on Face and Gesture Recognition*, pp. 267–272 (IEEE, New York, 2002).

[10]   S. Furui, "Cepstral analysis technique for automatic speaker verification," *IEEE Trans. Acoust. Speech Signal Process.*, **29**, (2), 254–272 (1981).

[11]   A. Elgammal, D. Harwood, and L. Davis, "Non-parametric model for background subtraction," *FRAME-RATE Workshop* (IEEE, New York, 1999).

[12]   H. Sakoe and S. Chiba, "Dynamic programming algorithm optimization for spoken word recognition," *IEEE Trans. Acoust. Speech Signal Process.*, **26** (1), 43–49, (1978).

[13]   J.M. Zacharia, "Text-independent speaker verification using segmental, suprasegmental and source features," M.S. thesis, IIT Madras Chennai, India, March 2002.

[14]   P.J. Philips, H. Moon, and S.A. Rizvi, "The feret evaluation methodology for face-recognition algorithms," *IEEE Trans. Pattern Anal. Machine Intell.*, **22** (10), 1090–1100 (2000).

[15]   L. Rabiner and H. Juang, *Fundamentals of Speech Recognition* (Prentice-Hall, Englewood, NJ, 1993).

[16]   W.I Scholhorn, B.M. Nigg, D.J. Stephanshyn, and W. Liu, "Identification of individual walking patterns using time discrete and time continuous data sets," *Gait Posture* **15**, 180–186 (2002).

2-D Periodic Patterns for Image Watermarking

Damien Delannay and Benoit Macq

Summary. The robustness of watermarking algorithms against common geometrical deformations has drawn the attention of many researchers for ten years already. Today, the design of communication systems that are able to recover from loss of synchronization is still an important challenge. In watermarking, the use of periodic structures has been proposed as a mean to fight such distortions. However, the security implications of this periodicity have not always been properly addressed. In this chapter, we first present a method that we have developed to generalize and introduce secrecy in the construction of periodic patterns for watermarking. The resistance against cropping and general affine transformations is presented and the related security and robustness issues are discussed.

15.1 Introduction

Watermarking can be described as a communication over noisy channel problem with specific security constraints. Digital communication systems over noisy channels use channel-coding techniques in order to achieve error-free transmissions. These techniques introduce a certain amount of redundancy in the transmitted message, which enables a diminution of the rate of error at the price of a lower transmitted information rate. Specific forms of redundancy can enhance the watermark detection performances. However, introducing redundancy can also have security implications for the watermarking system. These issues will be discussed in the present chapter.

The simplest way to introduce redundancy in a discrete signal is to perform repetition coding, where each symbol of the message is repeated n times and produces a code whose length is n times the original length of the message. In most circumstances, repetition coding is not the most appropriate way to introduce redundancy in a signal. Algebraic and convolutional error-coding codes are more complex but also much more efficient channel-coding techniques when the channel exhibits sufficiently low error probability. It has therefore been shown in [1] that the adequate channel-coding strategy relies on the combination of repetition and classical error-coding strategy.

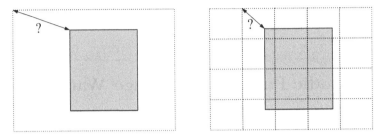

Fig. 15.1. (a) Searching for synchronization with reference marks, (b) tiling an elementary pattern to reduce search space.

In watermarking systems, the channel is characterized by a very low signal-to-noise ratio. Therefore, some kind of repetition coding will always be required before any other error-correcting coding can successfully be used. Moreover, as the communication channel is mostly unknown, one will have to consider the worst case—or most likely case—signal-to-noise ratio in the system's design.

Repetition of a set of symbols can be performed in many different ways. Some methods can bring more benefit than others. One specific form of repetition is the periodic repetition. One says that a sequence is periodic with period T if the same symbol occurrence can be observed at fixed-size intervals of samples.

In watermarking schemes, besides its channel-coding role, the use of periodicity in the embedded signal is motivated by two different but quite related objectives. The first objective is to limit the size of the searching space to one repetition period in methods performing an exhaustive search to recover from a loss of synchronization. As illustrated in Figure 15.1 the periodic tiling of an elementary rectangular watermark pattern permits us to operate the search for synchronization after a cropping over the area of the size of a single tile. This construction also guarantees that, provided the cropped image is still larger than the elementary tile, it contains an occurrence of every symbol of the hidden message.

The second objective is to benefit from the self-referencing structure of the signal. One can analyze the autocorrelation function of the periodic signal to determine which transformation the signal has undergone. This determination is achievable only when the number of periods represented in the signal is sufficiently large.

This chapter will present an original generalization method to build periodical n-dimensional discrete signals and expose its advantageous properties for watermarking schemes.

15.2 Weaknesses of Watermarking Schemes Using Periodicity

Periodicity, because of its very organized redundancy, can introduce security weakness in a watermarking scheme. The channel-coding role played by periodicity can also be exploited by an unauthorized user to compute an estimate of the embedded watermark through averaging of the signal over the different repetition periods.

Very early works [1] on watermarking techniques already discussed the choice of two-dimensional arrays with useful properties. However, they focused only on cross-correlation properties of rectangular arrays and did not research periodical structures. Existing periodic watermarking schemes [2–4] propose rigid watermark construction. Indeed, the periodicity results from the tiling of a fixed-size rectangular elementary pattern, as illustrated in Figure 15.1(b). The secrecy of the construction lies only in the nature of this elementary tile. Therefore, the repetition periods of the watermark are fixed and publicly known.

An important weakness issue arises in those schemes due to the absence of secrecy in the way the elementary pattern is repeated across the image. This enables any opponent to compute an estimate of the watermark by averaging the signal over the different periods' repetition. This opponent would then be able to remove important watermark power from the document or perform a copy attack [5] on other documents.

The first part of this chapter shows that it is possible to introduce secrecy through a generalization in the construction process of periodic structures. In the second part, we illustrate how one can benefit from a periodically structured watermark to estimate the amount of affine transformation of the media. We also discuss the security issues in those different usages of periodicity.

15.3 Construction of n-Dimensional Periodic Patterns: A Generalization

15.3.1 Problem Statement

In the following discussion, we consider finite-length discrete multidimensional signals that we refer to as patterns. Each element of the pattern is characterized by an index giving its position in the sequence and a value belonging to an alphabet of symbols M. In the multidimensional case, the indexes take values in the n-dimensional finite subset K of Z^n. The extent of the pattern is thus given by the set of indexes where a value for the signal is defined. We consider arrays as a category of patterns whose extend results from the inner product of index intervals in each dimension.

A pattern w is periodic with period T if the following relation is satisfied:

$$w(k) = w(k + T) \qquad \forall k \in K \mid (k + T) \in K. \tag{15.1}$$

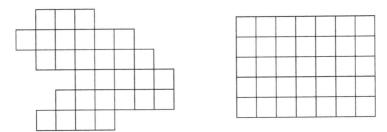

Fig. 15.2. (a) Two-dimensional pattern; (b) two-dimensional array.

When the signal has finite length, this relation must hold only for pairs $(k, k + T)$ belonging to the subset of indexes that determine the signal's extent.

Given C, a discrete sequence of symbols with length N, we aim at finding a method to construct an arbitrary-shaped pattern that contains all the symbols of C in such a way that the pattern is periodic. Moreover, the proportion of each symbol represented in the pattern should tend to the same proportion as in the sequence C when the pattern's extent tends to infinity.

15.3.2 Construction Scheme

In this section, we describe how one can construct different periodic patterns depending on the values chosen as construction keys. The construction method will be described for a two-dimensional pattern (see Figure 15.2), but its extension to n-dimensional pattern is straightforward.

Let S be a one-dimensional sequence of symbols with length N,

$$S[k], \qquad k = 0..N-1. \tag{15.2}$$

In general cases, no other information about this sequence is known. It could contain N different symbols values. No restriction exists on the length of S.

We start by choosing two keys, k_1 and k_2, which can be any integer comprised between 0 and $N-1$. We will explain below which combinations of keys should not be used. The pattern is filled with the elements of the sequence in such a way that there exists in the pattern a constant relation between elements in adjacent positions. If the position (x, y) in the pattern is filled with the ith symbol of the sequence, then we will take the symbol in position $(i + k_1) \bmod N$ in the sequence to fill position $(x + 1, y)$ of the pattern. The construction is illustrated in Figure 15.3. In the same way, the symbol at position (x, y) and the one at position $(x, y + 1)$ are k_2 cyclic positions away in the sequence.

Once a first element is chosen in the pattern, there exists only one way to complete the pattern given the two keys k_1 and k_2:

$$W[x, y] = S[i]; \tag{15.3}$$
$$W[x', y'] = S[(i + k_1 * (x' - x) + k_2 * (y' - y)) \bmod N]. \tag{15.4}$$

Sequence C

C[0]	C[1]	C[2]	C[3]	...	C[N–3]	C[N–2]	C[N–2]

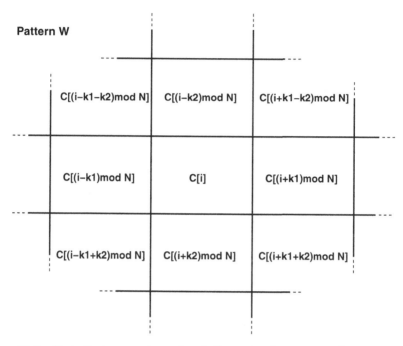

Pattern W

Fig. 15.3. Periodical expansion of a 1-dimensional sequence S onto a two-dimensional pattern

Figure 15.4 shows an example construction with initial sequence

$$S = ['P', 'A', 'T', 'T', 'E', 'R', 'N']$$

and construction keys $k_1 = 3$ and $k_2 = 2$ ($N = 7$).

A few precautions have to be taken while choosing the two keys k_1 and k_2. Indeed, we want all the symbols of the sequence to be represented in the pattern. In order to achieve this, the following condition has to be verified: the greatest common factor among k_1, k_2, and N must be equal to one. When this condition is not satisfied, and say the greatest common factor is m, only N/m out of the N symbols of S are represented in the pattern.

The number of different constructions depends on the length N of the symbol sequence S. If N is prime, then the number of allowed combinations of keys will be N^2. If N is not prime, some combinations will have to be ruled

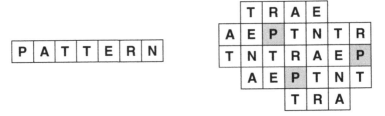

Fig. 15.4. Example pattern construction with construction keys $k_1 = 3$ and $k_2 = 2$ ($N = 7$)

out, and the total number will be slightly smaller. If the factorization of N in prime numbers is given by the expression,

$$N = f_1^{d_1} \times f_2^{d_2} \times \ldots \times f_n^{d_n} \tag{15.5}$$

with $f_1, f_2, \ldots, f_n$ being different prime numbers, then the total number of combinations that do not satisfy the condition stated above is

$$N_{\text{excluded}} = 1 + \sum_{i=1}^{n} \left[\left(\frac{N}{f_i} \right)^2 - 1 \right] - \frac{1}{2} \sum_{i,j=1 (i \neq j)}^{n} \left[\left(\frac{N}{f_i f_j} \right)^2 - 1 \right]. \tag{15.6}$$

The total number of allowed combinations of keys is given by

$$N_{\text{keys}} = N^2 - N_{\text{excluded}}. \tag{15.7}$$

Figure 15.5 shows the evolution of the total number of suitable combinations of keys as a function of the length N of the sequence. One can clearly observe the diminution of the number of combinations for lengths that are multiples of two, three, or five.

In the general case where all symbols in S are different, each permutation of symbols in the sequence produces a different pattern owning the same periodic structure. Therefore, the total number of different constructions of a periodic pattern whose repetition period contains strictly one occurrence of every symbols comprised in a sequence of length N is

$$N_{\text{patterns}} = N! \times N_{\text{keys}} \tag{15.8}$$

15.3.3 Reconstruction Property

A very interesting property of the pattern-construction scheme is the constant relation between symbols located in adjacent positions. It enables reverse processing on the pattern.

Let $S[k]$ be the initial ordering of symbols in the sequence before construction of the pattern; that is, after a possible permutation of the original

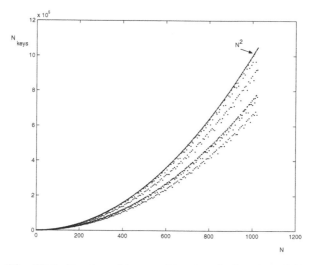

Fig. 15.5. Number of key combinations in function of N.

sequence. An inverse operation consisting in reconstruction of a sequence $S'[k]$ from the pattern can be performed. It follows from the construction scheme that the reconstructed sequence S' is a circularly shifted version of S whatever position is chosen as reference in the pattern. This is illustrated in Figure 15.6, where position $(1,2)$ was chosen as the starting index instead of $(2,3)$.

On the other hand, cyclic permutations of the sequence S produce shifted versions of the periodic pattern in the construction process.

Another consequence is that the one-dimensional cross-correlation properties of the sequence S are transferred to the two-dimensional constructed pattern. One might, for example, choose a maximum length sequence [6] as initial sequence S to generate a pattern with very good cross-correlation properties.

Note that the construction method does not guarantee that it is possible to obtain a rectangular pattern without truncating repetition periods. This means that 2-D cyclic crosscorrelation of an arbitrary-shaped rectangular pattern will not necessarily exhibit the same cross correlation as the sequence. It

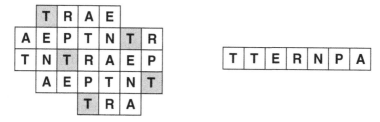

Fig. 15.6. Reconstruction of a circularly shifted sequence.

will, however, be the case for a few particular arrays: first, all arrays of size $N \times N$ whatever combinations of keys are used; second, arrays generated by a combination of keys that have common—but different—factors with N. With those keys, the adequate array size is

$$\frac{N}{\gcd(N, k_1)} \times \frac{N}{\gcd(N, k_2)} \tag{15.9}$$

15.3.4 Extension to N-Dimensional Periodic Structures

The construction presented can easily be extended to n-dimensional structures. For each dimension, one key has to be chosen. The construction rule is then given by the following expressions:

$$W(i_1, i_2, \ldots, i_n) = S[i]; \tag{15.10}$$
$$W(j_1, j_2, \ldots, j_n) = S[(i + k_1 * (j_1 - i_1) + k_2 * (j_2 - i_2)$$
$$+ \ldots + k_n * (j_n - i_n)) \bmod N] \tag{15.11}$$

The considerations concerning the choice of key values can easily be derived from the two-dimensional case.

15.4 A Spatial Watermarking Scheme Based on Generalized 2-D Periodic Patterns

In this section, we present a practical watermarking scheme using the previously described generalized periodic pattern structures. The proposed scheme is a spatial-domain additive scheme for a gray-scale image. The watermark-embedding procedure can be expressed as

$$I_w(x, y) = I(x, y) + \alpha(x, y) \, W(x, y), \tag{15.12}$$

where I_w is the watermarked image luminance, I the original image luminance array, α a perceptual weighting factor array, and W the watermark array. All arrays have the same size as the original image.

15.4.1 Watermark Construction

Let M be a binary sequence of length N_m representing the message to be hidden in the image. A succession of processing steps applied to this sequence M lead to the watermark array. The global procedure is illustrated in figure 15.7.

The first operation consists in performing convolutional error-correcting coding on the message sequence M to produce a coded message C with length N_c. The code rate N_c/N_m is limited by the channel bit-error probability and

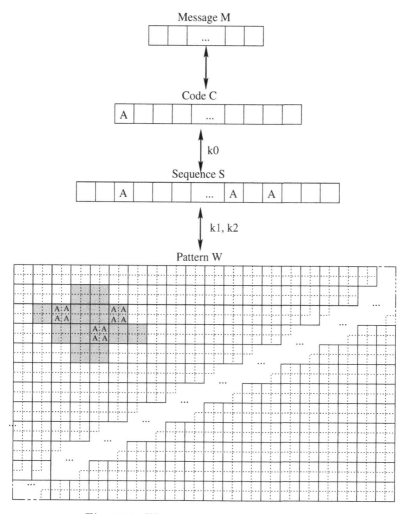

Fig. 15.7. Watermark pattern construction.

the capacity of the cover image. Indeed, the subsequent steps in the watermark construction perform pure repetition coding of the coded sequence C aiming at the achievement of a satisfying bit error probability on the coded message. Above a certain threshold, convolutional coding does not perform better than repetition coding.

In the second step, the bits of the coded message C are randomly repeated to form a new code sequence S with length N_s. This expansion is dependent on key k_0. One should take care that all the bits are equally repeated, such that N_s is a multiple of N_c. This operation can include bit inversion in such a way that an equal number of 1s and 0s is guaranteed in S and therefore that the insertion of the watermark pattern into the image does not change its mean luminance value.

The role of this expansion is to increase the size of S in order to increase the number of possible different pattern constructions and at the same time control the size of each periodic repetition in the pattern. One has to check that the random repetition does not produce a periodic sequence; otherwise, the pattern construction will produce the same structures as if sequence S were truncated to one repetition period.

The next operation is the periodic pattern construction as detailed in previous section. We will call S the generating sequence of the pattern. The pattern's dimensions are the same as for the cover image. One difference with the description from Section 15.3 is that each cell of the pattern covers an array of R_x by R_y pixels in the cover image. The repetition of the same bit value over an array of pixels permits us to limit the high-frequency components of the watermark. Most high frequencies do not survive common compression and interpolation operations.

The parameters $N_c, N_s, k_0, k_1, k_2, R_x$ and R_y will determine the watermark repetition structure. In a practical application, three parameters have fixed values: the watermark message length N_m, image expected size, and the worst channel characteristics. Image expected size means the minimum size of the image from which the message should be recovered. The rest of the parameters should be optimized for each situation.

15.4.2 Perceptual Masking

The design of the weighting factor α in Eq. (15.12) is motivated by the will to minimize the image degradation while maximizing the detection reliability. This leads to the shaping of the watermark signal to fit the image characteristics. Most approaches consist in optimizing watermark power distribution with perceptual sensitivity considerations. The easiest way to perform perceptual masking is to apply local scaling to the watermark proportional to a local activity measure of the cover image. An effective measure of local activity is the Laplacian high-pass filter:

$$L = \begin{bmatrix} -1 & -1 & -1 \\ -1 & 8 & -1 \\ -1 & -1 & -1 \end{bmatrix} / 8. \tag{15.13}$$

Another perceptual consideration is the lower sensitivity of the human eye to a change of luminance for darker luminance intensity levels. This effect is rendered by Weber's law.

A combination of Laplacian high-pass filtering and Weber's law-based weighting was chosen as scaling factor α for our method. Other more effective masking methods based, for example, on anisotropic wavelet decomposition analysis could be used. Many authors [7–9] have proposed human visual sensitivity models for watermarking applications. The comparison of existing perceptual masking models was not addressed in this work.

15.4.3 Detection Scheme

The present detection strategy considers the watermarking channel as begin linearly time-invariant with additive noise. The proposed detection scheme is essentially correlation-based. As the message bits are randomly distributed over the generating sequence S, the pattern elementary period can be considered as spectrally white. From detection theory, we know that correlation detectors are optimum in the presence of linear time-invariant channels with additive white Gaussian noise (AWGN). However, in watermarking channels, noise is due to the cover image, where neighboring pixels are highly correlated. Detection improvement can be achieved through whitening of the cover image spectrum prior to the correlation operation [10]. This necessitates a decorrelating filter. The following far-infrared (FIR) filter is appropriate for this operation:

$$F = \begin{bmatrix} 1 & -2 & 1 \\ -2 & 4 & -2 \\ 1 & -2 & 1 \end{bmatrix} /4. \qquad (15.14)$$

After the image has been filtered, one must proceed to the inverse construction of an estimated generating sequence S' according to the construction keys. The sequence results from the averaging of the different repetitions of the elementary pattern over the image. As pattern cells are spread over R_x by R_y pixels, as many different sequences S' corresponding to the possible cell shifts have to be considered. The sequence giving the best final correlation value will be retained. If the image was watermarked with the same keys, the inverse construction process will produce a sequence that is a circularly shifted estimate of the original sequence S used to construct the watermark pattern. One has to retrieve the amount shift. This is performed through the correlation of the estimated sequence S' with N_c different masks corresponding to each of the different bits in the coded message C. Those masks $m_1, m_2, \ldots, m_{N_c}$ are constructed thanks to the knowledge of key k_0 that was used to distribute the bits of C over the generating sequence S. They are constituted of 1, -1 and 0s according to the position of the different occurrences of each bit of C in sequence S. Let N be the number of occurrences of each bit of C in S,

$$N = \frac{N_s}{N_c}. \qquad (15.15)$$

For each circular shift, the N_c normalized correlation values of sequence S' with the different masks are squared and summed as expressed as follows:

$$d[k] = \sum_{i=1}^{N_c} \left[\frac{m_i^T S_k'}{N} \right]^2, \qquad (15.16)$$

where S'_k is a cyclic permutation of S'. The shift k leading to the maximum correlation value indicates the most probable amount of circular shift. This maximum correlation value $d_{\max}$ constitutes the decision criterion of the watermark detection process,

$$d_{\max} = \max_k d[k]. \qquad (15.17)$$

Considering this shift, an estimated coded message C' is generated, where each value results from the averaging of the different occurrences of each bit in S' according to k_0. The last operation consists in decoding the coded message C' through a viterbi soft decision algorithm and producing the decoded message M'.

In order to decide whether the image was watermarked, the decision criterion $d_{\max}$ has to be compared with a predefined threshold value. The threshold value T can be fixed according to the Neyman–Pearson strategy. This consists in setting a maximum false alarm probability that cannot be exceeded.

Two exclusive hypotheses have to be defined:

- H_0 The image was not watermarked.
- H_1 The image was watermarked.

The probability of a false alarm is defined as

$$P_{fa} = \text{Prob}[d_{\max} > T \mid H_0]. \qquad (15.18)$$

Under the hypothesis that the image does not contain a watermark, the elements of sequence S' are i.i.d. gaussian variables, $S_j \sim N(0, \sigma^2)$. Each correlation output with one of the mask m_i has therefore a gaussian distribution

$$m_i^T S'_k \sim N(0, \sigma^2/N) \qquad \forall i, k, \qquad (15.19)$$

which can be normalized to standard normal distribution $\sim N(0,1)$. Let $P_{d,k}$ be the probability density function associated with $d[k]$; that is, the detection value for a given shift k. According to Eq. (15.16) and provided correlation outputs are normalized, $P_{d,k}$ has the form of a χ^2 distribution with N_c degrees of freedom.

$$P_{d,k}(x) = \frac{x^{(N_c/2)-1}e^{-(x/2)}}{2^{(N_c/2)}\Gamma(\frac{N_c}{2})} \qquad \forall k. \qquad (15.20)$$

When N_c is large, χ^2 distributions can be reasonably approximated by a standard normal distribution such that

$$\text{Prob}[d[k] < T \mid H_0] \approx Q\left((T - N_c)(2N_c)^{-(1/2)}\right), \qquad (15.21)$$

where

$$Q(x) = \frac{1}{\sqrt{2\pi}} \int_{-\infty} x e^{(y^2/2)} dy. \qquad (15.22)$$

The probability of a false alarm for a fixed shift k and threshold value T is thus given by

$$P_{fa,k} = \text{Prob}[d[k] > T] \approx 1 - Q\left((T - N_c)(2N_c)^{-(1/2)}\right). \qquad (15.23)$$

Since $d_{\max}$ results from the maximum value of $d[k]$ over all k, the total probability of a false alarm given a threshold T is the probability that the maximum detection value is greater than T:

$$P_{fa,tot} = 1 - (1 - P_{fa,k})^{N_s}. \qquad (15.24)$$

Given a maximum acceptable false-alarm probability, a threshold value T for the decision criterion $d_{\max}$ can be derived. Notice that a different $d_{\max}$ is computed for every $R_x \times R_y$ subcell shift. Although these detection values are highly correlated, it will increase the total probability of a false alarm. One can also experimentally estimate $\text{Prob}[d_{\max} > x]$ and set the threshold value to an appropriate level.

15.4.4 Properties of the Watermarking Scheme

Whitening of the Cover Media

A first interesting property of the watermark structure is that the regular periodic repetition produces an efficient whitening of the cover-image power in the detection process. Samples originating from distant locations in the image are uncorrelated and ensure appropriate statistics of the detection sequence S'.

Invariance to Cropping

Another nice property of the detection scheme is that it is fully insensitive to spatial discrete shifts. As previously described, a loss of reference produces cyclic shifts in the correlation sequence. The detection scheme performs an exhaustive cross correlation to recover from this cyclic shift. If shifts are fractions of pixels, performances will be slightly degraded due to interpolation.

This property also permits boosting of the detection performance using an appropriate informed embedding procedure [11, 12]. Since the cover medium is known to the encoder, one can optimize the coding process to adapt the watermark to the channel characteristics. The procedure consists in looking for the shift of the watermark pattern to be embedded that leads to the best correlation value with the cover image. The number of different shifts that can be considered is equal to $N_s \times R_x \times R_y$. This increase in performance is illustrated in Figure 15.8.

One must keep in mind that an exhaustive search for synchronization tends to increase the false-positive alarm probability [13]. A benefit of the cyclic

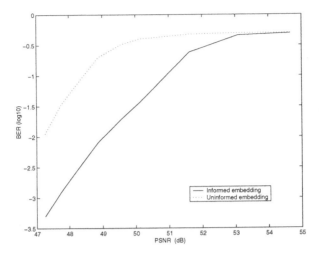

Fig. 15.8. Detection improvement using informed embedding.

pattern construction is that the exhaustive search for spatial shifts is always bounded by the size N_s of sequence S as expressed in Eq. (15.24). This ensures that the false alarm probability is not increased excessively. One should notice that $\text{Prob}[d_{\max} > x \mid H_1]$ also increases with N_s when an informed embedding procedure is applied.

Security Issues

The specificity that also motivated the design of the scheme is the close control on secrecy issues that it provides. Indeed, many different parameters influence the construction process, making the watermark repetition structure unknown to unauthorized users. The length of the generating sequence S plays an important role. First, it determines the total number of different repetition structures, which is linked to the computational complexity of an exhaustive search attack. Second, it represents the extent of one periodic repetition in the pattern which must be chosen carefully. Indeed, as we will describe in Section 15.5 periodic structures can be revealed through autocorrelation or Fourier analysis. However, it can be done only when a sufficient number of repetition periods are observable in the signal. Therefore, to ensure secrecy, one must take care that the elementary periodic structure is large enough with respect to the image size. The proposed scheme enables precise adjustments of the construction parameters depending on the image size and the watermark embedding power.

15.5 Using Periodical Structures to Estimate Undergone Geometrical Distortion

In communications, a classical approach to inform the receiver about channel characteristics is to use pilot signals. These are signals that have known and easily detectable features and therefore do not convey information. At the receiver, a correct registration of this received signal with the original uncorrupted pilot is required to perform channel estimation.

This technique is used in watermarking to design systems that are able to recover from geometrical distortions. Most approaches rely on the introduction or exploitation of robust singularities in the watermark signal, which can be detected even after various geometrical deformations. This pilot signal which is also embedded into the cover image, is often called the reference watermark as opposed to the informative watermark, which conveys a message.

Two different classes of methods exist. In the first class, the pilot signal is distinct from the informative watermark. It is usually called a template signal and often consists of a signal exhibiting peaks in the magnitude of the Fourier spectrum of the image. Different works [14–17] have addressed the use of such pilot signals in watermarking.

In the second class of methods, the pilot signal results from a particular structure of the informative watermark that exhibits exploitable singularities. Periodically structured watermarks have such characteristics. Indeed, the analysis of the autocorrelation function or magnitude of the Fourier spectrum of such structures reveals peaks organized on a well-defined grid alignment. These watermarks are said to be self-referencing. With such structures, the pilot signal does not introduce additional degradation to the image. Several authors [2, 18, 19, 4, 20, 21] have studied the implementation of such methods. The detection of periodicity in a two-dimensional signal will be the subject of this section.

The reference watermark used as a pilot signal must keep its characteristics even after geometrical distortions. One can realize that a two-dimensional periodic signal will remain periodic when the geometrical deformation can be expressed as an affine transform.

15.5.1 Detection of Periodicity in Autocorrelation Function Versus Fourier Magnitude Spectrum

The observation of the periodic nature of a finite signal depends on the number of periods that are represented. The periodization of a signal gives a discrete nature to its Fourier representation. When the period becomes large relative to the observation window, this discrete nature is hidden by the windowing effect and different peaks become harder to distinguish from one another. In the same way, the autocorrelation function of a periodic signal produces peak values for shifts that are multiples of the base period. If the period becomes large, fewer peaks will be observable. In both situations, peaks are organized along a grid that is a function of the repetition periods and directions.

When the periodic watermark signal is embedded in the image, detection of these characteristics is complicated by the noise resulting from the cover-image characteristics. Therefore, robust detection strategies have to be designed. Few works studied how one can perform detection with a low watermark-to-image power ratio. Recently, some authors [21] proposed a method to detect Fourier magnitude peaks using a Hough transform. Works proposing autocorrelation-based detection [2, 20] did not discuss robust detection strategies. In this section, we describe how one can perform robust detection of periodic watermark structure based on the autocorrelation function and illustrate results using the watermark construction described in Section 15.4.

15.5.2 Grid Alignment Detection in Autocorrelation Function

The subsequent steps leading to the grid determination will be illustrated with results coming from the example scenario in Figure 15.9. Image Lena (512×512) was watermarked using the method from Section 15.4 with a PSNR of 44.9 dB. The watermarked image contains 16 repetitions of the elementary pattern. In Figure 15.9(b), a scaling of 111% followed by a rotation of 7° and appropriate cropping was applied to the watermarked image. The periodic grid was recovered and the deformation was inverted to produce image 15.9(c). The 64 bit payload could successfully be extracted.

The first step in the detection process consists in the computation of the autocorrelation function of the estimated watermark. Subsequently, a decorrelating filter is applied to further reduce the contribution of the cover image. Peaks are extracted using a sliding observation window and a local threshold. The parameters are chosen to target a number of detected peaks to limit the complexity of subsequent steps. In our example, a total of 162 peaks were detected. The location of these detected peaks in the autocorrelation function is represented in Figure 15.10; expected locations are illustrated by circles.

(a) Watermarked image   (b) Distorted image   (c) Inverted image

Fig. 15.9. Rotation (7°) and scaling (111%) applied to the watermarked image.

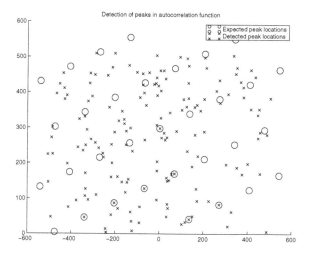

Fig. 15.10. Detection of peaks in the autocorrelation function.

The challenge is to identify the peaks that result from the autocorrelation of the periodic pattern. These peaks are organized along a regular grid that can be determined by two vectors, **a** and **b**,

$$\mathbf{v} = k\mathbf{a} + l\mathbf{b}, \qquad (15.25)$$

with k and l taking integer values. The grid is defined by the set of points **v** satisfying expression (15.25). This description supposes that the origin $(0,0)$ is part of the grid. The density $\mathcal{D}$ of the grid is defined as the norm of the cross product of **a** and **b**:

$$\mathcal{D} = \|\mathbf{a} * \mathbf{b}\|. \qquad (15.26)$$

The problem can be stated as follows: Given a set of points $\mathcal{G}$, find the largest subset of $\mathcal{G}$ such that all points belong to the same grid and $\mathcal{D} > \mathcal{D}_{\min}$. The restriction on the value of the grid's density must be set in order to avoid erroneous grid detection. The probability of finding a set of points belonging to the same grid increases as the density of the grid becomes smaller. When the grid's density tends to zero, any point becomes part of the grid. Therefore, $\mathcal{D}_{\min}$ must be set to the smallest possible size of the repetition pattern.

Considering all possible subsets of $\mathcal{G}$ becomes computationally unacceptable when the number of points in $\mathcal{G}$ gets larger than 30. A practical approach is proposed to perform a preselection among the detected peaks.

The preselection proceeds through the computation of a confidence value for each detected peak. All possible combinations of three points $\mathbf{v}_1$, $\mathbf{v}_2$ and $\mathbf{v}_3$ are analyzed. The confidence of each point of the combination is credited if one of the points can be expressed as an integer combination of the two other points:

$$\mathbf{v}_1 = k_1 \mathbf{v}_2 + l_1 \mathbf{v}_3 \qquad k_1, l_1 \in \mathbb{Z}$$

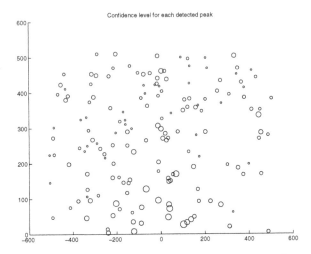

Fig. 15.11. Confidence measure for extracted peaks.

or

$$\mathbf{v}_2 = k_2\mathbf{v}_1 + l_2\mathbf{v}_3 \qquad k_2, l_2 \in \mathbb{Z}$$

or

$$\mathbf{v}_3 = k_3\mathbf{v}_1 + l_3\mathbf{v}_2 \qquad k_3, l_3 \in \mathbb{Z}. \qquad (15.27)$$

These relations can be verified through the projection of the first vector on each of the two other vectors. Approximations have to be considered because of the discrete form of the autocorrelation function.

The result of this preprocessing is illustrated in Figure 15.11. Each detected peak is represented by a circle whose radius is proportional to the confidence measure. The preselection consists in choosing the 15–25 points with the highest confidence measure.

One can now look for the largest subset of points belonging to the same grid with density $\mathcal{D} > \mathcal{D}_{\min}$. All possible subsets have to be considered, starting with the subsets containing the largest number of points. For a given subset of points, the determination of the grid's base vectors **a** and **b** can be performed by an iterative algorithm. Two points belonging to the set are chosen as starting base vectors. At each step, the biggest base vector is replaced by a smaller vector derived from a point not yet part of the intermediate grid. The iteration stops when all points are part of the grid or the density becomes smaller than $\mathcal{D}_{\min}$. At each step, in order to limit the accumulation of approximation errors, new estimations of the base vectors are computed using a least square error criterion. The first grid that is found with density greater than $\mathcal{D}_{\min}$ is selected. In our example, the subset of points leading to an acceptable density is represented by the black dots in Figure 15.12. One can realize that these points correspond with the expected autocorrelation peaks of Figure 15.10.

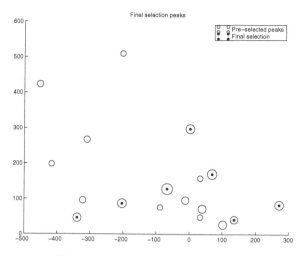

Fig. 15.12. Final selection of peaks.

15.5.3 Estimation of the Amount of Deformation and Informed Detection

Once a periodic grid has been detected, the base vectors can be compared with those from the grid that was used to watermark the image. Given the two original base vectors $\mathbf{a}_o$ and $\mathbf{b}_o$ and detected base vectors $\mathbf{a}$ and $\mathbf{b}$, there exists only one affine transform described by matrix A such that

$$\mathbf{a} = A\,\mathbf{a}_o,$$
$$\mathbf{b} = A\,\mathbf{b}_o. \tag{15.28}$$

Actually, one has to consider eight different possible affine transforms as vectors: $-\mathbf{a}$ and $-\mathbf{b}$ are also base vectors for the detected grid, and $\mathbf{a}_o$ and $\mathbf{b}_o$ can be permuted.

To determine which transform took place, one can try inverting exhaustively all eight possible transforms and perform watermark detection. One can also restrict the set of transforms to be considered to geometrically acceptable transforms. Indeed, most erroneous estimations will lead to transforms that induce severe perceptual nuisance. One can therefore rule out excessive aspect ratio change, shear, or rotations.

This hypothesis on the distortion severity can also be used to improve the grid's detection performance and avoid erroneous grid detection. Not only can a $\mathcal{D}_{\min}$ and a $\mathcal{D}_{\max}$ be derived but subsets of peaks leading to a grid with acceptable density $\mathcal{D}$ can be ruled out if they induce unacceptable geometrical distortion. It also enables us to distinguish from periodic structures that are present in the cover image. This approach consists in performing informed detection. When the embedded periodic structure is secret, this information is only available to an authorized user.

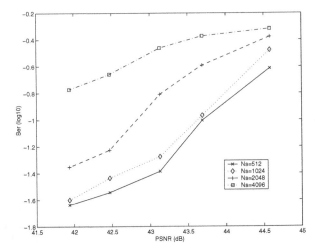

Fig. 15.13. Watermark detection performances after rotation of $7°$ as a function of the embedding strength for different pattern-repetition sizes ($R_x = R_y = 2$, image Lena 512×512).

As previously discussed, the ability to correctly detect the periodic structure depends on the embedding strength but also on the number of periods that are represented in the image signal. Figure 15.13 illustrates the detection performances for different pattern sizes after a rotation of $7°$ and an appropriate cropping were applied to image Lena. Results were obtained with one hundred different key combinations.

15.5.4 Weaknesses and Security Issues

One possible weakness of the grid detection scheme is that it can be fooled by periodic structures that are present in the cover image or added by an opponent. This problem can only partially be overcome through informed detection.

The main security issue relies on the ability of everyone to perform grid detection. It means that any opponent can estimate the watermark through averaging over the different repetitions and can effectively remove watermark power from the image. This is due to the fact that the general grid characteristic of our pilot signal is observable by everyone. A similar weakness appears [17] with template reference signals, where the pilot can often be erased from the watermarked image due to its publicly detectable characteristics. We proposed [22] an original approach to solve this security issue. The idea is to modulate the reference signal with a secret and content-dependent signal prior to its embedding into the image. This secret signal exhibits a white power spectrum and can be computed from the image even after rotation, scaling, and cropping operations. At the detection stage, the reference signal can only be detected using the secret modulating signal.

15.6 Conclusion

Using periodic structures in watermarking can help recover from loss of synchronization such as cropping or even affine transforms. However, to ensure the security of the system, one should carefully design the watermark. Watermarks with large repetition periods with respect to the image size can be made secure but do not provide means to resist against general affine deformations. Watermarks with smaller repetition periods can be designed to be robust against affine deformations but exhibit an important security weakness unless specific hiding strategies are adopted.

References

[1]   A. Tirkel, R. van Schyndel, and C. Osborne, "A two-dimensional digital watermark," in *Proceedings of Digital Image Computing, Technology and Applications (Dicta'95)*, University of Queensland, Brisbane, Australia, December 6–8, 1995, pp. 378–383.

[2]   M. Kutter, "Watermarking resistant to translation, rotation and scaling," *Proc. SPIE*, **3528** (1998).

[3]   M. Maes, T. Kalker, J. Linnartz, J. Talstra, G. Depovere, and J. Haitsma, "Digital watermarking for DVD video copy protection," *IEEE Signal Process. Mag.* **17**(5), 47–57 (2000).

[4]   S. Voloshynovskiy, F. Deguillaume, and T. Pun, "Multibit digital watermarking robust against local nonlinear geometrical distorsions," in *IEEE-ICIP'01*, Thessaloniki, Greece, October 7–10, 2001 (IEEE, New York, 2001).

[5]   M. Kutter, S. Volosjynovskiy, and A.Herrigel, "The watermark copy attack," *Proc. SPIE* **3971**, 371–380 (2000).

[6]   F. MacWilliams and N.J.A. Sloane, "Pseudo-random sequences and arrays," *Proc. IEEE*, **64**, 1715–1729 (1976).

[7]   J.F. Delaigle, C. DeVleeschouwer, and B. Macq, "Watermarking algorithm based on a human visual system," *Signal Process.* **66**, 319–336 (1998).

[8]   R.B. Wolfgang, C.I. Podilchuk, and E.J. Delp, "Perceptual watermarks for digital images and video," *Proc. IEEE* **87**(7), 1108–1126 (1999).

[9]   S.W. Martin Kutter, "A vision-based masking model for spread-spectrum image watermarking," *IEEE Trans. Image Process.* **11**, 16–25 (2002).

[10] G. Depovere, T. Kalker, and J. Linnartz, "Improved watermark detection reliability using filtering before correlation," in *IEEE-ICIP'98*, **I**, Chicago, IL, October 1998, pp. 430–434 (IEEE, New York, 1998).

[11] I.J. Cox, M.L. Miller, and A.L. McKellips, "Watermarking as communications with side information," *Proc. IEEE*, **87**, 1127–1141 (1999).

[12] S. Voloshynovskiy, F. Deguillaume, S. Pereira, and T. Pun, "Optimal adaptive diversity watermarking with channel state estimation," in E.W. Wong and

E.J. Delp, ed., *SPIE Photonics West, Electronic Imaging 2001, Security and Watermarking of Multimedia Contents III, No. paper 4314-74*, San Jose, CA, January 21-26, 2001, (SPIE, Bellingham, WA, 2001).

[13] J. Lichtenauer, I. Setyawan, T. Kalker, and R. Lagendijk, "Exhaustive geometrical search and false positive watermark detection probability," in *SPIE Electronic Imaging 2002, Security and Watermarking of Multimedia Contents V*, Santa Clara, CA, January 2003, (SPIE, Bellingham, WA, 2003).

[14] D. Fleet and D. Heeger, "Embedding invisible information in color images," in *IEEE-ICIP'97*, **1**, Santa Barbara, CA, 1997, pp. 532–535 (IEEE, New York, 1997).

[15] A. Herrigel, J.J.K.O. Ruanaidh, H. Petersen, S. Pereira, and T. Pun, "Secure copyright protection techniques for digital images," in David Aucsmith ed., *Information Hiding*, Second International Workshop IH'98, Portland, OR, April 15-17, 1998, Lecture Notes in Computer Science, **1525**, pp. 169–190, (Springer-Verlag, Berlin, 1998).

[16] S. Pereira and T. Pun, "Fast robust template matching for affine resistant image watermarking," in *International Workshop on Information Hiding, Dresden, Germany*, September 29–October 1, 1999, Lecture Notes in Computer Science, **1768**, pp. 200–210 (Springer-Verlag, Berlin, 1999).

[17] A. Herrigel, S. Volosjynovskiy, and Y. Rytsar, "The watermark template attack," in *Security and Watermarking of Multimedia Contents III*, San Jose, CA, January 2001, (SPIE, Bellingham, WA, 2001).

[18] C. Honsinger and M. Rabanni, "Data embedding using phase dispersion," *Proceedings of the International Conference on Information Technology*, 2000.

[19] S. Voloshynovskiy, F. Deguillaume, and T. Pun, "Content adaptive watermarking based on a stochastic multiresolution image modeling," in *Tenth European Signal Processing Conference (EUSIPCO'2000)*, Tampere, Finland, September 5-8, 2000.

[20] M. Alvarez-Rodriguez and F. Perez-Gonzalez, "Analysis of pilot-based synchronization algorithms for watermarking of still images," *Signal Process. Image Commun.*, 611–633, (2002).

[21] F. Deguillaume, S. Voloshynovskiy, and T. Pun, "A method for the estimation and recovers from general affine transforms in digital watermarking applications," in *SPIE Electronic Imaging 2002, Security and Watermarking of Multimedia Contents IV*, San Jose, CA, February 2002 (SPIE, Bellingham, WA, 2002).

[22] D. Delannay and B. Macq, "Method for hiding synchronization marks in scale and rotation resilient watermarking schemes," in *SPIE Electronic Imaging 2002, Security and Watermarking of Multimedia Contents IV*, San Jose, CA, February 2002 (SPIE, Bellingham, WA, 2002).

[23] D. Delannay and B. Macq, "Generalized 2-D cyclic patterns for secret watermark generation," in *IEEE-ICIP'2000*, **2**, 77–79, (2000).

Image Steganalysis

Rajarathnam Chandramouli and Nasir Memon

Summary. The past few years have seen an increasing interest in using images as cover media for steganographic communication. There have been a multitude of public-domain tools available for image-based steganography. Given this fact, detection of covert communications that utilize images has become an important issue. There have been several techniques for detecting stegoimages developed in the past few years. In this chapter, we review some fundamental notions related to steganography using image media, including security and capacity. We also describe in detail two steganalysis techniques that are representative of the different approaches that have been taken.

16.1 Introduction

Steganography refers to the science of "invisible" communication. Unlike cryptography, where the goal is to secure communications from an eavesdropper, steganographic techniques strive to hide the very presence of the message itself from an observer. Although steganography is an ancient subject, the modern formulation of it is often given in terms of the *prisoner's problem* [24], where Alice and Bob are two inmates who wish to communicate in order to hatch an escape plan. However, all communication between them is examined by the warden, Wendy, who will put them in solitary confinement at the slightest suspicion of the presence of secret messages. Specifically, in the general model for steganography, illustrated in Figure 16.1, we have Alice wishing to send a secret message m to Bob. In order to do so, she "embeds" m into a *cover object c* to obtain the *stego-object s*. The stego-object s is then sent through the public channel. In a *pure steganography* framework, the technique for embedding the message is unknown to Wendy and shared as a secret between Alice and Bob. However, it is generally not considered good practice to rely on the secrecy of the algorithm itself. In *private-key steganography*, Alice and Bob share a secret key that is used to embed the message. The secret key, for example, can be a password used to seed a pseudorandom generator to select pixel locations in an image cover object for embedding the secret message (possibly encrypted). Wendy has no knowledge about the secret key that Alice and Bob share, although she is aware of the algorithm that they could be employing for embedding messages. In *public key steganography*, Alice and Bob have private–public key pairs and know each other's public key. In this chapter, we restrict our attention to private-key steganography.

The warden, Wendy, who is free to examine all messages exchanged between Alice and Bob, can be passive or active. A *passive* warden simply examines the message and tries to determine if it potentially contains a hidden message. If it appears that it does, she suppresses the message and/or takes appropriate action; otherwise she lets the message through without any action. An *active* warden, on the other hand, can alter messages deliberately, even though she does not see any trace of a hidden message, to foil any secret communication that can nevertheless be occurring between Alice and Bob. The amount of change the warden is allowed to make depends on the model being used and the cover objects being employed. For example, with images, it would make sense that the warden is allowed to make changes as long as she does not alter significantly the subjective visual quality of a suspected stegoimage. In this chapter, we restrict our attention to the passive warden case and assume that no changes are made to the stego-object by the warden, Wendy.

It should be noted that the general idea of hiding some information in digital content has a wider class of applications that go beyond steganography. The techniques involved in such applications are collectively referred to as *information hiding*. For example, an image printed on a document could be annotated by metadata that could lead a user to its high-resolution version. In general, metadata provide additional information about an image. Although metadata can also be stored in the file header of a digital image, this approach has many limitations. Usually, when a file is transformed to another format

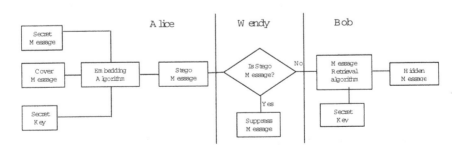

Fig. 16.1. Framework for secret-key passive warden steganography. Alice embeds a secret message in the cover image (left). Wendy the warden checks whether Alice's image is a stegoimage (center). If she cannot determine it to be so, she passes it on to Bob, who retrieves the hidden message based on a secret key (right) he shares with Alice.

(e.g., from TIFF to JPEG or to bmp), the metadata are lost. Similarly, cropping or any other form of image manipulation destroys the metadata. Finally, metadata can only be attached to an image as long as the image exists in the digital form, and it is lost once the image is printed. Information hiding allows the metadata to travel with the image regardless of the file format and image state (digital or analog).

A special case of information hiding is *digital watermarking*. Digital watermarking is the process of embedding information into digital multimedia content such that the information (the watermark) can later be extracted or detected for a variety of purposes, including copy prevention and control. Digital watermarking has become an active and important area of research, and development and commercialization of watermarking techniques is being deemed essential to help address some of the challenges faced by the rapid proliferation of digital content. The key difference between information hiding and watermarking is the absence of an active adversary. In watermarking applications, such as copyright protection and authentication, there is an active adversary that would attempt to remove, invalidate, or forge watermarks. In information hiding, there is no such active adversary, as there is no value associated with the act of removing the information-hidden in the content. Nevertheless, information hiding techniques need to be robust against accidental distortions.

Unlike information hiding and digital watermarking, the main goal of steganography is to communicate securely in a completely undetectable manner. That is, Wendy should not be able to distinguish in any sense between cover objects (objects not containing any secret message) and stego-objects (objects containing a secret message). In this context, *steganalysis* refers to the body of techniques that aid Wendy in distinguishing between cover objects and stego-objects. It should be noted that Wendy has to make this distinction without any knowledge of the secret key that Alice and Bob may be sharing and sometimes even without any knowledge of the specific algorithm that they might be using for embedding the secret message. Hence, steganalysis is inherently a difficult problem. However, it should also be noted that Wendy does not have to glean anything about the contents of the secret message m. Just determining the existence of a hidden message is enough. This fact makes her job a bit easier.

Given the proliferation of digital images, and given the high degree of redundancy present in a digital representation of an image (despite compression), there has been an increased interest in using digital images as cover objects for the purpose of steganography. For a good survey of image steganography techniques, the reader is referred to [19]. The development of techniques for image steganography and the widespread availability of tools for the same have led to an increased interest in steganalysis techniques for image data. The last two years have seen many new and powerful steganalysis techniques reported in the literature. Many such techniques are specific to different embedding methods and indeed have been shown to be quite effective in this regard. However, our intention here is not to present a comprehensive survey of different embedding techniques and possible ways to detect them. Instead, we focus on some general concepts and ideas that apply across different techniques and cover media. The rest of this chapter is organized as follows. In Section 16.2, we first establish a formal framework and define the notion of security for a steganographic system. We point out how conventional definitions do not really adequately cover image steganography (or steganography using any multimedia object for that matter) and provide alternate definitions. In Section 16.3, given our new definition of steganographic security, we then define the notion of steganographic capacity. In Section 16.4, we look at some practical approaches to steganalysis, focusing on the more general techniques that apply across different embedding algorithms and media formats. In Section 16.5, we present some universal steganalysis techniques and conclude with a discussion on future work in Section 16.6.

16.2 Steganographic Security

In this section, we explore the topic of steganographic security. Some of the earlier work on this topic can be found in [4, 26, 8]. Here, a steganographic system is considered to be insecure if the warden, Wendy, is able to prove the existence of a secret message; in other words, if she can distinguish between cover objects and stego-objects, assuming she has unlimited computing power. Let P_C denote the probability distribution of cover objects and P_S denote the probability distribution of stego-objects. Cachin [4] defines a steganographic algorithm to be ϵ-secure ($\epsilon \geq 0$) if the relative entropy between the cover object and the stego-object probability distributions (P_C and P_S, respectively) is, at most, ϵ; that is,

$$D(P_C||P_S) = \int P_C \cdot \log \frac{P_C}{P_S} \leq \epsilon. \qquad (16.1)$$

From this equation, we note that $D(.)$ increases with the ratio P_C/P_S, which in turn means that the reliability of steganalysis detection will also increase. A steganographic technique is said to be *perfectly secure* if $\epsilon = 0$ (i.e., $P_C = P_S$). In this case the probability distributions of the cover object and stego-objects are indistinguishable. Perfectly secure steganography algorithms (although impractical) are known to exist [4].

We observe that there are several shortcomings in the ϵ-secure definition presented in Eq. (16.1). Some of these are listed below.

1. The ϵ-secure notion as presented in [4] assumes that the cover object and stego-object are vectors of independent, identically distributed (i.i.d.) random variables. This is not true for many real-life cover signals such as images. One approach to rectify this problem is to put a constraint that the relative entropy computed using the nth order joint probability

distributions must be less than, say, ϵ_n and then force the embedding technique to preserve this constraint. But, it may then be possible, at least in theory, to use $(n+1)$st-order statistics for successful steganalysis. This line of thought clearly poses several interesting issues:

- Practicality of preserving nth order joint probability distribution during embedding for medium to large values of n.
- Behavior of the sequence $\{\epsilon_n\}$ depends on the cover message as well as the embedding algorithm. If this sequence exhibits a smooth variation, then, for a desired target value, say $\epsilon = \epsilon^*$, it may be possible to precompute a value of $n = n^*$ that achieves this target.

Of course, even if these nth-order distributions are preserved, there is no guarantee that embedding-induced perceptual distortion will be acceptable. If this distortion is significant, then it is not even necessary to use a statistical detector for steganalysis!

2. While the ϵ-secure definition may work for random bit streams (with no inherent statistical structure), for real-life cover objects such as audio, image, and video, it seems to fail. This is because real-life cover objects have a rich statistical structure in terms of correlation, higher-order dependence, etc. By exploiting this structure, it is possible to design good steganalysis detectors even if the first-order probability distribution is preserved (i.e., $\epsilon = 0$) during message embedding. If we approximate the probability distribution functions using histograms, then examples such as [16] show that it is possible to design good steganalysis detectors even if the histograms of the cover object and stego-object are the same.

3. Consider the following embedding example. Let X and Y be two binary random variables such that $P(X = 0) = P(Y = 0) = 1/2$, and let them represent the host and covert messages, respectively. Let the embedding function be given by the following:

$$Z = X + Y \bmod 2. \tag{16.2}$$

We then observe that $D(P_Z\|P_X) = 0$ but $E(X - Z)^2 = 1$. Therefore, the nonzero mean-squared error value may give away enough information to a steganalysis detector even though $D(.) = 0$.

Given these arguments, is there an alternative measure for stegosecurity that is perhaps more fundamental to steganalysis? In the rest of this section, we present an alternate definition of steganographic security. In our new definition, the *false-alarm probability* ($\alpha = P$(detect message present|message absent)) and the *detection probability* ($\beta = P$(detect message present|message present) play important roles. A steganalysis detector's receiver operating characteristic (ROC) is a plot of α versus β. Points on the ROC curve represent the achievable performance of the steganalysis detector. The average error probability of steganalysis detection is given by

$$P_e = (1 - \beta)P(\text{message embedded}) + \alpha P(\text{message not embedded}). \tag{16.3}$$

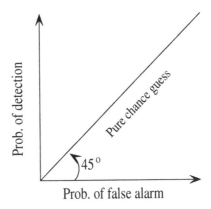

Fig. 16.2. Detector ROC plane.

If we assume P(message embedded) $= P$(message not embedded), then, from Eq. (16.3),

$$P_e = \frac{1}{2}\left[(1 - \beta) + \alpha\right]. \tag{16.4}$$

Note that α and β are detector-dependent values. For example, for a chosen value of α, β can be maximized by using a Neyman–Pearson statistical detector [7] or both α and β can be fixed and traded off with the number of observations required for detection by using Wald's sequential probability ratio test [9]. Observe from Eq. (16.4) that if $\alpha = \beta$, then $P_e = 1/2$, as shown in Figure 16.2. That is, the detector makes purely random guesses when it operates or is forced to operate on the 45° line in the ROC plane. This means that the detector does not have sufficient information to make an intelligent decision. Therefore, if the embedder forces the detector to operate on the 45° ROC line by employing appropriate algorithms and/or parameters, then we say that the stegomessage is secure and obtain the following definitions.

Definition 1. *A stego-embedding algorithm is said to be $\gamma_{\mathcal{D}}$-secure w.r.t. a steganalysis detector $\mathcal{D}$ if $|\beta_{\mathcal{D}} - \alpha_{\mathcal{D}}| \leq \gamma_{\mathcal{D}}$, where $0 \leq \gamma_{\mathcal{D}} \leq 1$.*

Definition 2. *A stego-embedding algorithm is said to be perfectly secure w.r.t. a steganalysis detector $\mathcal{D}$ if $\gamma_{\mathcal{D}} = 0$.*

Clearly, from these definitions we can think of embedding and steganalysis as a zero sum game where the embedder attempts to minimize $|\beta - \alpha|$ while the steganalyst attempts to maximize it.

16.3 Steganographic Capacity

One of the key performance measures used to compare different data-hiding algorithms is *capacity*. In a broader sense, capacity refers to the maximum message size that can be embedded in a cover object, subject to certain constraints. These constraints are chosen based on the specific application. For example, in digital watermarking systems, a power constraint on the watermark message is placed so that it can survive certain types of attacks. For covert communications, placing an upper bound on the amount of embedding-induced statistical changes in the host message is important. If this upper bound is small enough, then it is fair to say that steganalysis detection of the covert message will be difficult. Perceptual distortion constraints are also popular. Intuitively, there is an inherent trade-off between the embedding capacity and the strictness of the constraint. The capacity can be improved if the constraint is relaxed and vice versa.

There have been several attempts at computing the data-hiding embedding capacity for different mathematical models (e.g., see [5, 21, 20, 22, 23, 6, 10, 4] and references therein). The majority of these use tools from information theory. Here, data embedding is viewed as a communication channel, and capacity estimates for different source and channel models are derived. Then, the data-hiding capacity is defined as the maximum achievable embedding rate for which the error probability of message reception asymptotically vanishes. This definition is consistent with the goal of data-hiding and watermarking systems, where message reliability is of concern (i.e., it is necessary to detect the (possibly corrupted) embedded message at the receiver). Depending on the nature of the attack channel—compression, scaling, rotation, etc.—a capacity value can be computed.

Whereas data-hiding systems attempt to maximize the message reliability in the presence of malicious/nonmalicious attacks, the goal of steganographic techniques is to convey the message securely. That is, the primary goal of steganographic systems is covertness rather than reliability or robustness to signal-processing attacks. Attacks in a passive warden steganography framework are attempts to detect or estimate the embedded message rather than destroy or corrupt it. Therefore, it can be argued that Shannon's information-theoretic definition of capacity may not be the right one for steganographic systems. Intuitively, steganographic capacity must measure the maximum message size that can be embedded so that a steganalysis detector is only able to make a perfectly random guess about the presence/absence of a covert message. This raises the following important question: in steganographic embedding, what is the security trade-off with respect to message size? This question can be answered in several different ways. In image steganalysis it is shown that [7] when the embedded message size is larger than a threshold, it becomes easier for a steganalysis algorithm to detect it. Clearly, this defeats the primary purpose of steganography. Therefore, it can be observed that the definition of steganographic capacity must involve parameters of the embedding function as well as those of the steganalysis detector as justified below.

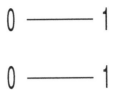

Fig. 16.3. Simple stegochannel.

- On average, a large message size will significantly perturb the host message statistics. This in turn will improve the steganalysis success probability.
- A steganalysis algorithm that is designed to detect one specific embedding algorithm may not be able to detect other ones even if the message size is significant.

These arguments motivate a definition of steganographic capacity from a detection-theoretic perspective rather than an information-theoretic one. In the following, we present some of these ideas in a more rigorous manner.

Recall the definition of γ-security defined above. We shall build on this notion to arrive at an alternative notion of steganographic capacity. However, before doing this, we present a simple numerical example to illustrate why it is useful to define a steganographic capacity measure from a steganalysis perspective. This example illustrates why Shannon's capacity definition may not be applicable directly in computing steganographic capacity. Consider the simple stegochannel, as shown in Figure 16.3. The inputs and outputs of the stegochannel are shown in this figure. These could be the least significant bits (LSB) of an image, for example. We could interpret the stego channel shown in Figure 16.3, as message embedding when the LSB and the secret message bit agree – a modified LSB embedding. Suppose that every LSB carries a message bit. Let us assume that the steganalysis detector knows that LSB embedding is used but does not know that the modified embedding algorithm has been used. Now, consider the following two steganalysis detectors, $\mathcal{D}_1$ and $\mathcal{D}_2$:

"Believe what you see detector," $\mathcal{D}_1$: decode 0 as 0 and decode 1 as 1

"Detect always 1 detector," $\mathcal{D}_2$: decode 0 as 1 and decode 1 as 1

It is quite obvious that $\alpha_{\mathcal{D}_1} = P(1|0) = 0$ and $\beta_{\mathcal{D}_1} = P(1|1) = 1$ and $\alpha_{\mathcal{D}_2} = 1$ and $\beta_{\mathcal{D}_2} = 1$. If we take the inequality in Definition 16.1 to be an equality, then $\gamma_{\mathcal{D}_1} = 1$, meaning that the embedding algorithm is totally insecure w.r.t. $\mathcal{D}_1$, and $\gamma_{\mathcal{D}_2} = 0$, implying that it is perfectly secure w.r.t. $\mathcal{D}_2$. Therefore, we can expect the steganographic capacity to be zero if $\mathcal{D}_1$ is employed since every embedded message bit can be decoded perfectly by the steganalysis detector. However, we note that the Shannon capacity of the stegochannel in Figure 16.3 is 1 bit/symbol, independent of the steganalysis detector used.

Before giving a formal definition of stegocapacity in the presence of steganalysis, we present another numerical example. Consider an embedding algorithm that changes the mean value of a Gaussian$(0,\sigma^2)$ cover signal by $m > 0$ to embed a message bit. Let the length of the message symbols be N. Let the steganalysis detector employ the minimum probability of error criterion. Then it is not difficult to show that

$$\alpha = \frac{1}{2\sqrt{2}}\left[1 - \operatorname{erf}\left(\frac{-m\sqrt{N}}{2\sigma}\right)\right], \tag{16.5}$$

$$\beta = \frac{1}{2\sqrt{2}}\left[1 - \operatorname{erf}\left(\frac{m\sqrt{N}}{2\sigma}\right)\right] \tag{16.6}$$

for this detector. If we assume that $m^2 \ll \sigma^2$ (i.e., the message-to-noise ratio is small due to perceptual considerations, etc.) and $m\sqrt{N}/2\sigma \ll 1$, then, to satisfy $|\beta - \alpha| \leq \gamma$, the embedder has to choose

$$N \leq \frac{2\pi\sigma^2\gamma^2}{m^2}. \tag{16.7}$$

We observe from this formula that the number of symbols that can be used for embedding and still satisfy the γ-security constraint is inversely proportional to the message-to-noise ratio. Therefore, if the message symbol strength is increased, the number of message-carrying symbols must decrease to maintain the same level of security. Also note that for other types of steganalysis detectors the rate of change of N with respect to message-to-noise ratio may be different. This means that the embedding *capacity* in the presence of steganalysis varies with respect to the steganalysis detector, thereby leading us to the following definition.

Definition 3. *[8] Let the number of message-carrying symbols be N, and let $\alpha_{\mathcal{D}}^{(N)}$ and $\beta_{\mathcal{D}}^{(N)}$ be the corresponding false-alarm and detection probabilities for a steganalysis detector $\mathcal{D}$. Then, define the stegocapacity as*

$$N_\gamma^* = \{\max N \text{ subject to } |\beta_{\mathcal{D}}^{(N)} - \alpha_{\mathcal{D}}^{(N)}| \leq \gamma_{\mathcal{D}}\} \text{ symbols.} \tag{16.8}$$

Based on the ideas above, Chandramouli and Memon [7] do a theoretical analysis of the capacity of LSB steganography and investigate the conditions under which an observer can distinguish between stegoimages and cover images. They derive a closed-form expression of the probability of false detection in terms of the number of bits that are hidden. They then formulate the steganalysis problem as a multiple-hypothesis testing problem and pool results at individual pixels to arrive at a global detection rule. This gives an analytical bound on the number of bits we can hide in an image using LSB-based techniques without causing statistically significant modifications. Although this work is interesting from an academic point of view, it results in loose upper bounds for steganographic capacity of LSB encoding in general. Some of the specific techniques described below for LSB steganalysis give better results in practice with real images.

16.4 Some Practical Steganalysis Techniques

16.4.1 Detection of LSB Steganography

The simplest image-steganography techniques essentially embed the message in a subset of the LSB plane of the image, possibly after encryption. It is well-known that an image is generally not visually affected when its least-significant-bit plane is changed. Popular steganographic tools based on LSB-like embedding vary in their approach for hiding information. Some methods use LSB embedding in the spatial domain, while others embed in the frequency. For brevity, we collectively refer to these techniques as LSB steganography.

Perhaps some of the earliest work in this regard was reported by Johnson and Jajodia [18]. They mainly look at palette tables in GIF images and anomalies caused therein by common stegotools that perform LSB embedding in GIF images. Since pixel values in a palette image are represented by indices into a color look-up table that contains the actual color RGB value, even minor modifications to these indices can result in annoying artifacts. Visual inspection or simple statistical analysis of such stegoimages can yield enough tell-tale evidence to discriminate between stego-images and cover images.

Another early approach to LSB steganalysis was presented in [25] by Westfeld and Pfitzmann. They note that LSB embedding induces a partitioning of image pixels into pairs of values (PoVs) that get mapped to one another. For example, the value 2 gets mapped to 3 on LSB flipping, and likewise 3 gets mapped to 2, so (2, 3) forms a PoV. Now LSB embedding causes the frequency of individual elements of a PoV to flatten out with respect to one another. So, for example, if an image has 50 pixels that have a value 2 and 100 pixels that have a value 3, then after LSB embedding of the entire LSB plane the expected frequencies of 2 and 3 are 75 and 75, respectively. This of course is when the entire LSB plane is modified. However, as long as the embedded message is large enough, there will be a statistically discernible flattening of PoV distributions, and this fact is exploited by their steganalysis technique. The length constraint, on the other hand, turns out to be the main limitation of their technique. LSB embedding can only be reliably detected when the message length becomes comparable with the number of pixels in the image. In the case where the message placement is known, shorter messages can be detected. But requiring knowledge of message placement is too strong an assumption, as one of the key factors playing in the favor of Alice and Bob is the fact that the secret message is hidden in a location unknown to Wendy.

Another steganalysis tool on similar lines but with a higher detection rate, even for short messages, was proposed by Fridrich, Du, and Meng [14]. They define pixels that are close in color intensity to be a difference of not more than one count in any of the three color planes. They then show that the ratio of close colors to the total number of unique colors increases significantly when a new message of a selected length is embedded in a cover image as opposed to when the same message is embedded in a stegoimage (that is, an image already carrying an LSB-encoded message). It is this difference that enables them to distinguish cover images from stegoimages for the case of LSB steganography.

A more direct approach that analytically estimates the length of an LSB-embedded message in an image was proposed by Dumitrescu et al. [11]. Their technique to detect LSB steganography in continuous-tone natural images is based on an important statistical identity related to certain sets of pixels. This identity is very sensitive to LSB embedding, and the change in the identity can quantify the length of the embedded message. In the rest of this subsection, we describe this technique in more detail. Our description is adopted from [11].

Consider the partition of an image into pairs of horizontally adjacent pixels. Let $\mathcal{P}$ be the set of all these pixel pairs. Define the subsets X, Y, and Z of $\mathcal{P}$ as follows:

- X is the set of pairs $(u, v) \in \mathcal{P}$ such that v is even and $u < v$, or v is odd and $u > v$.
- Y is the setof pairs $(u, v) \in \mathcal{P}$ such that v is even and $u > v$, or v is odd and $u < v$.
- Z is the subset of pairs $(u, v) \in \mathcal{P}$ such that $u = v$.

After having made the definitions above, the authors of [11] make the assumption that statistically we will have

$$|X| = |Y|. \tag{16.9}$$

This assumption is true for natural images, as such images are isotropic in terms of the gradient of intensity function. In other words, the gradient in any direction has an equal probability of being to be positive or negative.

Furthermore, they partition the set Y into two subsets, W and V, with W being the set of pairs in $\mathcal{P}$ of the form $(2k, 2k + 1)$ or $(2k + 1, 2k)$ and $V = Y - W$. Then $\mathcal{P} = X \cup W \cup V \cup Z$. They call sets X, V, W, and Z *primary sets*.

When LSB embedding is done pixel values get modified and so does the membership of pixel pairs in the primary sets. More specifically, given a pixel pair (u, v), they identify the following four situations:

00) both values u and v remain unmodified;
01) only v is modified;
10) only u is modified;
11) both u and v are modified.

The corresponding change of membership in the primary sets is shown in Figure 16.4.

Now consider a subset A of $\mathcal{P}$ and denote by $\rho(\pi, A)$ the probability that the pixel pairs of A are modified with pattern $\pi \in \{00, 10, 01, 11\}$.

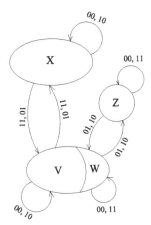

Fig. 16.4. State-transition diagram for sets X, V, W, Z under LSB flipping.

The authors make the critical assumption that for each modification pattern $\pi \in \{00, 10, 01, 11\}$ and each primary set $A \in \{X, V, W, Z\}$

$$\rho(\pi, A) = \rho(\pi, \mathcal{P}).$$

The assumption above essentially implies that message embedding is done by flipping LSB bits that are randomly scattered throughout the image. In fact, the entire LSB steganalysis technique proposed relies on this assumption and the technique can fail if pixel locations are selected judiciously for embedding. In any case, it is now easy to see the following facts:

i) $\rho(00, \mathcal{P}) = (1 - p/2)^2$;
ii) $\rho(01, \mathcal{P}) = \rho(10, \mathcal{P}) = p/2(1 - p/2)$;
iii) $\rho(11, \mathcal{P}) = (p/2)^2$.

These facts are all that are needed to derive a steganalysis system for LSB embedding (conditioned on the two assumptions that were stated earlier). Let p be the length of the embedded message in bits divided by the total number of pixels. Then the fraction of the pixels modified by the LSB embedding is $p/2$. For each $A \in \{X, Y, V, W, Z\}$, denote by A' the set defined in the same way as the set A but considering the pixel values after the embedding. This results in the following equations:

$$|X'| = |X| \cdot (1 - p/2) + |V| \cdot p/2, \tag{16.10}$$
$$|V'| = |V| \cdot (1 - p/2) + |X| \cdot p/2, \tag{16.11}$$
$$|W'| = |W|(1 - p + p^2/2) + |Z|p(1 - p/2). \tag{16.12}$$

Relations (16.10) and (16.11) imply that

$$|X'| - |V'| = (|X| - |V|)(1 - p). \tag{16.13}$$

By some simple manipulations, the authors finally arrive at the equation

$$0.5\gamma p^2 + (2|X'| - |\mathcal{P}|)p + |Y'| - |X'| = 0, \qquad (16.14)$$

where $\gamma = |W| + |Z| = |W'| + |Z'|$. The equation above allows one to estimate p, (i.e., the length of the embedded message, based on X', Y', W', Z', which can all be measured from the image being examined for possible steganography). Of course, it should be noted that we cannot have $\gamma = 0$, the probability of which for natural images is very small.

In fact, the pairs-based steganalysis described above was inspired by an effectively identical technique, although from a very different approach, called RS steganalysis by Fridrich et al. in [15], that had first provided remarkable detection accuracy and message-length estimation even for short messages. However, RS steganalysis does not offer a direct analytical explanation that can account for its success. It is based more on empirical observations and their modeling. It is interesting to see that the pairs-based steganalysis technique essentially ends up with exactly the same steganalyzer as RS steganalysis.

16.5 Universal Steganalysis Techniques

The steganalysis techniques described above were all specific to a particular family of algorithms (i.e., LSB embedding). A more general class of steganalysis techniques pioneered independently by Avcibas et al. [1–3] and Farid [12] are designed to work with any steganographic embedding algorithm, even an unknown algorithm. Such techniques have subsequently been called *universal steganalysis* techniques. Such techniques essentially design a classifier based on a training set of cover objects and stegoobjects arrived at from a variety of different algorithms. Classification is done based on some inherent "features" of typical natural images that can get violated when an image undergoes some embedding process. In the rest of this section, we describe in detail one such technique, proposed by Avcibas et al. [3].

In [3], Avcibas et al. develop a discriminator for cover images and stegoimages using an appropriate set of image quality metrics (IQMs). Objective image-quality measures are values based on image features, a function of which should correlate well with subjective judgment; that is, the degree of (dis)satisfaction of an observer. Such measures have been utilized in coding artifact evaluation, performance prediction of vision algorithms, quality loss due to sensor inadequacy and other applications. In [3], they are used not as predictors of subjective image quality or algorithmic performance but specifically as a steganalysis tool; that is, as features in distinguishing cover objects from stego-objects.

Typically, a good IQM should be accurate, consistent, and monotonic in predicting quality. In the context of steganalysis, prediction accuracy can be interpreted as the ability of the measure to detect the presence of a hidden message with minimum error on average. Similarly, prediction monotonicity signifies that IQM scores should ideally be monotonic in their relationship to the embedded message size. Finally, prediction consistency relates to the quality measure's ability to provide consistently accurate predictions for a large set of steganography techniques and image types. This implies that the spread of quality scores due to factors of image variety should not eclipse the score differences arising from message-embedding artifacts. In this context, the authors use analysis of variance (ANOVA) techniques to determine whether a metric's response was consistent with a change in the image or whether it was a random effect. They arrive at a ranking of IQMs based on their F-scores in the ANOVA tests to identify the ones that responded most consistently and strongly to message embedding. The idea is to seek IQMs that are sensitive specifically to steganography effects, that is, those measures for which the variability in score data can be explained better because of some treatment rather than as random variations due to the image set. The rationale of using several quality measures is that different measures respond with differing sensitivities to artifacts and distortions. For example, measures such as mean-square error respond more to additive noise, whereas others such as spectral phase or mean-square HVS-weighted (human visual system) error are more sensitive to pure blur and the gradient measure reacts to distortions concentrated around edges and textures. Similarly, embedding techniques affect different aspects of images.

The idea behind detection of watermark or hidden message presence is to obtain a consistent distance metric for images containing a watermark or hidden message vis-à-vis those without with respect to a common reference. The reference processing should possibly recover the original unwatermarked image and for this purpose they use low-pass filtering based on a Gaussian kernel. Other approaches such as denoising and Wiener filtering are also possible. In fact, they report that a Wiener filtering approach gave better results, for example, in the case of certain steganographic algorithms, while denoising proved more effective in the other cases. The Gaussian filtering approach gave uniformly good results across all steganographic techniques. The reason why Gaussian blurring works fine as a common reference is that it gives us the local mean, which is also the maximum-likelihood (ML) estimate of the image under the Gaussian assumption. Under a Laplacian distribution assumption, the median would have been the ML estimate. Therefore, the blurred image minus the original image yields the maximum-likelihood estimate of the additive watermark.

The authors consistently obtained statistically different quality scores from embedded and filtered images and from filtered but not embedded sources. For example, Figure 16.5 gives an instance of the cover object and stego-object class separability based on a scatter diagram of three image quality-metrics. The steganalysis detector they then develop is based on regression analysis of a number of relevant IQMs. In the design phase of the steganalyzer, they regress normalized IQM scores to, respectively, −1 and 1, depending upon

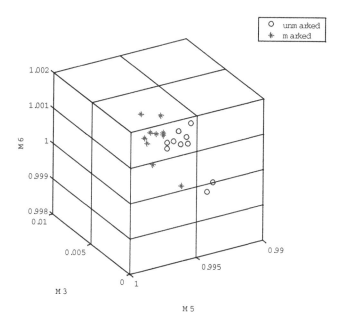

Fig. 16.5. Scatter plot of three image-quality measures showing separation of marked and unmarked images.

whether an image did not or did contain a covert message. Similarly, IQM scores are calculated between the original images and their filtered versions. In the regression model, they expressed each decision label y in a sample of N observations as a linear function of the IQM scores, denoted as x's, plus a random error, ϵ:

$$y_1 = \beta_1 x_{11} + \beta_2 x_{12} + \ldots + \beta_q x_{1q} + \epsilon_1$$
$$y_2 = \beta_1 x_{21} + \beta_2 x_{22} + \ldots + \beta_q x_{2q} + \epsilon_2$$
$$\vdots$$
$$y_n = \beta_1 x_{N1} + \beta_2 x_{N2} + \ldots + \beta_q x_{Nq} + \epsilon_N$$

In the first expression, x_{ij} denotes the IQM score, where the first index indicates the i'th image $i = 1, \ldots, N$ and the second one the quality measure, $j = 1, \ldots, q$, so that the total number of quality measures considered is q. The βs denote the regression coefficients. The complete statement of the standard linear model is

$$y = \mathbf{X}_{nxq}\beta + \epsilon.$$

In the second expression, the $N \times q$ data matrix has rank q and ϵ is a zero-mean Gaussian noise. The corresponding optimal MMSE linear predictor β

can be obtained by

$$\hat{\beta} = \left(\mathbf{X}^T\mathbf{X}\right)^{-1}\left(\mathbf{X}^T\mathbf{y}\right).$$

In the third expression, the prediction coefficients are obtained in the training phase, these coefficients can be used in the testing phase. Given an image in the test phase, first it is filtered and the q IQM scores are obtained using the image and its filtered version. Then, using the prediction coefficients, these scores are regressed to the output value. If the output exceeds the threshold 0 then the decision is that the image is embedded; otherwise the decision is that the image is not embedded. That is,

$$\hat{y} = \hat{\beta}_1 x_1 + \hat{\beta}_2 x_2 + \ldots + \hat{\beta}_q x_q,$$

for $\hat{y} \geq 0$ the image contains a watermark, and for $\hat{y} < 0$ it does not.

Simulation results with the chosen feature set and well-known watermarking and steganographic techniques indicate that their approach is able with reasonable accuracy to distinguish between cover images and stegoimages. The selected IQMs form a multidimensional feature space whose points cluster well enough to do a classification of marked and nonmarked images. The classifier is also able to do a classification when the tested images come from an embedding technique unknown to it, indicating that it has a generalizing capability of capturing the general intrinsic characteristics of steganographic techniques.

16.6 Conclusions

The past few years have seen an increasing interest in using images as cover media for steganographic communication. There have been a multitude of public-domain tools, albeit many being adhoc and naive, available for image-based steganography. Given this fact, detection of covert communications that utilize images has become an important issue. There have been several techniques for detecting stegoimages that have been developed in the past few months. In this chapter, we have have reviewed some fundamental notions related to steganography using image media, including security and capacity. We also described in detail two steganalysis techniques that are representative of the different approaches that have been taken.

Acknowledgment.

This work was sponsored by Air Force Research Laboratory under agreement #F306020-02-2-0193 and NSF DAS 0242417.

References

[1] I. Avcibas, N. Memon, and B. Sankur, "Steganalysis using Image Quality Metrics," in *Security and Watermarking of Multimedia Contents*, San Jose, CA, February 2001.

[2] I. Avcibas, N. Memon, and B. Sankur, "Image steganalysis with binary similarity measures," *IEEE International Conference on Image Processing*, Rochester, New York, September 2002.

[3] I. Avcibas, N. Memon, and B. Sankur, "Steganalysis using image quality metrics," *IEEE Trans. Image Process.*, **12**, 221–229 (2003).

[4] C. Cachin, "An information-theoretic model for steganography," *2nd International Workshop on Information Hiding*, Lecture Notes in Computer Science, **1525**, pp. 306–318 (Springer-Verlag, Berlin, 1998).

[5] R. Chandramouli, "Data hiding capacity in the presence of an imperfectly known channel," *SPIE Proceedings of Security and Watermarking of Multimedia Contents II*, vol. **4314**, pp. 517–522 (SPIE, Bellingham, WA, 2001).

[6] R. Chandramouli, "Watermarking capacity in the presence of multiple watermarks and partially known channel," *SPIE Multimedia Systems and Applications IV*, vol. **4518**, pp. 210–215 (SPIE, Bellingham, WA, 2001).

[7] R. Chandramouli and N. Memon, "Analysis of LSB image steganography techniques," *IEEE Int. Conf. Image Process.*, **3**, 1019–1022 (2001).

[8] R. Chandramouli and N. Memon, "Steganography capacity: A steganalysis perspective," *SPIE Security and Watermarking of Multimedia Contents V*, vol. **5020** (SPIE, Bellingham, WA, 2003).

[9] R. Chandramouli and N. Memon, "On sequential watermark detection," *IEEE Trans. Signal Process.*, **51**(4), 1034–1044 (2003).

[10] A. Cohen and A. Lapidoth, "On the Gaussian watermarking game," *International Symposium on Information Theory*, p. 48, June 2000.

[11] S. Dumitrescu, X. Wu, and N. Memon, "On steganalysis of random LSB embedding in continuous-tone images," *IEEE International Conference on Image Processing*, Rochester, New York, September 2002.

[12] H. Farid, "Detecting steganographic messages in digital images," Technical Report, TR 2001-412, Dartmouth College, Hanover, NH, 2001.

[13] J. Fridrich and Rui Du, "Secure steganographic methods for palette images," *Proceeding of the 3rd Information Hiding Workshop*, **1768**, pp. 47–60 (Springer-Verlag, New York, 2000).

[14] J. Fridrich, R. Du, and L. Meng, "Steganalysis of LSB encoding in color images," *ICME 2000*, New York, July 31–August 2, 2000.

[15] J. Fridrich, M. Goljan, and R. Du, "Detecting LSB steganography in color and gray-scale images," *IEEE Multimedia Spec. Issue Security*, pp. 22–28, October–November 2001.

[16] J. Fridrich, M. Goljan and D. Hogea, "Attacking the OutGuess," *Proceedings of the ACM Workshop on Multimedia and Security 2002*, Juan-les-Pins, France, December 6, 2002.

[17] J. Fridrich, M. Goljan, and D. Hogea, "Steganalysis of JPEG images: Breaking the F5 algorithm," *5th Information Hiding Workshop,* Noordwijkerhout, The Netherlands, October 7–9, 2002, pp. 310–323.

[18] N.F. Johnson and S. Jajodia, "Steganalysis of images created using current steganography software", in David Aucsmith, ed., *Information Hiding*, Lecture Notes in Computer Science, **1525**, pp. 32–47 (Springer-Verlag, Berlin, Heidelberg, 1998).

[19] N.F. Johnson and S. Katzenbeisser, "A survey of steganographic techniques," in S. Katzenbeisser and F. Petitcolas, eds., *Information Hiding*, pp. 43–78 (Artech House, Norwood, MA, 2000).

[20] P. Moulin and M. Mihcak, "The data hiding capacity of image sources," preprint available at http://www.ifp.uiuc.edu/~moulin/paper.html.

[21] P. Moulin and J. Sullivan, "Information theoretic analysis of information hiding," To appear in *IEEE Trans. Information Theory* (2003).

[22] M. Ramkumar and A. Akansu, "Information theoretic bounds for data hiding in compressed images," *IEEE 2nd Workshop on Multimedia Signal Processing*, December 1998, pp. 267–272.

[23] M. Ramkumar and A. Akansu, "Theoretical capacity measures for data hiding in compressed images," *SPIE Multimedia Systems and Application*, vol. **3528**, pp. 482–492 (SPIE, Bellingham, WA, 1998).

[24] G. Simmons, "The prisoners problem and the subliminal channel," in D. Chaum, ed., *Advances in Cryptology: Proceedings of CRYPTO '83*, pp. 51–67 (Plenum, New York, 1983).

[25] A. Westfeld and A. Pfitzmann, "Attacks on steganographic systems," in A. Pfitzmann, ed., *Information Hiding*, Lecture Notes in Computer Science, **1768**, pp. 61–76 (Springer-Verlag, Berlin, Heidelberg, 1999).

[26] J. Zollner, H. Federrath, H. Klimant, A. Pfitzman, R. Piotraschke, A. Westfeld, G. Wicke, and G. Wolf, "Modeling the security of steganographic systems," *2nd Information Hiding Workshop*, April 1998, pp. 345–355.

Public-Key Cryptography: An Overview of some Algorithms

Franck Leprévost

Summary. The security of communication systems requires more and more tehnical tools and approaches. This involves in particular cryptographic aspects. We describe here an overview of some algorithms used for public-key cryptography, and digital signatures. The focus here is mainly on the description of the approaches, and does not address an exhaustive description of the available publicc-key cryptosystems.

17.1 Introduction

Since the mid-1970s, the security of communication systems has no longer belonged strictly to the pure military or governmental area but has become a field of interest in R&D departments of companies and in the academic world. Since then, this area of activity has become more and more important as the ways of communication have changed and secure communications are required for financial transactions, protection of copyrights and privacy, and other concerns (see [10] and [13] for some aspects related to electronic surveillance and the Echelon network, [14] for tasks related to counterterrorism, and [11] and [12] for tasks related to the protection of digital copyrights in Europe). At the base of all the solutions aiming at secure communications, one finds one or many *cryptographic modules*.

The word *cryptology* concerns two aspects: on the one hand, *cryptography*, which covers the algorithms allowing confidentiality, integrity, authenticity, and nonrepudiation of data; on the other hand, *cryptanalysis*, which covers the attacks on cryptosystems. Cryptology itself splits again into two subdisciplines: secret-key cryptosystems and public-key cryptosystems.

We shall not describe here secret-key cryptosystems (the new standard is the Advanced Encryption Standard, AES). Rather, we explain their general principle. A key K is generated and is used both for enciphering and deciphering messages. This key should be kept secret and is shared only by the two communicating entities (Alice and Bob, say). Moreover, in a network of

n communicating entities, it is necessary for the keys to be pairwise different. This implies that one needs $n(n-1)/2$ different keys.

One sees that secret-key cryptosystems state the problems of key generation and distribution. These problems are mainly solved when one relies on public-key infrastructure (PKI). Such an infrastructure requires public-key algorithms. We shall avoid the aspects of PKI (see [17] and [25]) and concentrate on the algorithms.

A public-key cryptosystem requires two keys per communicating entity. Each communicating entity publishes (as in a telephone directory) a public key corresponding to its name, say. If Bob wants to communicate with Alice, he recovers the public key of Alice, uses it to encipher his message, and sends the encrypted message to Alice. Only Alice can decipher the received message, using her secret key. The secret key is of course mathematically related to the public key. However, it should not be possible to recover the secret key knowing the public key (some other more technical conditions are necessary, but here is the principle). This security condition is measured by the difficulty in solving a mathematical problem in practice.

One can use public-key cryptosystems in order to agree on a common value, say K, on an insecure channel. This value can then be used as the secret key of a secret-key cryptosystem such as AES, for instance. When this procedure is achieved, the two corresponding entities can forget the public-key cryptosystem and use only the secret-key algorithm with this key for their communications (say of the day). This is a good way to handle the problem because, in general, public-key cryptosystems are about 1000 times slower than secret-key cryptosystems. Public-key cryptosystems in general may be used for digital signatures as well.

In Section 17.2, I describe some selected public-key cryptosystems and their applications: the encryption scheme RSA, the Diffie–Hellman key-exchange protocol, and the digital signature algorithm. The security of these systems is based on two mathematical problems: the integer factorization problem and the discrete logarithm problem. These problems are introduced in Section 17.2 and more closely considered in Sections 17.3 and 17.4. In Section 17.5, we outline some aspects that are not considered in this chapter (because their understanding requires a background that goes beyond the scope of this chapter) but are definitely of high interest.

Let us move on to Alice and Bob, our two friends in this story.

17.2 Public-Key Cryptosytems and their Applications

The aim of this section is to illustrate, with some examples, the uses of public-key cryptosystems. First, we describe the encryption scheme RSA. Its security is based on the factorization problem, and this will be considered later in this chapter. Then, we explain the Diffie–Hellman key-exchange protocol and the digital signature algorithms. The security of these algorithms is based on the discrete logarithm problem, which we state in this section and consider more closely later in this chapter.

17.2.1 The RSA Cryptosystem

The cryptosystem RSA ([3]), after the names of the authors Rivest, Shamir, and Adleman, is the most used in the world. Its security is mainly based on the integer factorization problem: given an integer n, a product of two primes p and q, find p and/or q. This problem turns out to be quite difficult if the primes are of length ≥ 256 bits.

Generation of the Keys

Alice randomly chooses two prime numbers p and q (see [18], [20], [22]) and computes $n = pq$. Alice randomly chooses an integer e (odd) such that $1 < e < \varphi(n) = (p-1)(q-1)$ and $\gcd(e, \varphi(n)) = 1$. Alice then computes the integer $1 < d < \varphi(n)$ such that

$$ed \equiv 1 \bmod \varphi(n).$$

The public key of Alice is (n, e), and her secret key is d. One calls e the RSA exponent and n the RSA module.

Encryption

We describe here the basic RSA algorithm. Bob recovers the public key of Alice and wants to send her the text m such that $0 \leq m < n$. Bob computes

$$c = m^e \bmod n$$

Bob sends the quantity c to Alice.

Decryption

The decryption uses the following result: let (n, e) be a public key for the RSA algorithm and d the corresponding secret key. Then, for $0 \leq m < n$, the following relation holds:

$$(m^e)^d \bmod n = m.$$

As a consequence, when Alice receives c, she computes $c^d \bmod n$ and recovers the message m sent by Bob.

Security

The goal of the pirate is to recover the clear text m. The best way to do this, currently, is to factorize the modulus n. However, it is not proved whether this is the only way. The current limits of the factorization methods are around 160 digits. The record is now in Germany, where F. Bahr, J. Franke, and T. Kleinjung (University of Bonn) factorized on January 18th, 2002, a number of 158 digits, which is a divisor of $2^{953} - 1$.

17.2.2 Diffie–Hellman's Key Exchange Protocol

In this section, we describe a method, due to Diffie and Hellman, which allows Alice and Bob to agree on a common secret key (to be used as the secret key of AES for instance) on a nonsecure channel. The security of this method is deeply related to the discrete logarithm problem, which we explain first.

DLP: The Discrete Logarithm Problem

A group G is a set G together with an internal binary operation satisfying some natural axioms: associativity, neutral element, and inverse element. For instance, $\mathbf{Z}$ together with the operation $+$ is a group, but $\mathbf{N}$ with $+$ is not because, except for 0, no element has an inverse in $\mathbf{N}$ (they are in $\mathbf{Z}$!). In cryptography, we need finite groups (i.e., the set G is finite). Moreover, we need cyclic groups (generated by one element), and it is very often useful to denote these groups multiplicatively (so the binary operation is the multiplication of elements). Let $G = \langle g \rangle$ be a cyclic group generated by an element g. Let $h \in G$. In this setting, the discrete logarithm problem is: knowing G, g, h, find $x \in \mathbf{Z}$ (denoted $x = \log_g h$ or also the logarithm of h in basis g) such that

$$h = g^x.$$

Because g is a generator of G, such an x exists.

The Key-Exchange Protocol of Diffie–Hellman

With these notations, supposing G and g are public, the Diffie–Hellman key-exchange protocol in G is the following procedure: let $n = \operatorname{Card} G$ (public). Alice chooses a random integer $1 \leq a \leq n - 1$ and computes

$$A = g^a.$$

Alice send A to Bob. Bob does the same: he randomly chooses an integer $1 \leq b \leq n - 1$ and computes
$$B = g^b.$$

Bob sends B to Alice. Alice can compute B^a, and Bob can compute A^b. The common key is
$$K = g^{ab} = A^b = B^a.$$

Security

An observer, looking at the exchanges between Alice and Bob, faces the following mathematical problem (the so-called Diffe–Hellman mathematical problem): knowing G, g, g^a, and g^b, compute g^{ab}. If he can solve this problem in practice he can recover the common secret key K, and the protocol is unsecure.

This is for instance the case if he can efficiently solve the discrete logarithm problem in G. Currently, the best known way to solve the Diffe–Hellman problem is to solve first the discrete logarithm problem in the underlying group. However, it is not proved that these two problems are algorithmically equivalent to each other.

So which group G should we choose? Two conditions must be satisfied. First, one should be able to compute efficiently in the group; in particular, the rise of the generator g to a given power should be fast. On the other hand, the discrete logarithm problem should be hard to solve in the group G. With regard to the current knowledge, these conditions are satisfied in the multiplicative group of a finite field $G = \mathbf{F}_p^*$ for convenient choices of primes p and of a generator of G (this group is simply the set $\mathbf{Z}/p\mathbf{Z}-\{0\}$ of the nonzero remainders of the division of integers by p together with the multiplication). There are other choices (see the last part of this chapter).

Other attacks, related to the public-key infrastructure (such as the *man in the middle* attack), are not considered here.

17.2.3 Digital Signatures and the DSA Standard

We describe here one of the main applications of public-key cryptography: digital signatures. We first explain the general principle and then outline the standard DSA (digital signature algorithm), whose security is based on the discrete logarithm problem.

General Idea

It is becoming more and more important to be able to sign digital data. The main difference with the signatures on paper documents is that the signature does not only depend on the person who signs but also on the document she wants to sign. If this was not the case, using a simple cut-and-paste procedure, one would be able to falsify a signature. Procedures deeply related to public-key cryptography allow this problem to be solved and provide good solutions for digital signatures.

Alice, with her secret key d, signs a digital document m by producing a signature $s(d, m)$. With the public key e of Alice, anybody, in particular a judge, can check the validity of the signature of the document without the knowledge of the secret key. The signature algorithm is considered secure if it is not possible in practice to falsify a signature $s(d, m)$ of a document m without knowing the secret key d (other conditions are necessary, but we prefer here to keep it simple). How does such an algorithm work? Let C and D be, respectively, the enciphering and deciphering algorithms. The public key is e, and the secret key is d. Suppose that the cryptosystems satisfy the following equation for every document m to sign:

$$C(D(m, d), e) = m.$$

The signature s of the document m is simple $s = D(m, d)$. The verification's procedure consists in checking the validity of the equation above. If the equation is satisfied, the signature is accepted.

There are many algorithms for digital signatures, such as the RSA signature algorithm, the Rabin signature algorithm, and the El Gamal signature algorithm, for instance. The following section outlines the DSA.

The Standard DSA

DSA means digital signature algorithm and is a standard of the NIST (National Institute for Standards and Technology).

Generation of the Parameters

Alice generates a prime number q of 160 bits:

$$2^{159} < q < 2^{160}.$$

Alice then chooses a prime number p of length between 512 and 1024 bits, satisfying the following conditions:

- $2^{511+64t} < p < 2^{512+64t}$ for an integer t such that $0 \leq t \leq 8$,
- the prime number q divides $p - 1$.

The second condition implies that $\mathbf{F}_p^*$ contains a subgroup of order q. Let $\tilde{g}$ be a primitive root modulo p (a generator of $\mathbf{F}_p^*$), and let us denote

$$g = \tilde{g}^{(p-1/q)} \bmod p,$$

a generator of the subgroup of $\mathbf{F}_p^*$ of order q. Alice chooses a random element $a \in \{1, \ldots, q-1\}$ and computes

$$A = g^a \bmod p.$$

The public key of Alice is (p, q, g, A). The underlying discrete logarithm problem is in the group of order q.

Generation of a Signature

Alice wants to sign $m \in \{1, \ldots, q-1\}$. Alice chooses an integer $k \in \{1, \ldots, q-1\}$ and computes

$$r = (g^k \bmod p) \bmod q$$

and

$$s = k^{-1}(m + ar) \bmod q,$$

where k^{-1} is the inverse of k modulo q. The signature of m is then (r, s).

Verification of a Signature

Bob wants to check the validity of the signature (r, s) of m sent by Alice. He recovers first the public key (p, q, g, A) of Alice. He then checks the formats of the parameters, in particular that the following holds:

$$1 \leq r, \ s \leq q - 1.$$

If it is not the case, the signature is considered as nonvalid. If the conditions on the formats of the parameters are satisfied, Bob checks the following equation:

$$r = ((g^{s^{-1}m \bmod q} A^{rs^{-1} \bmod q}) \bmod p) \bmod q.$$

The signature is valid if and only if this equation is satisfied.

17.3 The Integer Factorization Problem and the Quadratic Sieve Method

The security of the cryptosystems RSA (or of Rabin) is deeply related to the number-theorectic problems of factorization. There are several methods for proving that a given number n is composite. The fastest of them (such as the Rabin–Miller algorithm) actually do not provide a divisor of n, but just output the information that the number n is composite or probably prime (with a high probability). So, it is necessary to rely on factorization methods. Many approaches are possible such as the method of Fermat and the $p-1$ and the ρ methods of Pollard (see [6], [7], [8], [9], [20]). In this chapter, we only describe the quadratic sieve method of Pomerance, which is very efficient for numbers with less than 129 digits. There is a natural generalization of it, called the general number field sieve method (GNFS), which in practice provides the best results for numbers with more than 130 digits, but the necessary background is beyond the scope of this chapter.

17.3.1 Description of the Method

Let n be a composite integer. One looks for two integers x and y such that

$$x^2 \equiv y^2 \bmod n.$$

and

$$x \not\equiv \pm y \bmod n.$$

If these conditions are satisfied, then n divides $x^2 - y^2 = (x + y)(x - y)$; however, n does not divide $x \pm y$. As a consequence, $d = \gcd(x - y, n)$ is a strict divisor of n. Let $m = \lfloor \sqrt{n} \rfloor$ and

$$f(X) = (X + m)^2 - n.$$

Obviously, one has

$$(t + m)^2 \equiv f(t) \bmod n$$

for all integers t. The core of the method is to find enough integers t providing such congruences that their product provides the relation $x^2 \equiv y^2 \bmod n$ we are looking for. In order to achieve that, let B be a nonnegative number, and $F(B) = \{p \in \mathbf{P}; p \leq B\} \cup \{-1\} = \{p_1, p_2, \ldots, p_k\}$, where $p_1 = -1$ and p_2, $p_3, \ldots, p_k$ are prime numbers $\leq B$. Suppose first that we have found $r > k$ values $t_1, \ldots, t_r$ such that, for $1 \leq i \leq r$, $f(t_i)$ is B-smooth. In other words,

$$f(t_i) = \prod_{j=1}^{k} p_j^{\alpha_{i,j}}.$$

Let $a_{i,j} = \alpha_{i,j} \bmod 2$.

Because $r > k$, the vectors v_i of the matrix $a_{i,j}$ satisfy a linear relation in $\mathbf{F}_2^k$, and one gets

$$v_{i_1} + \cdots + v_{i_s} = 0,$$

where $s \leq r$.

In particular,

$$f(t_{i_1}) \ldots f(t_{i_s}) = \left(\prod_{j=1}^{k} p_j^{(\alpha_{i_1,j} + \cdots + \alpha_{i_s,j}/2)} \right)^2$$

is a square. We then let

$$x = (t_{i_1} + m) \ldots (t_{i_s} + m) \bmod n,$$

and

$$y = \prod_{j=1}^{k} f_j^{(\alpha_{i_1,j} + \cdots + \alpha_{i_s,j}/2)} \bmod n$$

How do we get the values t for which $f(t)$ is B-smooth? A first approach is to check whether $f(t)$ is B-smooth for $t = 0, \pm 1, \pm 2, \ldots$. It is quite time-consuming because one has to divide $f(t)$ by all prime numbers $p \leq B$ in order to check that $f(t)$ is not B-smooth. The following table provides some information for the time needed to achieve the computation with this approach:

Number of ciphers of n	50	60	70	80	90	100	110	120
Cardinality of the factor basis (in thousands)	3	4	7	15	30	51	120	245

For this reason, one prefers to use a sieve method. We describe here a simple version of it: one fixes a sieve interval

$$T = \{-c, -c + 1, \ldots, 0, 1, \ldots, c\}.$$

One then searches the values $t \in T$ such that $f(t)$ is B-smooth. First, one systematically computes $f(t)$ for all the values $t \in T$. Then, for each prime number p of the basis factor $F(B)$, one divides $f(t)$ by the highest power of p dividing $f(t)$. Finally, $f(t)$ is B-smooth if, at the end of the process, the value obtained equals ± 1.

Example. Let $n = 8633$, and let us try to factorize n by the quadratic sieve method. One has $m = \lfloor \sqrt{8633} \rfloor = 92$, and

$$f(X) = (X + 92)^2 - 8633.$$

Let us choose $B = 11$, the factor basis $\{-1, 2, 3, 5, 7, 11\}$, and the sieve interval $\{-7, -6, -5, -4, -3, -2, -1, 0, 1, 2, 3, 4, 5, 6, 7\}$. The sieve is provided in the following table:

t	-7	-6	-5	-4	-3
$(t + 92)^2 - 8633$	-1408	-1237	-1064	-889	-712
Sieve with 2	-11		-133		-89
Sieve with 3					
Sieve with 5					
Sieve with 7			-19	-127	
Sieve with 11	-1				

t	-2	-1	0	1	2
$(t + 92)^2 - 8633$	-533	-352	-169	16	203
Sieve with 2		-11		1	
Sieve with 3					
Sieve with 5					
Sieve with 7					29
Sieve with 11	-1				

t	3	4	5	6	7
$(t + 92)^2 - 8633$	392	583	776	971	1168
Sieve with 2	49		97		73
Sieve with 3					
Sieve with 5					
Sieve with 7	1				
Sieve with 11		53			

One finds

$$\begin{aligned}
f(-7) &= 85^2 - 8633 = -1408 = -2^7 \times 11 \\
f(-1) &= 91^2 - 8633 = \ -352 = -2^5 \times 11 \\
f(1) &= 93^2 - 8633 = \quad 16 = 2^4 \\
f(3) &= 95^2 - 8633 = \quad 392 = 2^3 7^2.
\end{aligned}$$

The first two equations provide

$$85^2 \equiv -2^7 \times 11 \bmod 8633$$
$$91^2 \equiv -2^5 \times 11 \bmod 8633.$$

In particular, if one sets $x = 85.91 \bmod 8633 = 7735$ and $y = 2^6 \times 11 \bmod 8633 = 704$, one gets $x^2 \equiv y^2 \bmod 8633$ and

$$\gcd(x - y, 8633) = 89, \quad \gcd(x + y, 8633) = 97,$$

and the factorization of n is $8633 = 89 \cdot 97$.

17.4 The Discrete Logarithm Problem and the Pohlig–Hellman Reduction

Let $G = \langle g \rangle$ be a group generated by an element g of order n. Let $h \in G$. In this setting, the discrete logarithm problem (DLP) is: knowing G, g, h, find $x \in \mathbf{Z}$ such that

$$h = g^x.$$

There are several methods for solving this problem. By general, we mean that these methods do not use particular extra information on the group (see the conclusion for more on this subject). Such methods are, for instance, the baby-step giant-step (BSGS) method of Shanks, the ρ method of Pollard, and the index calculus method (see [19], [24], [26]). Here we describe the Pohlig–Hellman reduction method. The goal of this method is to reduce the DLP in the group G of order n into many DLPs in groups of orders p, for all prime divisors p of n. Hence, this approach needs the knowledge of the factorization of the order n.

Suppose that we know the prime decomposition of

$$n = \operatorname{Card} G = \prod_{p \mid n} p^{v_p(n)}.$$

We simply write v_p for $v_p(n)$ here. The Pohlig–Hellman reduction needs two steps, which we explain in Sections 17.4.1 and 17.4.2 below.

17.4.1 Reduction for n to p^{v_p} for all p dividing n

One denotes $n_p = n/p^{v_p}$, $g_p = g^{n_p}$, and $h_p = h^{n_p}$. The order of g_p is hence exactly p^{v_p}, and the following equation holds:

$$h_p = g_p^x.$$

The element g_p generates a subgroup of G of order p^{v_p}, and this subgroup contains h_p. It follows that there is a discrete logarithm of h_p in the basis g_p, denoted by $x(p)$. Then, the solution $x \in \{0, \ldots, n-1\}$ of all the congruences

$$x \equiv x(p) \bmod p^{v_p}$$

for all prime numbers p dividing n, obtained with the Chinese remainder theorem, is the solution to the discrete logarithm problem we are trying to solve in G. One needs to compute the quantities $x(p)$. One may obtain them with a general method such as those of Shanks or Pollard. One may as well, if $v_p \geq 2$, proceed to a new reduction before applying these methods.

17.4.2 Reduction for p^{v_p} to p

We have seen above how to reduce the discrete logarithm problem in a cyclic group G of order n into a set of discrete logarithm problems in cyclic groups of order p^{v_p}, where the ps are the prime numbers dividing n, and v_p their exponent in the decomposition of n into its prime factors. To simplify the notations, suppose now that G is a cyclic group of order p^v, where p is a prime number and v is an integer ≥ 2. Let g be a generator of G. The discrete logarithm problem we study in this section is: given $h \in G$, find $x \in \{0, \dots, p^n - 1\}$ such that

$$h = g^x.$$

We can write the integer x in basis p: there exist integers $(x_i)_{0 \leq i \leq v-1}$ such that

$$x = x_0 + x_1 p + \cdots + x_{v-1} p^{v-1}, \ 0 \leq x_i < p.$$

Finding x amounts to finding these integers x_i for $0 \leq i \leq v - 1$. The second step of the Pohlig–Hellman reduction algorithm shows, that these x_i are solutions of a discrete logarithm problem in a group of order p. Let use consider the equation

$$p^{v-1} x = x_0 p^{v-1} + p^v (x_1 + x_2 p + \cdots + x_{v-1} p^{v-1}).$$

Thanks to Fermat's small theorem, one deduces from the rise to the p^{v-1}-th power of the equation $h = g^x$ the following relation:

$$h^{p^{v-1}} = g^{p^{v-1} x} = g^{p^{v-1} x_0} = \left(g^{p^{v-1}}\right)^{x_0}.$$

But the element $g^{p^{v-1}}$ is of order p; hence x_0 is the solution of a DLP in a group of order p. It is then sufficient to apply this procedure recursively. Suppose that the integers $x_0, x_1, \dots, x_{i-1}$ are already known. The right-hand side of the following equation, denoted by h_i, is known as well:

$$g^{x_i p^i + \cdots + x_{v-1} p^{v-1}} = h g^{-(x_0 + x_1 p + \cdots + x_{i-1} p^{i-1})} = h_i.$$

The rise of this relation to the p^{v-i-1}-th power leads to

$$(g^{p^{v-1}})^{x_i} = h_i^{p^{v-i-1}} \quad \text{for } 0 \leq i < v.$$

For the same reason as previously, one sees that x_i is the solution of a DLP in a group of order p. Finally, this second step requires solving v discrete logarithm problems in groups of order p.

17.4.3 Complexity of the Pohlig–Hellman Reduction Algorithm

With the notation of the preceding sections, and assuming that the factorization of the order $n = \prod_{p|n} p^{v_p}$ of the group G is known, the Pohlig–Hellman algorithm requires the computation of h_p and g_p for all prime factors p of n and the knowledge of the coefficients $x_i(p)$ for $0 \leq i < v_p$. These coefficients, solutions of DLPs in groups of order p, can be obtained, for example, with the BSGS algorithm of Shanks or with the ρ method of Pollard. One shows that the Pohlig–Hellman reduction algorithm solves the DLP in the group G in $O(\Sigma_{p|n} v_p(\log n + \sqrt{p}))$ operations in the group G.

Example 1. We want to compute the discrete logarithm problem of 5 in basis 3 in $(\mathbf{Z}/2161\mathbf{Z})^*$ i.e., find the integer $1 \leq x \leq 2160$ such that $847 \equiv 23^x \bmod 2161$ using the Pohlig–Hellman method. Computation shows that the order of $\mathbf{F}^*_{2161}$ is $n = 2160 = 2^4 \times 3^3 \times 5$. One computes first $x(2) = x \bmod 2^4$ as the solution of the congruence

$$847^{3^3 \times 5} \equiv (23^{3^3 \times 5})^{x(2)} \bmod 2161.$$

This may be rewritten as

$$1934 \equiv 934^{x(2)} \bmod 2161.$$

The decomposition of $x(2)$ in basis 2 is

$$x(2) = x_0(2) + 2x_1(2) + 2^2 x_2(2) + 2^3 x_3(2).$$

It happens that $x_0(2)$ is the solution of the congruence

$$2160 \equiv 2160^{x_0(2)} \bmod 2161,$$

and $x_0(2) = 1$. The next step leads to the congruence

$$1 \equiv (-1)^{x_1(2)} \bmod 2161,$$

and one finds that $x_1(2) = 0$. The next computation gives

$$-1 \equiv (-1)^{x_2(2)} \bmod 2161,$$

and hence $x_2(2) = 1$. The last congruence is

$$-1 \equiv (-1)^{x_3(2)} \bmod 2161,$$

and one has $x_3(2) = 1$. Finally,

$$x(2) = 1 + 2^2 + 2^3 = 13.$$

Let us now compute $x(3) = x \bmod 3^3$ as the solution of the congruence

$$847^{2^4 \times 5} \equiv (23^{2^4 \times 5})^{x(3)} \bmod 2161.$$

This is equivalent to

$$1755 \equiv 2065^{x(3)} \bmod 2161.$$

One writes $x(3)$ in basis 3:

$$x(3) = x_0(3) + 3x_1(3) + 3^2 x_2(3).$$

It turns out that $x_0(3)$ is the solution of the relation

$$1 \equiv 593^{x_0(3)} \bmod 2161,$$

and one finds $x_0(3) = 0$. The next step leads to

$$593 \equiv 593^{x_1(3)} \bmod 2161,$$

and hence $x_1(3) = 1$. The last relation is

$$1567 \equiv 593^{x_2(3)} \bmod 2161,$$

and hence $x_2(3) = 2$. Finally,

$$x(3) = 0 + 1.3 + 2.3^2 = 21.$$

Lastly, one computes $x(5) = x \bmod 5$ as the solution of the relation

$$847^{2^4 \times 3^3} \equiv (23^{2^4 \times 3^3})^{x(5)} \bmod 2161.$$

In other words,

$$1161 \equiv 953^{x(5)} \bmod 2161.$$

Computation shows that $x(5) = 4$. We have the following relations:

$$\begin{cases} x \equiv 13 \mod 16, \\ x \equiv 21 \mod 27, \\ x \equiv 4 \mod 5. \end{cases}$$

The Chinese remainder theorem provides the solution:

$$x = 669.$$

At this stage, we invite the reader to have a look again at the DSA and to understand some of the choices made for the parameters.

17.5 Conclusion

This short chapter presents some selected aspects of public-key cryptography. We have tried to keep the presentation quite elementary, without going into all details. For several reasons, choices have been made, and we would like to mention here some other aspects that should be considered by the interested reader. First, there are a lot of algorithmic aspects that should be considered: how to compute a gcd efficiently, how to show that a given number is (probably) prime or composite, and others. We have also presented just a few cryptosystems. Of course there are plenty of other ones, such as Rabin, El Gamal, and ECDSA, to name a few. We have explained one factorization method. There are several other ones that are very useful, such as Fermat's method or Pollard's $p-1$ method, culminating with the GNFS. Again, just one method for solving the DLP has been described here. There are other general methods, such as the BSGS of Shanks and Pollard's ρ method, for example. A more modern approach consists in using elliptic curves in cryptography (see [27], [4], [5], [16]). These are quite sophisticated mathematical objects, providing very secure and very fast cryptosystems and digital signature schemes, but their study and presentation are definitely beyond the scope of this chapter (see [2] and [15]). Moreover, one uses generalizations of these objects, such as the Jacobian of hyperelliptic curves defined over finite fields. More recently, my research group succeeded in using Drinfeld elliptic modules in cryptography: the story keeps going on. Finally, a very important aspect of cryptology concerns public-key infrastructures (SSL, SSH, X.509 certificates, and others). This should also be studied by the interested reader.

References

[1]   AES Home Page. http://csrc.nist.gov/encryption/aes/.

[2]   IEEE-P1363 standard on public-key cryptology.

[3]   L.M. Adelman, R.L. Rivest, and A. Shamir, "A method for obtaining digital signatures and public-key cryptosystems." *Commun. ACM* 21, 120–126 (1978).

[4]   A.O.L. Atkin and F. Morain, "Elliptic curves and primality proving," *Math. Comput.* **61**, 29–68 (1993).

[5]   I. Blake, G. Seroussi, and N. Smart, "Elliptic curves in cryptography," *London Math. Soc.* LNS **265** (1999).

[6]   H. Cohen, *A Course in Algorithmic Algebraic Number Theory*, Graduate Texts in Mathematics vol. **138**, (Springer-Verlag, New York, 1993).

[7]   N. Koblitz, *A Course in Number Theory and Cryptology*, Graduate Texts in Mathematics **114**, (Springer-Verlag, New York, 1994).

[8]   H.W. Lenstra, Jr., "Factoring integers with elliptic curves," *Ann. Math.* **126**, 649–673 (1987).

[9]   A.K. Lenstra and H.W. Lenstra, Jr., *The Development of the Number Field Sieve*, A.K. Lenstra and H.W. Lenstra, Jr., eds., Lecture Notes in Mathematics **1554** (Springer-Verlag, Berlin, 1993).

[10] F. Leprévost, "Encryption and cryptosystems in electronic surveillance: A survey of the technology assessment issues," European Parliament (1998). Global Project No. EP/IV/B/SSSTOA/98/111401/01: Development of surveillance technology and risk of abuse of economic information.

[11] F. Leprévost and B. Warusfel, "Security technologies for digital media," European Parliament (2001). Study No. EP/IV/A/STOA/2000/06/01.

[12] F. Leprévost and B. Warusfel, "Protection and implementation of intellectual property rights in security technologies for digital media," European Parliament (2002–2003). Study No. EP/IV/STOA/2002/13/02.

[13] F. Leprévost and B. Warusfel, *Echelon: origines et perspectives d'un débat transnational.* Annulaire Franḉis des Relations Internationales, Vol. 2 (2001).

[14] R. Leprévost, *Cryptographie et lutte contre le terrorisme: éviter les fausses solutions sécuritaires.* Revue Droit et Défense (Publications du Centre de Recherche Droit et Défense de l'Université Paris V, Paris, 2002).

[15] F. Leprévost, Les standards cryptographiques du XXI-eme siecle: AES et IEEE-P1363," *Gazette Math.*

[16] A.J. Menezes, *Elliptic Curve Public-Key Cryptosystems* (Kluwer Academic Publishers, Dordrecht, 1993).

[17] A.J. Menezes, P.C. van Oorschot, and S.A. Vanstone, *Handbook of Applied Cryptography*, (CRC Press, Boca Raton, 1997)

[18] J.-L. Nicolas, "Tests de primalité," *Expo. Math.* **2**, 223–234 (1984).

[19] A. Odlyzko, "Discrete logarithms in finite fields and their cryptographic significance," In *Advances in Cryptology-Eurocrypt '84*, Lecture Notes in Computer Science, **209**, 224–314 (Springer-Verlag, Berlin, 1985).

[20] J. Pollard, "Theorems on factorization and primality testing," *Proc. Cambridge Philos. Soc.* **76**, 521–528 (1974).

[21] J. Pollard, "A Monte-Carlo method for factorization," *Bit* **15**, 331–334 (1975).

[22] C. Pomerance, J.L. Selfridge, and S.S. Wagstaff, Jr., "The pseudoprimes to $25 \cdot 10^9$," *Math. Comput.*, **35**, 1003–1026 (1980).

[23] P. Ribenboim, *The New Book of Prime Number Records*, 3rd ed., (Springer-Verlag, New York, 1996).

[24] O. Schirauker, D. Weber, and T. Denny, "Discrete logarithm: The effectiveness of the index calculus method," in H. Cohen, ed., *Algorithmic Number Theory Symposium*, Proceedings of the Second International Symposium ANTS II, Talence, France, May 1996, Lecture Notes in Computer Science, **1122** (Springer-Verlag, Berlin, 1996).

[25] B. Schneier, *Applied Cryptography*, 2nd ed. (John Wiley and Sons, New York, 1996).

[26] D. Shanks, "Class number, a theory of factorization, and genera." in Proceedings of Symposia in Pure Mathematics, **20**, pp. 415–440 (American Mathematical Society, Providence, RI, 1971).

[27] J. Silverman, *The Arthmetic of Elliptic Curves*, Graduate Texts in Mathematics **106**, (Springer, New York, 1992).

Index